Injection Molding of Polymers and
Polymer Composites

Injection Molding of Polymers and Polymer Composites

Editors

Andrew N. Hrymak
Shengtai Zhou

Basel • Beijing • Wuhan • Barcelona • Belgrade • Novi Sad • Cluj • Manchester

Editors
Andrew N. Hrymak
Western University
London
Canada

Shengtai Zhou
Sichuan University
Chengdu
China

Editorial Office
MDPI AG
Grosspeteranlage 5
4052 Basel, Switzerland

This is a reprint of articles from the Special Issue published online in the open access journal *Polymers* (ISSN 2073-4360) (available at: https://www.mdpi.com/journal/polymers/special_issues/injec_mold).

For citation purposes, cite each article independently as indicated on the article page online and as indicated below:

Lastname, A.A.; Lastname, B.B. Article Title. *Journal Name* **Year**, *Volume Number*, Page Range.

ISBN 978-3-7258-1767-2 (Hbk)
ISBN 978-3-7258-1768-9 (PDF)
doi.org/10.3390/books978-3-7258-1768-9

Contents

About the Editors

Andrew N. Hrymak

Andrew N. Hrymak (P. Eng, FCIC, FCAE, Ph. D. Chem. Eng., Carnegie Mellon University; B.Eng., Chem Eng., McMaster University) is a Professor of Chemical and Biochemical Engineering and co-Director of the Fraunhofer Innovation Platform for Composites Research at Western University, London, Ontario, Canada. In recent years, he has also served as Provost and Vice-President Academic and Dean of Engineering at Western University. His research interests include the modeling, design, and optimization of material processing systems, with a current emphasis on polymer composites. He previously served in the Department of Chemical Engineering at McMaster University from 1985 to 2009, where he was Department Chair (2000–2009), Director of the McMaster Manufacturing Research Institute (2001–05), and Director of the Walter G. Booth School of Engineering Practice (2004–09). Hrymak has served as an Associate Editor of *Computers and Chemical Engineering*, the Editor-in-Chief of *International Polymer Processing*, is President of the International Society of Coating Science and Technology (ISCST), and Chair of the Board of Directors of the Chemical Institute of Canada.

Shengtai Zhou

Shengtai Zhou (Ph.D., Chem. Biochem. Eng., Western University; M.Sc., Compos. Mater., Sichuan University; B.Sc., Polym. Mater. Eng., Qingdao University) is an Associate Professor at the State Key Laboratory of Polymer Materials Engineering and at the Polymer Research Institute of Sichuan University. He received his Doctoral degree in 2018, under the joint supervision of Prof. Andy Hrymak (Western University, London, Canada) and Prof. Musa R. Kamal (McGill University, Montreal, Canada). His primary research interest concerns the processing of polymer composites, with a special focus on conductive polymer composites and self-lubricating polymer composites. He conducts studies on porous materials, including, for instance, the preparation of rigid/flexible polyimide foams, polyimide aerogels, and polyurethane foams. Aside from the above-mentioned areas, he also conducts research on other topics that relate to polymer processing. Dr. Zhou has co-authored one book chapter and more than 120 peer-reviewed journal articles. He was awarded a Graduate Travel Award by the Polymer Processing Society in 2018.

Preface

Injection molding is a versatile polymer processing method that is widely used to prepare polymers and polymer composites for industrial product applications. The properties of injection-molded parts are affected by the prevailing shearing and extensional flow fields, as well as temperature gradients. These processing factors determine the state of orientation of polymer chains and functional fillers (especially for those with very high aspect ratios), the development of crystalline structures for semi-crystalline polymers, as well as the state of the distribution of functional fillers. As a result, the optimization of the processing parameters and the design of functional composites play a crucial role in fabricating injection-molded parts for targeted applications.

In recent years, advances in injection molding technology have promoted the development of polymeric parts for application in various industrial fields, such as automation and optical and thermal energy storage. This reprint provides a collection of research studies that address the process–structure–properties of injection molded polymers and polymer composites. This reprint gathers 10 important contributions from researchers around the globe. In addition, we express our sincere gratitude to the reviewers and members of the Editorial Office who aided in the development of this Special Issue.

Andrew N. Hrymak and Shengtai Zhou
Editors

Editorial

Injection Molding of Polymers and Polymer Composites

Shengtai Zhou [1] and Andrew N. Hrymak [2,*]

[1] The State Key Laboratory of Polymer Materials Engineering, Polymer Research Institute of Sichuan University, Chengdu 610065, China; szhou@scu.edu.cn or qduzst@163.com

[2] Department of Chemical and Biochemical Engineering, The University of Western Ontario, London, ON N6A 5B9, Canada

* Correspondence: ahrymak@uwo.ca

Citation: Zhou, S.; Hrymak, A.N. Injection Molding of Polymers and Polymer Composites. *Polymers* **2024**, *16*, 1796. https://doi.org/10.3390/polym16131796

Received: 24 May 2024
Accepted: 21 June 2024
Published: 25 June 2024

Injection molding technology has been widely adopted to fabricate multifunctional polymeric components or structural parts for applications in fields such as automotives, electronics, packaging, aerospace, and many others. It is also widely accepted that the properties of injection moldings are greatly affected by the development of crystalline structure, the distribution and orientation of functional fillers which are incurred by the prevailing shearing and extensional flow fields as well as the cooling effect during injection molding. In addition, the properties of polymeric parts are determined by the types of fillers and host matrices, part geometry, and processing parameters. Therefore, elucidating the relationship of processing–structure–properties in injection molding is particularly important for both the academic and industrial spheres.

The Special Issue, "Injection Molding of Polymers and Polymer Composites", serves as a suitable platform for the state-of-the-art research progress in injection molding. This Special Issue collates 10 research articles, with contributions from Germany (1), China (4), the United States of America (1), Japan (2), Vietnam (1) and the Czech Republic (1), which covered a very broad range of topics relating to injection molding technology. The contributions to this Special Issue are listed below:

Contribution 1: Zhang, J.; Zhang, Y.; Li, Y.; Luo, M.; Zhang, J. Influence of Strong Shear Field on Structure and Performance of HDPE/PA6 In Situ Microfibril Composites. Polymers 2024, 16, 1032. https://doi.org/10.3390/polym16081032

Contribution 2: Bednarik, M.; Pata, V.; Ovsik, M.; Mizera, A.; Husar, J.; Manas, M.; Hanzlik, J.; Karhankova, M. The Modification of Useful Injection-Molded Parts' Properties Induced Using High-Energy Radiation. Polymers 2024, 16, 450. https://doi.org/10.3390/polym16040450

Contribution 3: Minh, P.S.; Nguyen, V.-T.; Uyen, T.M.T.; Huy, V.Q.; Le Dang, H.N.; Nguyen, V.T.T. Enhancing Amplification in Compliant Mechanisms: Optimization of Plastic Types and Injection Conditions. Polymers 2024, 16, 394. https://doi.org/10.3390/polym16030394

Contribution 4: Tian, J.; Wang, C.; Wang, K.; Xue, R.; Liu, X.; Yang, Q. Flexible Polyolefin Elastomer/Paraffin Wax/Alumina/Graphene Nanoplatelets Phase Change Materials with Enhanced Thermal Conductivity and Mechanical Performance for Solar Conversion and Thermal Energy Storage Applications. Polymers 2024, 16, 362. https://doi.org/10.3390/polym16030362

Contribution 5: Jiang, Q.; Takayama, T.; Nishioka, A. Impact Energy Dissipation and Quantitative Models of Injection Molded Short Fiber-Reinforced Thermoplastics. Polymers 2023, 15, 4297. https://doi.org/10.3390/polym15214297

Contribution 6: Takayama, T.; Shibazaki, R. Mechanical Anisotropy of Injection-Molded PP/PS Polymer Blends and Correlation with Morphology. Polymers 2023, 15, 4167. https://doi.org/10.3390/polym15204167

Contribution 7: Myers, M.; Mulyana, R.; Castro, J.M.; Hoffman, B. Experimental Development of an Injection Molding Process Window. Polymers 2023, 15, 3207. https://doi.org/10.3390/polym15153207

Contribution 8: Yao, H.; Xue, R.; Wang, C.; Chen, C.; Xie, X.; Zhang, P.; Zhao, Z.; Li, Y. High-Temperature Response Polylactic Acid Composites by Tuning Double-Percolated Structures. Polymers 2023, 15, 138. https://doi.org/10.3390/polym15010138

Contribution 9: Zhou, C.; Bai, Y.; Zou, H.; Zhou, S. Improving Thermal Conductivity of Injection Molded Polycarbonate/Boron Nitride Composites by Incorporating Spherical Alumina Particles: The Influence of Alumina Particle Size. Polymers 2022, 14, 3477. https://doi.org/10.3390/polym14173477

Contribution 10: Maertens, R.; Liebig, W.V.; Weidenmann, K.A.; Elsner, P. Development of an Injection Molding Process for Long Glass Fiber-Reinforced Phenolic Resins. Polymers 2022, 14, 2890. https://doi.org/10.3390/polym14142890

The following provides an overview of the articles published in this Special Issue:

Zhang et al. (contribution 1) adopted a multi-flow vibration injection molding (MFVIM) technology to improve the mechanical properties of high-density polyethylene (HDPE) by tuning the crystalline structure of HDPE and creating polyamide 6 (PA) microfibers in situ. An HDPE/PA6 blend with a component mass ratio of 90:10 was used as the model system, and it was subjected to conventional injection molding (CIM, without vibration) and MFVIM, respectively. The tensile strength and tensile modulus of HDPE/PA6, which was prepared by implementing six vibration times during the packing stage, were 66.5 and 981.4 MPa, figures which are 91% and 32% higher than those for CIM pure HDPE, which were 83% and 27% higher than their CIM counterparts, respectively. The in situ formation of PA6 microfibers, as well as the numerous shish-kebab and hybrid shish-kebab structures of HDPE induced by the multiple shear zones generated by vibration, were considered as contributing factors.

Bednarik et al. (contribution 2) adopted a high energy radiation (β radiation) technique to modify the properties of injection-molded HDPE (representative commodity thermoplastic) and glass fiber (GF)-reinforced PA66 composites (GF content: 30 wt%, representative technical plastic). Their results show that the free surface energy of samples was altered, which was likely caused by oxidation, and this greatly affected the adhesive properties of the tested materials. In addition, the tensile and bending strength of both samples were enhanced after radiation treatment, which was attributed to the radiation induced cross-linking process. In the case of HDPE, an optimal dose was reported in a range from 145 to 150 kGry by taking both the surface and mechanical properties into account, and the optimal dose for PA66 was 128~135 kGry. They proposed that a well-chosen radiation dose leads to the improvement of both the mechanical and surface properties of injection-molded products that can broaden their practical applications.

Nguyen et al. (contribution 3) investigated the effect of process parameters such as filling time, filling pressure, filling speed, packing time, packing pressure, cooling time and melt temperature on the amplification ratio of the compliant mechanism injection molded flexure hinges made from ABS, PP, and HDPE. Their results demonstrated a linear relationship between the input and output data of ABS, PP, and HDPE flexure hinges at different process parameters. The packing pressure had the greatest impact on the amplification ratio of the ABS flexure hinge, filling time had the highest effect with a PP flexural hinge, and packing time had the greatest effect with HDPE flexural hinges. This work provides some insights to broaden the application of plastic flexure hinges by optimizing plastic types and injection-molding process parameters.

Yang et al. (contribution 4) prepared electrically insulative and thermally conductive polyolefin elastomer/paraffin wax (POE/PW) phase-change materials (PCMs) with spherical alumina (Al_2O_3) particles and graphene nanoplatelets (GNPs), using injection molding technique. The hybrid addition of Al_2O_3 and GNPs was found to be helpful for establishing three-dimensional thermal conductive pathways, thus improving thermal conductivity. The in-plane thermal conductivity of the POE/PW/GNPs 5 wt%/Al_2O_3 40 wt% composite reached as high as 1.82 W/mK, which is approximately 269.5% higher than that of unfilled POE/PW. In addition, the POE/PW/GNPs 5 wt%/Al_2O_3 40 wt% composite demonstrated outstanding electrical insulation, mechanical performance, and efficient solar energy conversion. This study showcased developing flexible PCMs for solar conversion and thermal storage applications.

Takayama et al. (contribution 5) proposed a mechanical model to explain the notched impact strength of injection-molded short glass-fiber-reinforced thermoplastics (SGFRTP). The model showed a good agreement ($R^2 > 0.95$) with the experimental values obtained from GF-reinforced polypropylene (PP) and polystyrene (PS) composites. The authors suggested that the model could be applied to different fiber orientation angles and a range of fiber lengths in the molded products, provided that the fiber length was sufficiently shorter than the critical fiber length. They also stated that the universality of the proposed model needed to be verified if the fibers have weaker interfacial strength or less susceptibility to fracture during injection molding.

Takayama and Shibazaki (contribution 6) adopted a short-beam shear testing method to evaluate mechanical anisotropy as the stress concentration factor. The correlation between the evaluation results and the phase structure of PP/PS blends was clarified. They found that the yield condition under uniaxial tensile testing was interface debonding for the continuous-phase PP with a sea–island structure; the phase structure was dispersed and elongated in the flow direction for continuous-phase PS. Unlike continuous-phase PP, the structure of continuous-phase PS was greatly altered with the addition of styrene–ethylene–butadiene–styrene (SEBS). The yielding condition under uniaxial tensile loading was shear yielding. The development of the phase structure was affected by the types of PP, and the addition of SEBS to PS/H-PP (i.e., homo-type PP) resulted in a phase morphology with a cylindrical dispersed phase with relatively small diameter, and a dispersed phase arranged in a network for PS/B-PP (i.e., block-type PP). Moreover, the mechanical anisotropy of PP/PS blends was correlated with the aspect ratio of the dispersed phase. The higher the aspect ratio of the dispersed phase, the greater mechanical anisotropy for corresponding blends.

Castro et al. (contribution 7) studied the relationship between some key machine settings, which were classified as primary control variables (mold temperature, melt temperature, packing pressure), secondary control variables (injection screw speed, packing/cooling time), and tertiary control variables (shot size, clamping force) with the successful operation of injection molding. Their study provided a more standardized and thorough procedure for experimentally developing injection molding process windows to obtain injection-molded parts with acceptable appearance and part qualities. Additionally, it was proposed that testing the mechanical properties is a need for determining process windows for semi-crystalline polymers.

Zhao et al. (contribution 8) prepared electrically conductive polymers (CPCs) with a double-percolation structure consisting of poly(lactide acid) (PLA), poly(butylene adipate terephthalate) (PBAT), and GNPs. The results show that adding 5 wt% PBAT resulted in an electrical conductivity, which is about two orders of magnitude higher than the PLA/GNP 3.5 wt% counterpart. PBAT significantly reduced the action time from 14.15 to 2.19 min during temperature-response measurements. Moreover, PLA/PBAT/GNP samples showed more sensibility and stability during the cyclic temperature-response tests, which demonstrate potential applications for fabricating temperature-sensing devices.

Zhou et al. (contribution 9) systematically studied the effect of Al_2O_3 particle size and filling content on the properties of polycarbonate (PC)/boron nitride (BN) composites. They reported that both the in-plane (i.e., parallel to the flow direction) and through-plane (i.e., perpendicular to flow direction) thermal conductivities of injection-molded PC/BN/Al_2O_3 composites were significantly enhanced with the addition of Al_2O_3 particles. In addition, thermal conductivity was greatly improved with increasing Al_2O_3 concentration and particle size. The change in the orientation state of BN platelets that was induced by the added Al_2O_3 was crucial to improving through-plane thermal conductivity. Additionally, PC/BN/Al_2O_3 composites exhibit exceptional electrical insulation and reasonable mechanical properties that may provide potential applications in industrial sectors.

Maertens et al. (contribution 10) developed a long fiber direct thermoset injection molding process to prepare GF-reinforced phenolic resin composites with a higher proportion of long fibers, aiming to improve its mechanical properties. A novel screw-mixing element was adopted with considerations of balancing the desired mixing action, an undesired preliminary curing of phenolic resin, and a reduction in fiber length. The results show that a weighted average fiber length of 571 μm (initial fiber length: 5000 μm) was achieved in molded parts, which is twice that observed for a short fiber-reinforced phenolic resin under comparable processing conditions. In addition, a homogeneous distribution of fibers was found to outweigh the disadvantages of reduced fiber length because samples prepared with the highest mixing energy input had a tensile strength of 57 MPa, and the samples prepared with the lowest mixing energy input was only 21 MPa.

Finally, we would like to thank all researchers and anonymous reviewers who contributed to the production of this Special Issue. In addition, we express our sincere gratitude to the editorial team, especially Ms. Shelly Gu who contributed to the development and subsequent launch of this Special Issue.

Author Contributions: S.Z. and A.N.H. contributed to writing the original draft as well as reviewing and editing. All authors have read and agreed to the published version of the manuscript.

Funding: S.Z. is appreciated for the support from the National Natural Science Foundation of China (52103040) and the China Postdoctoral Science Foundation (2020M673217).

Institutional Review Board Statement: Not applicable.

Data Availability Statement: Data sharing is not applicable to this article.

Conflicts of Interest: The authors declare no conflicts of interest.

Article

Influence of Strong Shear Field on Structure and Performance of HDPE/PA6 In Situ Microfibril Composites

Junwen Zhang, Yiwei Zhang, Yanjiang Li, Mengna Luo and Jie Zhang *

College of Polymer Science and Engineering, State Key Laboratory of Polymer Materials Engineering, Sichuan University, Chengdu 610065, China; wenjunzh11@outlook.com (J.Z.); z403084973@163.com (Y.Z.); 13881505715@163.com (Y.L.); z1907747660@163.com (M.L.)
* Correspondence: zhangjie@scu.edu.cn

Abstract: As one of the most widely applied general-purpose plastics, high-density polyethylene (HDPE) exhibits good comprehensive performance. However, mechanical strength limits its wider application. In this work, we introduced the engineering plastic PA6 as a dispersed phase to modify the HDPE matrix and applied multiple shears generated by vibration to the polymer melt during the packing stage of injection molding. SEM, 2D-WXRD and 2D-SAXS were used to characterize the morphology and structure of the samples. The results show that under the effect of a strong shear field, the dispersed phase in the composites can form in situ microfibers and numerous high-strength shish-kebab and hybrid shish-kebab structures are formed. Additionally, the distribution of fibers and high-strength oriented structures in the composites expands to the core region with the increase in vibration times. As a result, the tensile strength, tensile modulus and surface hardness of VIM-6 can reach a high level of 66.5 MPa, 981.4 MPa and 72, respectively. Therefore, a high-performance HDPE product is successfully prepared in this study, which is of great importance for expanding the application range of HDPE products.

Keywords: HDPE; in situ microfibril composites; strong shear field; shish-kebab; hybrid shish-kebab

Citation: Zhang, J.; Zhang, Y.; Li, Y.; Luo, M.; Zhang, J. Influence of Strong Shear Field on Structure and Performance of HDPE/PA6 In Situ Microfibril Composites. *Polymers* **2024**, *16*, 1032. https://doi.org/10.3390/polym16081032

Academic Editors: Andrew N. Hrymak and Shengtai Zhou

Received: 12 March 2024
Revised: 4 April 2024
Accepted: 8 April 2024
Published: 10 April 2024

1. Introduction

High-density polyethylene (HDPE), as a high-crystallinity, non-polar thermoplastic resin, has good comprehensive properties such as low toxicity, excellent cold resistance, good mechanical performance, great chemical resistance and ease of processing [1–6]. Therefore, it has become one of the most widely applied general-purpose plastics. However, with the advancement of science and technology, the mechanical strength of existing HDPE can no longer meet the demand for high-strength plastic products [7]. Therefore, it is of great significance to further improve the mechanical properties of HDPE to expand its application range.

High-performance polymer composites have been widely researched and applied owing to their excellent mechanical properties [8,9]. Adding fibers to the polymer matrix to prepare polymer composites is one of the common methods of polymer reinforcement [10,11]. At present, the commonly used reinforcing fibers are mainly high-strength inorganic fibers, such as glass fibers (GFs) [12,13], carbon fibers (CFs) [14,15], etc. But during the process of manufacturing these composites, the fibers may break as a result of shearing and agitation by the processing equipment, which reduces the length-to-diameter (L/D) ratio of the fibers and even leads to a decrease in the mechanical properties of the material [16]. Fortunately, the in situ microfibril composite (MFC) has stood out from the traditional polymer composites and attracted the attention of many researchers due to its great potential in mechanical enhancement and huge advantage in manufacturing over the past few decades [17,18]. Compared with common inorganic fiber polymer composites, the fibers of MFCs are formed during processing, which prevents fibers breaking caused by screw agitation. Traditional methods of preparing MFCs are usually divided into three steps: melt blending, stretching into fibers, processing and molding [19–21]. MFCs obtained by this method usually need to undergo a secondary process after the formation of in situ microfibers, which may lead to the aggregation of fibers, thereby affecting the final properties of the products [8]. Therefore, if it is possible for the dispersed phase to form in situ fibers directly during the molding process, it will not only ensure the quality of the fibers but also greatly simplify the processing. In recent years, several studies [8,22–27] have proven the feasibility of this method. For example, Qin et al. [22] obtained high-performance HDPE/PET MFCs through volume-pulsatile injection molding (VPIM). The results demonstrated that under the effect of the vibration force field, the PET dispersed phase formed rigid microfibers and the strength as well as the toughness of the products greatly improved. The tensile strength and impact strength of the material reached 31.1 MPa and 41.4 MPa, with an increase of 29% and 658%, respectively. Jiang et al. [27] successfully prepared PP/PS in situ microfibril composites by fused filament fabrication (FFF) and observed that the tensile strength of the product increased by 43.6%. Shen et al. [24] combined in situ fiber formation with the stacked extrusion technique to prepare PP/PA6 in situ microfiber composites. After the formation of in situ microfibers, the tensile strength of the composites significantly improved from 19 MPa to 47 MPa, increasing by 147%.

In addition to adding fibers, introducing external fields during processing to form self-reinforcement structures is also a common method of polymer reinforcement. Shear field is one of the popular external fields in polymer processing. In general, polymer materials tend to form highly oriented structure (shish-kebab) under a strong shear field [28]. Therefore, some new processing technologies (such as VPIM [29], dynamic packing injection molding (DPIM) [30–32], loop oscillatory push–pull molding (LOPPM) [33–35], multi-flow vibration injection molding (MFVIM) [36,37], etc.) have been developed to provide a strong shear field, which could introduce plenty of shish-kebab structures into the polymer products, thereby significantly improving the strength and modulus of the materials. For example, Liu et al. [38] prepared simultaneously self-reinforced and self-toughened HDPE products using LOPPM. The results indicated that the tensile strength and modulus of the samples improved by about 1.8 and 1.2 times, respectively. Liu et al. [35] used LOPPM technology to prepare high-performance HDPE/ultra-high-molecular-weight polyethylene (UHMWPE) samples. The results showed that under the effect of the dynamic shear field provided by LOPPM, a large number of shish-kebab structures were introduced into the products. The tensile strength, modulus and impact strength of the materials were increased by 2.8, 4.9 and 5.8 times, respectively. Mi et al. [39] prepared iPP samples with different thicknesses of the shear layer by using MFVIM and investigated the effect of the content of the shish-kebab on the mechanical properties of iPP materials. The results showed that the content of the shish-kebab was related to the duration of the shear field; the tensile strength and modulus significantly improved with the content of the shish-kebab.

In this work, we chose PA6 to modify HDPE. This is because the mechanical strength of PA6 is much higher than that of HDPE [40,41]. Meanwhile, we applied the MFVIM technique to prepare injection-molded parts. On the one hand, we would like the PA6 dispersed phase to form in situ microfibers under the effect of a strong shear field. On the other hand, it is expected that plenty of shish-kebab and hybrid shish-kebab structures can be formed to achieve the purpose of enhancing the mechanical properties of HDPE. The results show that the content of fibers as well as (hybrid) the shish-kebab improves with the increase in vibration times (the duration of the shear field). The tensile strength and modulus of products are significantly improved, increasing by 91% and 32%, respectively.

2. Materials and Methods

2.1. Materials

HDPE (4902T) was purchased from Yangzi Petroleum Chemical Co., Ltd., Nanjing, China and the melt flow rate (MFR) is 0.3 g/10 min (190 °C/5 kg). PA6 (M2500I) was purchased from Guangdong Xinhui Meida Nylon Co., Ltd., Jiangmen, China with an MFR of 26.4 g/10 min (190 °C/2.16 kg).

2.2. Sample Preparation

PA6 was dried at 80 °C for 12 h under vacuum to prevent degradation during processing. Then, HDPE/PA6 pellets with a mass blending ratio of 90:10 were fully melted and mixed through a co-rotating twin-screw extruder with L/D = 40 (CPE 20 plus, Coperion, Nanjing, China). The temperatures from the hopper to the die head were between 225 and 240 °C, and the screw rotation was maintained at 140 rpm. The extruded strand was then pelletized, and the obtained blend granules were dried at 80 °C for 12 h again. In this experiment, the dried blended granules were processed into injection-molded parts by the self-developed MFVIM equipment, whose mechanism was described in detail in our previous studies [37,42]. The mold temperature was 40 °C, and the melt temperatures from the hopper to the nozzle were 180 °C, 190 °C, 210 °C, 220 °C and 220 °C, respectively. During the packing stage, vibration times were set to 2, 4 and 6, respectively. And the specific parameter settings are shown in Figure 1 and Table 1. For the purpose of comparison, conventional injection molding (CIM) samples of HDPE/PA6 blends were also prepared in this experiment. The prepared samples are denoted as CIM and VIM-x, where CIM and VIM represent the samples molded by CIM and MFVIM, while x means the vibration times. The sampling methods for various characterizations are shown in Figure 2.

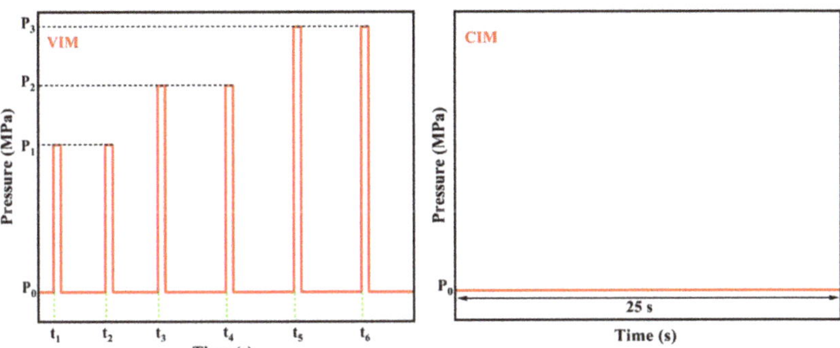

Figure 1. The relationship between the vibration pressure and time of VIM and CIM samples during the packing stage.

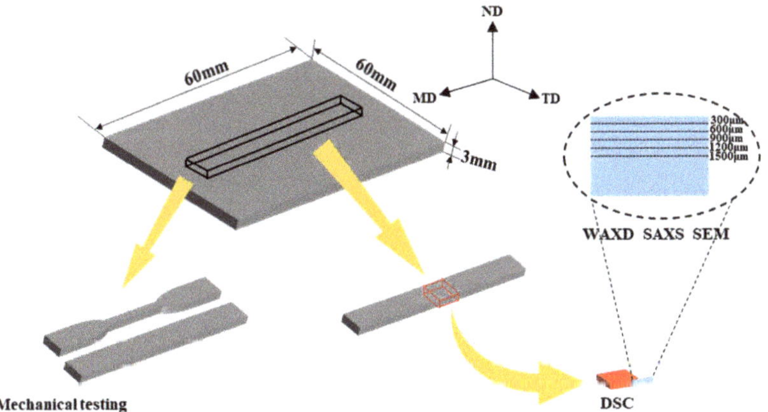

Figure 2. Schematic diagram of sampling method for characterizations. FD: flow direction, TD: transverse direction, ND: Normal direction.

Table 1. Processing parameters during packing stage. (P: pressure/MPa, t: start time/s).

	Packing Pressure (P_0)	1st		2nd		3rd		4th		5th		6th	
		P_1	t_1	P_1	t_2	P_2	t_3	P_2	t_4	P_3	t_5	P_3	t_6
VIM-2		80	1	80	4.5								
VIM-4	30	80	1	80	4.5	100	9	100	13.5				
VIM-6		80	1	80	4.5	100	9	100	13.5	120	18	120	22.5

2.3. Differential Scanning Calorimetry (DSC)

The thermal behavior and crystallinity of different samples were analyzed using a differential scanning calorimetric instrument (TA Q250, TA Instruments, New Castle, DE, USA). All measurements were performed under a nitrogen atmosphere. About 3–5 mg of the samples was heated from 40 °C to 240 °C at a heating rate of 10 °C/min. The crystallinity (X_c) of each sample can be calculated according to the following equation:

$$X_c = \frac{\Delta H_m}{w_f \Delta H_m^0} \times 100\%$$ (1)

where ΔH_m indicates the measured value of the enthalpy of melting obtained from the DSC experiment. ΔH_m^0 means the enthalpy of the melting of completely crystallized PLA, which is 293 J/g [43]. w_f represents the mass fraction of HDPE in the mixture, which is 0.9 in this work.

2.4. Scanning Electron Microscopy (SEM)

In order to examine the distribution and morphology of the dispersed phase in different samples, we etched the specimens by using formic acid to remove the PA6 dispersed phase. To further observe the crystal morphology, the samples were placed in a mixed acid solution for etching to remove the amorphous phase. The etched surfaces were cleaned with distilled water in an ultrasonic bath and dried afterwards. After gold sputtering treatment, the etched surfaces were examined by a field-emission scanning electron microscope (Nova Nano SEM450, FEI company, Hillsboro, OR, USA) along the MD-ND plane with an accelerating voltage of 10 kV.

2.5. 2D-WAXD and 2D-SAXS Measurements

Two-dimensional wide-angle X-ray diffraction (2D-WAXD) and two-dimensional small-angle X-ray scattering (2D-SAXS) measurements were carried out on the BL16B1 beamline of the Shanghai Synchrotron Radiation Facility (SSRF), Shanghai, China. The size of the rectangular beam was 0.5×0.8 mm^2, and the wavelength of the light was 0.124 nm. The detector-to-sample distances for WAXD and SAXS were 88.5 mm and 1000 mm, respectively. The rectangular beam, perpendicular to the MD-ND plane in Figure 2, was moved from the upper surface to the center region of the samples and irradiated at five positions with an interval distance of 300 µm. The distances between the upper surface and each position are about 300, 600, 900, 1200 and 1500 µm, respectively.

The crystal orientation is calculated by using Herman's orientation function, which is defined as

$$f = \frac{3 < \cos^2\varphi >_{hkl} - 1}{2}$$ (2)

where $\langle \cos^2\varphi \rangle_{hkl}$ is the orientation factor, which is defined as

$$\cos^2\varphi = \frac{\int_0^{\pi/2} I(\varphi)\sin(\varphi)\cos^2\varphi d\varphi}{\int_0^{\pi/2} I(\varphi)\sin\varphi d\varphi}$$ (3)

where φ is the angle between the molecular chain direction and the melt flow direction, and $I(\varphi)$ is the scattering intensity at angle φ. When f is equal to 0, the molecular chains are randomly arranged. When f is −0.5 or 1.0, the c axes of all the crystals are exactly perpendicular or parallel to the flow direction, respectively.

2.6. Mechanical Test

The mechanical strength of injection-molded products is tested using an electronic universal testing machine (68TM-10, Instron, Boston, MA, USA). According to ASTM D-638-V, the samples for the tensile test were cut into dumbbell bars with the dimensions of 65 mm × 4 mm × 3 mm (length × width of narrow part × thickness). All tests were carried out at room temperature with a cross-head speed of 50 mm/min. Five specimens of each group were tested, and the results were averaged.

A LX-D Shore Hardness Tester (HANDPI, Yueqing, China) was used to measure the surface hardness of the injection-molded parts. According to ASTM D2240, the hardness test was performed at room temperature with a specimen thickness of 6 mm. The hardness of the samples was tested at five different locations on the surface of each specimen, and the results were averaged.

3. Results and Discussion

3.1. Phase Morphology

In order to investigate the morphology of the dispersed phase at different locations in the samples, the etched surfaces were observed by SEM, and the results are shown in Figure 3. For CIM, although the dispersed phase can be stretched along the flow direction due to the shear effect introduced by the filling process, it is limited around 300 μm that oriented fibers can be observed. Meanwhile, at 600 μm, the PA6 dispersed phase gradually transforms from fibers to rod-like particles with a low length–diameter (L/D) ratio. Moreover, there is a typical sea-island structure in the core region, in which the dispersed phase is mainly ellipsoidal or spherical. However, after the introduction of a strong shear field, the situation changes dramatically. For VIM-2, in situ microfibers with a large L/D ratio can be observed at both 300 μm and 600 μm. In addition, the dispersed phase can still form oriented short rods at 900 μm and 1200 μm. Meanwhile, for VIM-4 and VIM-6, highly oriented fibers can be found from 300 μm to 1200 μm. Furthermore, there are still a small number of fibers remaining at 1500 μm. From the above results, it can be concluded that in situ microfibers can be formed under the effect of a strong shear field provided by the MFVIM technology, and with the increase in vibration times, the distribution of fibers gradually expands to the core region.

3.2. Thermal Behavior

Figure 4a shows the DSC curves of the composites prepared under different processing conditions, and the crystallinity calculated according to the curves is displayed in Figure 4b. In order to illustrate the role of the PA6 dispersed phase in the crystallization process of the HDPE matrix, we also performed DSC analysis on CIM samples of pure HDPE (donated as CIM-H). As can be seen from the figure, the crystallinity of CIM-H is 62.1%, while the crystallinity of the HDPE/PA6 composites are all around 65.5%. The crystallinity of the HDPE matrix increases by about 3.4% after the addition of the PA6 phase. Therefore, in this experiment, the improvement in the crystallinity of the HDPE matrix is mainly due to the heterogeneous nucleation of the PA6 dispersed phase, and the difference in the morphology of the dispersed phase has little effect on the crystallinity of the matrix.

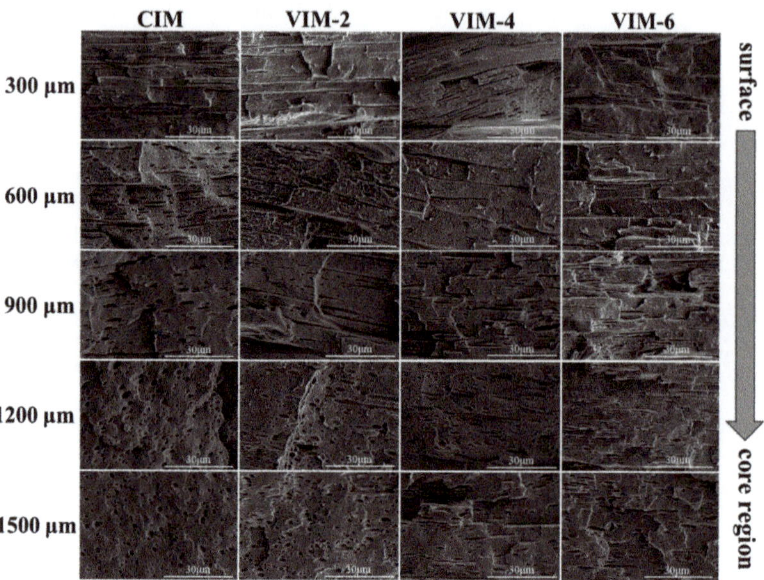

Figure 3. SEM micrographs of CIM, VIM-2, VIM-4 and VIM-6 at different positions with a certain distance away from the surface.

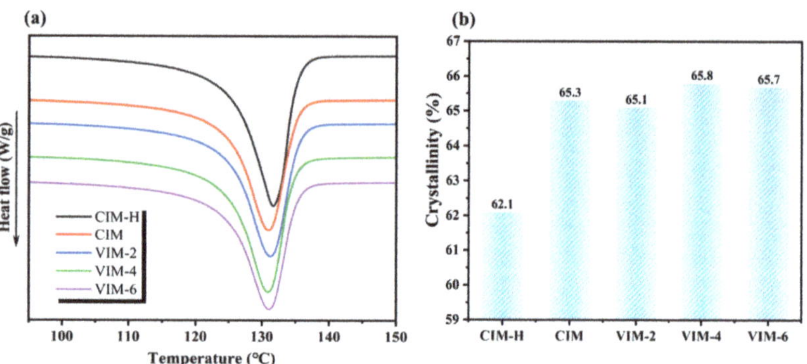

Figure 4. (**a**) DSC melting curves and (**b**) crystallinity of CIM-H, CIM, VIM-2, VIM-4 and VIM-6, respectively.

3.3. Crystalline Structure

In order to obtain detailed information about the crystal structure of the samples, the amorphous phase was etched away using a mixed acid solution, and the etched surfaces were characterized by SEM. The SEM images taken at different locations of different samples are shown in Figure 5. For CIM, some lamellae oriented along the flow direction can be observed at 400 μm, while at other locations, the lamellae are randomly aligned. After the introduction of a strong shear field, although the obtained samples are all dominated by oriented structures at 400 μm and 900 μm, there is still a difference in the crystal structures. For VIM-2, some shish-kebabs can be seen at 400 μm, while the crystal is mainly the oriented lamellae at 900 μm. For VIM-4 and VIM-6, it can be found that plenty of shish-kebab and hybrid shish-kebab structures oriented along the flow direction can be observed at 400 μm and 900 μm, and the lamellae still remains oriented at the core region (1500 μm). Moreover, there are more shish-kebab and hybrid shish-kebab structures formed in VIM-6 compared to VIM-4. It indicates that the content of shish-kebabs and hybrid shish-kebabs in the samples grows with the increase in vibration times.

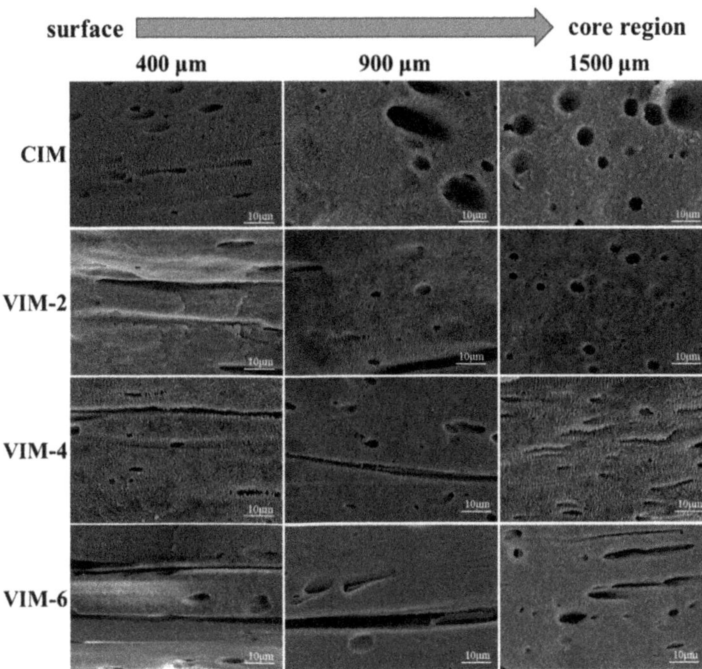

Figure 5. The crystalline structure of CIM, VIM-2, VIM-4 and VIM-6 at different positions with a certain distance away from the surface.

The enlarged micrographs of shish-kebab and hybrid shish-kebab structures in VIM-6 at 900 μm are shown in Figure 6. From Figure 6a, it can be found that the molten molecular chains are highly oriented along the flow direction under the effect of a strong shear field. Moreover, during the cooling process, the HDPE lamellar grows in attachment to the oriented molecular chains to form shish-kebab structures. Meanwhile, for hybrid shish-kebab structures, it can be seen from Figure 6b that the PA6 in situ microfibers replace the oriented HDPE molecular chains as the "shish" and the oriented HDPE lamellar grows periodically on the surface of the fibers.

The 2D-WAXD patterns of different samples at different locations are given in Figure 7. The diffraction pattern consists of two diffraction rings, where the inner one and outer one represent the (110) and (200) lattice planes of HDPE, respectively. The diffraction patterns of CIM are arc-like at 300 μm and 600 μm, indicating that there are oriented structures formed at these locations. Meanwhile, at other locations, the diffraction patterns exhibit typical isotropic diffraction rings. This suggests that during the process of CIM, the formation of oriented structures is limited in thickness due to the weak shear effect. However, for VIM-2, it displays arc-like diffraction patterns at the positions of 300 μm to 900 μm. Meanwhile, at 1200 μm and 1500 μm, the diffraction patterns are isotropic diffraction rings. In addition, the diffraction patterns all show arc-like patterns from 300 μm to 1500 μm for VIM-4 and VIM-6, indicating that the oriented structures are generated from the surface layer to the core layer in these two samples. In general, the distribution of oriented structures in the samples is consistent with the results of SEM.

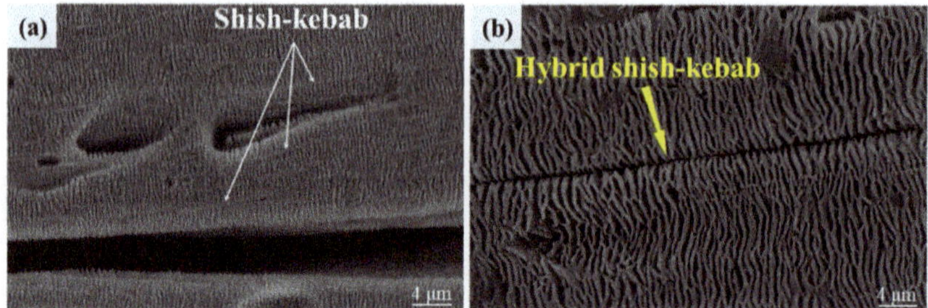

Figure 6. An enlarged micrograph (the magnification is 50,000×) of shish-kebab (**a**) and hybrid shish-kebab (**b**) structures in VIM-6. The distance from the surface of the samples is 900 μm.

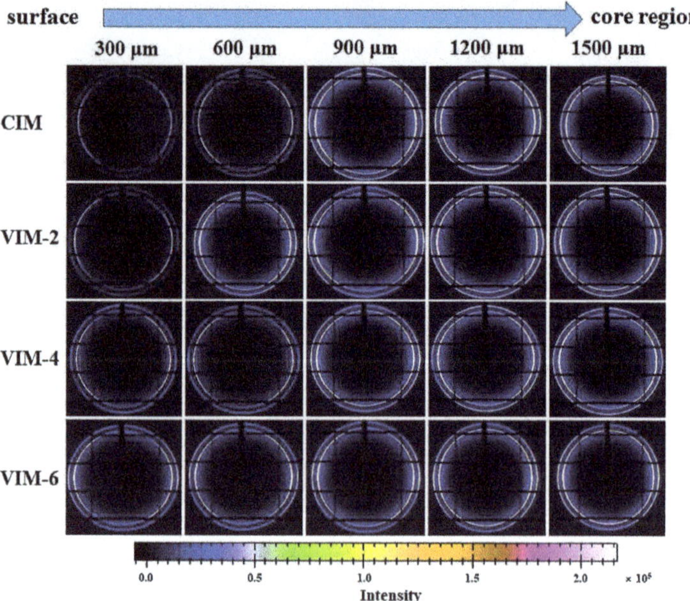

Figure 7. 2D-WAXD patterns of CIM, VIM-2, VIM-4 and VIM-6 at different positions with a certain distance away from the surface.

In order to quantitatively investigate the orientation degree of different layers in different samples, the calculated results of the orientation degree are summarized in Figure 8. The orientation degree of CIM is kept at 0.2–0.4 (except 300 μm) and always lower than that of the MFVIM samples. For MFVIM samples, as the distance from the surface increases, the orientation degree of all the composites rises sharply and then falls. Moreover, the orientation degree of VIM-2 is obviously lower than the other two samples, while the orientation degree of VIM-4 and VIM-6 shows little difference. This suggests that the orientation degree improves dramatically with the increase in vibration times. However, when the number of vibrations is greater than four, continuing to increase the vibration times has little effect on the improvement in the orientation degree.

To further study the difference in the crystal structure of HDPE/PA6 composites prepared by different processing conditions, the 2D-SAXS patterns of CIM, VIM-2, VIM-4 and VIM-6 samples at different positions are shown in Figure 9. For CIM, orientation signals appear in the equatorial direction of the patterns at 300 μm and 600 μm, indicating that there are oriented fibers or crystals formed in the sample at these locations. On the contrary, other positions of CIM show typical isotropic scattering patterns, meaning that both the dispersed phase and crystal at these positions are randomly distributed. As for the MFVIM samples, typical shish-kebab signals are observed at 300 μm and 600 μm in all three samples, including a "shish" signal along the equatorial direction and a "kebab" signal along the meridian direction. This suggests that under the effect of a strong shear field, the surface layer of the samples all form highly oriented shish-kebab structures. In addition, shish-kebab signals also appear at 900 μm and 1200 μm of VIM-4 and VIM-6, whereas only conventional orientation signals appear at the same positions of VIM-2. Surprisingly, weak shish-kebab signals can be observed at 1500 μm in VIM-4 and VIM-6, suggesting that small amounts of shish-kebab structures form in the core region of the two samples. Furthermore, it is evident that the shish-kebab signals at 300–900 μm are stronger with the increase in vibration times.

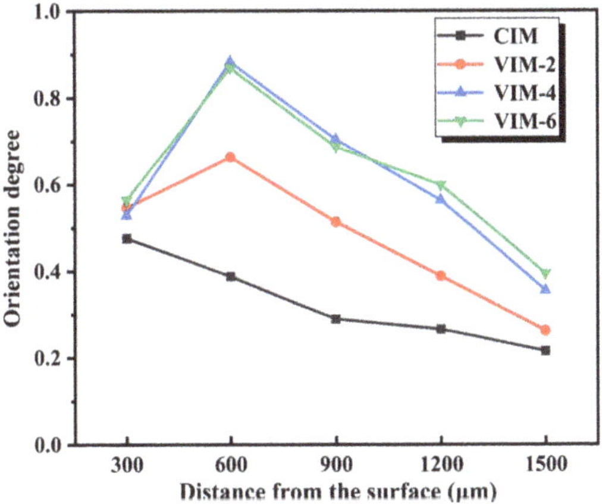

Figure 8. Orientation degree of different samples.

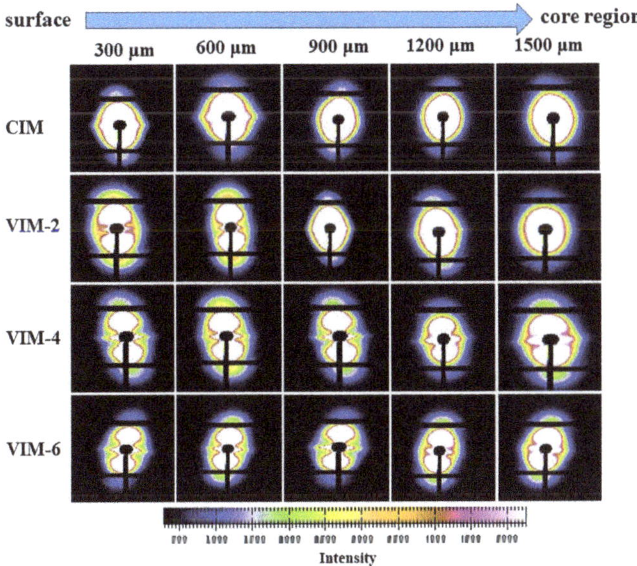

Figure 9. 2D-SAXS patterns of CIM, VIM-2, VIM-4 and VIM-6 at different positions with a certain distance away from the surface.

3.4. Mechanical Properties

The tensile strength, tensile modulus and elongation at break of different samples are given in Figure 10a–c, respectively. To explain the role of the PA6 dispersed phase in composites for the enhancement of mechanical properties, we also performed tensile tests on CIM-H. The tensile strength, tensile modulus and elongation at break of CIM-H (not shown in the figure) are 34.8 MPa, 744.3 MPa and 162.2%, respectively. From Figure 10a,b, it can be seen that after the addition of the PA6 dispersed phase, the tensile strength and tensile modulus of CIM increase to 36.4 MPa and 774.5, respectively. The improvement is very limited, which is due to the incompatibility of the HDPE matrix with the PA6 dispersed phase, although PA6 is much stronger than HDPE [44]. However, after the introduction of a strong shear field, the tensile strength and tensile modulus of the samples improve significantly and increase with increasing vibration times. VIM-6 exhibits the maximum tensile strength and tensile modulus, which are 66.5 MPa and 981.4 MPa, respectively. Compared with CIM-H, the tensile strength and tensile modulus increase by 91% and 32%, respectively. Meanwhile, compared to CIM, they improve by 83% and 27%, respectively. The reason for the enhancement of the mechanical strength is the massive generation of in situ microfibrils, shish-kebab and hybrid shish-kebab structures under the effect of the strong shear field [8,31].

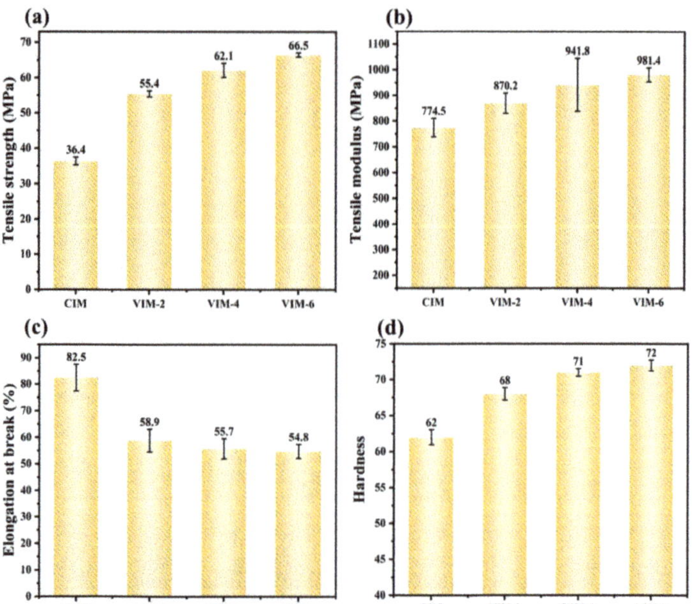

Figure 10. Mechanical performance of different samples prepared by different conditions: (**a**) tensile strength, (**b**) tensile modulus, (**c**) elongation at break and (**d**) hardness.

Meanwhile, for tensile toughness, it can be noticed from Figure 10c that the sample CIM possesses the maximum elongation at break. However, compared to the CIM-H sample, there is a considerable decrease in the tensile toughness of CIM after the addition of the PA6 dispersed phase, from 162.2% to 82.5%, which decreases by nearly 50%. After the introduction of the strong shear field, the elongation at break continues to decrease. But the elongation at break of all MFVIM samples remains above 50%, and the difference among them is small. It indicates that the formation of microfibers and oriented structures decreases the elongation at break of the material [35,39], while increasing vibration times has little effect on the tensile toughness. In addition, the MFVIM samples can maintain an acceptable level of toughness (>50%) while the mechanical strength is significantly improved. This suggests that products with balanced strength and toughness can be prepared through this method.

In order to investigate the ability to resist the localized plastic deformation of the material, the surface hardness of the material was tested [45]. The results are shown in Figure 10d. As can be seen from the figure, CIM has the lowest surface hardness of 62. After introducing the strong shear field, the surface hardness of the samples increases. The surface hardness of VIM-2 reaches 68, improving by 9.6% compared to CIM. Moreover, the surface hardness of the material further increases with increasing vibration times. VIM-6 possesses the maximum surface hardness of 72, which is 16% higher compared to CIM. This suggests that the formation of in situ microfibers as well as shish-kebab structures within the material facilitates the surface hardness of the HDPE material.

4. Conclusions

In this work, high-performance HDPE/PA6 microfibril composites were successfully fabricated by MFVIM technology. Under the effect of the strong shear field, the PA6 dispersed phase forms in situ microfibers and a large number of shish-kebab and hybrid shish-kebab structures are formed in the composites. Moreover, the distribution of microfibers gradually expands to the core region with the increase in vibration times. As a result, the mechanical performance of the products improves dramatically. The tensile strength, tensile modulus and surface hardness of VIM-6 reach up to 66.5 MPa, 981.4 MPa and 72, respectively. Compared with CIM-H, the tensile strength and tensile modulus increase by 91% and 32%, respectively. This study proposes an effective method for the fabrication of high-performance HDPE-based composites, which is of great significance for the preparation and application of HDPE products.

Author Contributions: Conceptualization, J.Z. (Junwen Zhang) and Y.Z.; formal analysis, J.Z. (Junwen Zhang); investigation, J.Z. (Junwen Zhang), Y.Z. and Y.L.; methodology, M.L.; resources, J.Z. (Jie Zhang); supervision, J.Z. (Jie Zhang); visualization, J.Z. (Junwen Zhang); writing—original draft, J.Z. (Junwen Zhang); writing—review and editing, J.Z. (Jie Zhang). All authors have read and agreed to the published version of the manuscript.

Funding: This research was funded by the National Natural Science Foundation of China grant number [No. 21627804].

Institutional Review Board Statement: Not applicable.

Data Availability Statement: Data are contained within the article.

Acknowledgments: The authors gratefully acknowledge the technical support of the Shanghai Synchrotron Radiation Facility (SSRF, Shanghai, China) for help with X-ray measurements.

Conflicts of Interest: The authors declare no conflicts of interest.

References

1. Amjadi, M.; Fatemi, A. Tensile behavior of high-density polyethylene including the effects of processing technique, thickness, temperature, and strain rate. *Polymers* **2020**, *12*, 1857. [CrossRef]
2. Khanal, S.; Zhang, W.P.; Ahmed, S.; Ali, M.; Xu, S. Effects of intumescent flame retardant system consisting of tris (2-hydroxyethyl) isocyanurate and ammonium polyphosphate on the flame retardant properties of high-density polyethylene composites. *Compos. Part A Appl. Sci. Manuf.* **2018**, *112*, 444–451. [CrossRef]
3. Hu, J.; Wang, Z.W.; Yan, S.M.; Gao, X.Q.; Deng, C.; Zhang, J.; Shen, K.Z. The morphology and tensile strength of high density polyethylene/nano-calcium carbonate composites prepared by dynamic packing injection molding. *Polym. Plast. Technol. Eng.* **2012**, *51*, 1127–1132. [CrossRef]
4. Feng, J.; Zhang, R.Y.; Wu, J.J.; Yang, W.; Yang, M.-B. Largely enhanced molecular orientation and mechanical property of injection-molded high-density polyethylene parts via the synergistic effect of polyamide 6 in situ microfibrillar and intense shear flow. *Colloid Polym. Sci.* **2014**, *292*, 3033–3044. [CrossRef]
5. Zhang, Y.; Xie, Z.X.; Zhong, G.J.; Lei, J.; Gao, X.; Jiang, L. Effect of cellulose nanocrystals and hot stretching on shish-kebab structures of high-density polyethylene. *Ind. Eng. Chem. Res.* **2023**, *62*, 15018–15028. [CrossRef]

6. Salakhov, I.; Shaidullin, N.M.; Chalykh, A.E.; Matsko, M.A.; Shapagin, A.V.; Batyrshin, A.Z.; Shandryuk, G.A.; Nifant'ev, I.E. Low-temperature mechanical properties of high-density and low-density polyethylene and their blends. *Polymers* **2021**, *13*, 1821. [CrossRef] [PubMed]

7. Costa, I.L.M.; Zanini, N.C.; Mulinari, D.R. Thermal and mechanical properties of hdpe reinforced with al2o3 nanoparticles processed by thermokinectic mixer. *J. Inorg. Organomet. Polym. Mater.* **2021**, *31*, 220–228. [CrossRef]

8. Jiang, Y.X.; Mi, D.S.; Wang, Y.X.; Wang, T.; Shen, K.; Zhang, J. Composite contains large content of in situ microfibril, prepared directly by injection molding: Morphology and property. *Macromol. Mater. Eng.* **2018**, *303*, 1800270. [CrossRef]

9. Zhu, J.; Ou, Y.C.; Feng, Y.P. Studies on the mechanical-properties and crystallization behavior of polyethylene composites. *Chin. J. Polym. Sci.* **1995**, *13*, 218–227.

10. Ma, L.L.; Liu, F.; Liu, D.Y.; Liu, Y. Review of strain rate effects of fiber-reinforced polymer composites. *Polymers* **2021**, *13*, 2839. [CrossRef]

11. Ansari, M.T.A.; Singh, K.K.; Azam, M.S. Fatigue damage analysis of fiber-reinforced polymer composites-a review. *J. Reinf. Plast. Compos.* **2018**, *37*, 636–654. [CrossRef]

12. Cech, V.; Palesch, E.; Lukes, J. The glass fiber-polymer matrix interface/interphase characterized by nanoscale imaging techniques. *Compos. Sci. Technol.* **2013**, *83*, 22–26. [CrossRef]

13. Thieme, M.; Boehm, R.; Gude, M.; Hufenbach, W. Probabilistic failure simulation of glass fibre reinforced weft-knitted thermoplastics. *Compos. Sci. Technol.* **2014**, *90*, 25–31. [CrossRef]

14. Hu, C.; Liao, X.W.; Qin, Q.H.; Wang, G. The fabrication and characterization of high density polyethylene composites reinforced by carbon nanotube coated carbon fibers. *Compos. Part A Appl. Sci. Manuf.* **2019**, *121*, 149–156. [CrossRef]

15. Kumar, S.; Doshi, H.; Srinivasarao, M.; Park, J.O.; Schiraldi, D.A. Fibers from polypropylene/nano carbon fiber composites. *Polymer* **2002**, *43*, 1701–1703. [CrossRef]

16. Tseng, H.-C.; Chang, R.-Y.; Hsu, C.H. Numerical predictions of fiber orientation and mechanical properties for injection-molded long-glass-fiber thermoplastic composites. *Compos. Sci. Technol.* **2017**, *150*, 181–186. [CrossRef]

17. Li, Z.M.; Yang, M.B.; Huang, R.; Yang, W.; Feng, J.M. Poly(ethylene terephthalate)/polyethylene composite based on in-situ microfiber formation. *Polym. Plast. Technol. Eng.* **2002**, *41*, 19–32. [CrossRef]

18. Jiang, Y.X.; Mi, D.S.; Wang, Y.X.; Wang, T.; Shen, K.; Zhang, J. Insight into understanding the influence of blending ratio on the structure and properties of high-density polyethylene/polystyrene microfibril composites prepared by vibration injection molding. *Ind. Eng. Chem. Res.* **2019**, *58*, 1190–1199. [CrossRef]

19. Jayanarayanan, K.; Thomas, S.; Joseph, K. Morphology, static and dynamic mechanical properties of in situ microfibrillar composites based on polypropylene/poly (ethylene terephthalate) blends. *Compos. Part A Appl. Sci. Manuf.* **2008**, *39*, 164–175. [CrossRef]

20. Rizvi, A.; Tabatabaei, A.; Barzegari, M.R.; Mahmood, S.H.; Park, C.B. In situ fibrillation of co2-philic polymers: Sustainable route to polymer foams in a continuous process. *Polymer* **2013**, *54*, 4645–4652. [CrossRef]

21. Kuzmanovic, M.; Delva, L.; Cardon, L.; Ragaert, K. The effect of injection molding temperature on the morphology and mechanical properties of pp/pet blends and microfibrillar composites. *Polymers* **2016**, *8*, 355. [CrossRef] [PubMed]

22. Qin, S.; Lu, X.; Lv, S.-Y.; Xu, W.-H.; Zhang, H.-H.; Tan, L.-C.; Qu, J.-P. Simultaneously toughening and reinforcing high-density polyethylene via an industrial volume-pulsatile injection molding machine and poly (ethylene terephthalate). *Compos. Part B Eng.* **2020**, *198*, 108243. [CrossRef]

23. Zheng, G.Q.; Yang, W.; Huang, L.; Li, Z.-M.; Yang, M.-B.; Yin, B.; Li, Q.; Liu, C.-T.; Shen, C.-Y. The role of gas penetration on morphological formation of polycarbonate/polyethylene blend molded by gas-assisted injection molding. *J. Mater. Sci.* **2007**, *42*, 7275–7285. [CrossRef]

24. Shen, J.; Wang, M.; Li, J.; Guo, S. In situ fibrillation of polyamide 6 in isotactic polypropylene occurring in the laminating-multiplying die. *Polym. Adv. Technol.* **2011**, *22*, 237–245. [CrossRef]

25. Zhao, Z.; Yang, Q.; Kong, M.; Tang, D.; Chen, Q.; Liu, Y.; Lou, F.; Huang, Y.; Liao, X. Unusual hierarchical structures of micro-injection molded isotactic polypropylene in presence of an in situ microfibrillar network and a β-nucleating agent. *RSC Adv.* **2015**, *5*, 43571–43580. [CrossRef]

26. Liu, W.; Nie, M.; Wang, Q. Biaxial reinforcements for polybutene-1 medical-tubes achieved via flow-design controlled morphological development of incorporated polystyrene: In-situ microfibrillation, alignment manipulation and performance optimization. *Compos. Sci. Technol.* **2015**, *119*, 124–130. [CrossRef]

27. Jiang, Y.; Wu, J.; Leng, J.; Cardon, L.; Zhang, J. Reinforced and toughened pp/ps composites prepared by Fused Filament Fabrication (FFF) with in-situ microfibril and shish-kebab structure. *Polymer* **2020**, *186*, 121971. [CrossRef]

28. Zhou, D.; Yang, S.-G.; Lei, J.; Hsiao, B.S.; Li, Z.-M. Role of stably entangled chain network density in shish-kebab formation in polyethylene under an intense flow field. *Macromolecules* **2015**, *48*, 6652–6661. [CrossRef]

29. Qin, S.; Jiang, H.-W.; Zhang, H.-H.; Huang, Z.-X.; Qu, J.-P. Simultaneously achieving super toughness and reinforcement of immiscible high-density polyethylene/poly (ethylene terephthalate) composite via oriented spherical crystal structure. *Compos. Part A Appl. Sci. Manuf.* **2022**, *163*, 107186. [CrossRef]

30. Zhou, W.-C.; Xie, Z.-X.; Gao, N.; Zhong, G.-J.; Deng, C.; Gao, X.-Q. Regulating crystalline morphology jointly by dynamic shearing and solid phase stretching endows polyethylene high modulus, strength and heat-resistance. *Polymer* **2023**, *283*, 126219. [CrossRef]

31. Liang, S.; Wang, K.; Tang, C.; Zhang, Q.; Du, R.; Fu, Q. Unexpected molecular weight dependence of shish-kebab structure in the oriented linear low density polyethylene/high density polyethylene blends. *J. Chem. Phys.* **2008**, *128*, 174902. [CrossRef] [PubMed]
32. Wang, Y.; Na, B.; Fu, Q. Super polyolefin blends achieved via dynamic packing injection molding: Morphology and properties. *Chin. J. Polym. Sci.* **2003**, *21*, 505–514.
33. Li, L.; Li, W.; Geng, L.; Chen, B.; Mi, H.; Hong, K.; Peng, X.; Kuang, T. Formation of stretched fibrils and nanohybrid shish-kebabs in isotactic polypropylene-based nanocomposites by application of a dynamic oscillatory shear. *Chem. Eng. J.* **2018**, *348*, 546–556. [CrossRef]
34. Geng, L.; Li, L.; Mi, H.; Chen, B.; Sharma, P.; Ma, H.; Hsiao, B.S.; Peng, X.; Kuang, T. Superior impact toughness and excellent storage modulus of poly (lactic acid) foams reinforced by shish-kebab nanoporous structure. *ACS Appl. Mater. Interfaces* **2017**, *9*, 21071–21076. [CrossRef] [PubMed]
35. Liu, T.; Huang, A.; Geng, L.H.; Lian, X.-H.; Chen, B.-Y.; Hsiao, B.S.; Kuang, T.-R.; Peng, X.-F. Ultra-strong, tough and high wear resistance high-density polyethylene for structural engineering application: A facile strategy towards using the combination of extensional dynamic oscillatory shear flow and ultra-high-molecular-weight polyethylene. *Compos. Sci. Technol.* **2018**, *167*, 301–312. [CrossRef]
36. Mi, D.; Hou, F.; Zhou, M.; Zhang, J. Improving the mechanical and thermal properties of shish-kebab via partial melting and re-crystallization. *Eur. Polym. J.* **2018**, *101*, 1–11. [CrossRef]
37. Hu, M.L.; Deng, C.J.; Gu, X.B.; Fu, Q.; Zhang, J. Manipulating the strength-toughness balance of poly(l-lactide) (plla) via introducing ductile poly(ε-caprolactone) (PCL) and strong shear flow. *Ind. Eng. Chem. Res.* **2020**, *59*, 1000–1009. [CrossRef]
38. Liu, T.; Li, W.; Li, L.; Peng, X.; Kuang, T. Effect of dynamic oscillation shear flow intensity on the mechanical and morphological properties of high-density polyethylene: An integrated experimental and molecular dynamics simulation study. *Polym. Test.* **2019**, *80*, 106122. [CrossRef]
39. Mi, D.; Xia, C.; Jin, M.; Wang, F.; Shen, K.; Zhang, J. Quantification of the effect of shish-kebab structure on the mechanical properties of polypropylene samples by controlling shear layer thickness. *Macromolecules* **2016**, *49*, 4571–4578. [CrossRef]
40. Dencheva, N.V.; Oliveira, M.J.; Pouzada, A.S.; Kearns, M.P.; Denchev, Z.Z. Mechanical properties of polyamide 6 reinforced microfibrilar composites. *Polym. Compos.* **2011**, *32*, 407–417. [CrossRef]
41. Sang, L.; Wang, C.; Wang, Y.Y.; Hou, W. Effects of hydrothermal aging on moisture absorption and property prediction of short carbon fiber reinforced polyamide 6 composites. *Compos. Part B Eng.* **2018**, *153*, 306–314. [CrossRef]
42. Min, J.; Hu, M.; Yan, Z.; Wang, T.; Fu, Q.; Gao, X.; Zhang, J. Preparation of ppr/ps blends by multiflow vibration injection molding: Influence of shish–kebab structures and in situ fibers. *Ind. Eng. Chem. Res.* **2024**, *63*, 2769–2778. [CrossRef]
43. Stack, G.M.; Mandelkern, L.; Kröhnke, C.; Wegner, G. Melting and crystallization kinetics of a high molecular-weight n-alkane—c192h386. *Macromolecules* **1989**, *22*, 4351–4361. [CrossRef]
44. Cappello, M.; Strangis, G.; Cinelli, P.; Camodeca, C.; Filippi, S.; Polacco, G.; Seggiani, M. From waste vegetable oil to a green compatibilizer for hdpe/pa6 blends. *Polymers* **2023**, *15*, 4178. [CrossRef]
45. Saha, A.; Kumar, S. Effects of graphene nanoparticles with organic wood particles: A synergistic effect on the structural, physical, thermal, and mechanical behavior of hybrid composites. *Polym. Adv. Technol.* **2022**, *33*, 3201–3215. [CrossRef]

Article

The Modification of Useful Injection-Molded Parts' Properties Induced Using High-Energy Radiation

Martin Bednarik [1],*, Vladimir Pata [1], Martin Ovsik [1], Ales Mizera [2], Jakub Husar [2], Miroslav Manas [2], Jan Hanzlik [1] and Michaela Karhankova [2]

[1] Faculty of Technology, Tomas Bata University in Zlin, Vavreckova 5669, 760 01 Zlin, Czech Republic; pata@utb.cz (V.P.); ovsik@utb.cz (M.O.); j_hanzlik@utb.cz (J.H.)
[2] Faculty of Applied Informatics, Tomas Bata University in Zlin, Nad Stranemi 4511, 760 05 Zlin, Czech Republic; mizera@utb.cz (A.M.); husar@utb.cz (J.H.); manas@utb.cz (M.M.); m_karhankova@utb.cz (M.K.)
* Correspondence: mbednarik@utb.cz

Abstract: The modification of polymer materials' useful properties can be applicable in many industrial areas due to the ability to make commodity and technical plastics (plastics that offer many benefits, such as processability, by injection molding) useful in more demanding applications. In the case of injection-molded parts, one of the most suitable methods for modification appears to be high-energy irradiation, which is currently used primarily for the modification of mechanical and thermal properties. However, well-chosen doses can effectively modify the properties of the surface layer as well. The purpose of this study is to provide a complex description of high-energy radiation's (β radiation) influence on the useful properties of injection-molded parts made from common polymers. The results indicate that β radiation initiates the cross-linking process in material and leads to improved mechanical properties. Besides the cross-linking process, the material also experiences oxidation, which influences the properties of the surface layer. Based on the measured results, the main outputs of this study are appropriately designed regression models that determine the optimal dose of radiation.

Keywords: polymers; beta radiation; injection molding; cross-linking; oxidation; regression; mechanical properties; surface properties

Citation: Bednarik, M.; Pata, V.; Ovsik, M.; Mizera, A.; Husar, J.; Manas, M.; Hanzlik, J.; Karhankova, M. The Modification of Useful Injection-Molded Parts' Properties Induced Using High-Energy Radiation. *Polymers* **2024**, *16*, 450. https://doi.org/10.3390/polym16040450

Academic Editors: Andrew N. Hrymak and Shengtai Zhou

Received: 17 November 2023
Revised: 23 January 2024
Accepted: 26 January 2024
Published: 6 February 2024

1. Introduction

The ever-increasing consumption of polymer materials in the entire engineering industry continually places pressure on the rising requirements put on base materials. One of the options which can meet high technical requirements is the use of special, although expensive, and difficult-to-process polymer materials [1–3]. Another alternative could be the use of cheaper, commodity, and technical plastics in combination with a suitable type of modification, which can enhance the material throughout its volume and improve other mechanical and thermal properties [3,4]. These specific applications often require the joining of individual parts to larger assemblies. One of the most important methods for this kind of application is bonding. Unlike mechanical methods of connecting (welding, riveting), bonding introduces no additional tension, dampens vibrations, increases rigidity and buckling strength, and can be used for water- and gas-tight applications. Furthermore, mechanical joints come with the drawback of being heavier [5]. The bonding of commodity and technical plastics is usually preceded by suitable modifications [6], which target the material's surface layer properties, such as the wettability of joined surfaces, and increase its free surface energy and adhesive properties [7].

If an application requires commodity and technical plastics, which can be polyethylene (PE), polyamide (PA), etc., to satisfy the aforementioned criteria with regard to the mechanical properties and quality of adhesive bonds, it is necessary to choose a suitable type of modification for both the mechanical and surface properties [3,4,8]. Current studies describe a wide spectrum of methods that primarily focuses on either the modification of mechanical and thermal properties [9–11] or on the alteration of surface properties [12–14]. Due to these modifications, many common polymers find their use not only in numerous engineering applications but also outside of these, for example, in the biomedical field [15] or energy industry [16].

Regarding the modification of a polymer's surface, the most widely used methods are plasma treatment [8,13,15,17], corona treatment [12,14], or chemical etching [12]. The aforementioned methods are quite effective in the modification of surface layer properties; however, their potential in the adjustment of mechanical properties is significantly limited. It would be beneficial for industrial practice if one method could modify both the mechanical [18,19] and surface layer properties of injection-molded materials [13,14,17]. From this point of view, irradiation appears to be the most suitable method, as some recent studies suggest that its use with a correctly chosen radiation dose could lead to an improvement of not only the mechanical properties [18,19] but also lead to an effective modification of surface layer properties, such as wettability or free surface energy [5].

The process of radiation cross-linking of injection-molded parts manufactured from polymer materials is performed, exposing them to radiation (most commonly the electron beams from electron accelerators). As mentioned by Makuuchi et al. [20] and other authors [21,22], the primary interactions of accelerated electrons with polymer material are the ionization, excitation, stabilization, neutralization, and generation of free radicals. Free radicals can be created either due to the scission of a main chain or due to the dissociation of a side chain. Following the primary reactions are secondary reactions, mainly hydrogen abstraction, recombination (cross-linking or branching), chain scission, oxidation, and grafting. All of these processes and reactions evoke either a positive or a negative change in the targeted group of properties. One of the main desired reactions is, above all, cross-linking, especially when an improvement in the material's characteristics is the primary objective [18,19]. The cross-linking process is underway when cross-linking prevails over degradation, i.e., the number of cross-linked chains is higher than the number of chain scissions. This observation was expanded upon by Gheysari et al. [23], who found that tested polymers demonstrated varying values of useful properties dependent upon the absorbed radiation dose (tensile strength rose with the increasing dose while elongation at break decreased). The actuator of these changes was the cross-linking process, which resulted in the creation of free radicals (breakup of C-H bonds) that subsequently recombined into a spatial network due to the linking of two free radicals together (C-C bond) with neighboring chains. As a result, mechanical properties were changed, which corresponds with the conclusions of other authors [18,24–26]. In the case of some polymers, an addition of a polyfunctional monomer is necessary to increase the initiation and recombination tendency of its radicals. For injection-molded parts, this additive is mixed into the polymer blend before the injection molding process [27]. This problem was studied by Malinowski [28], who reached the conclusion that exposing polybutylene adipate terephthalate (PBAT) filled by polyfunctional monomer triallyl isocyanurate (TAIC) to high-energy radiation led to cross-linking, which resulted in the creation of a significant gel fraction and, thus, the increase in tensile strength. Comparable results with regard to free radical recombination in polymers with polyfunctional monomers were observed in other studies as well [27,29].

Besides the cross-linking, irradiated polymers also face degradation processes, which can result in the worsening of useful properties. As shown in studies by Hama et al. [30] and other authors [31,32], one of these degradation processes is oxidation, which can cause the decline of the mechanical properties. On the other hand, degradation can also lead to the more frequent creation of carbonyl functional groups on the polymer surface [5,6], which can be useful for the improvement of adhesive properties and wettability of the surface. As can be seen, the potential of radiation cross-linking is quite high. Due to its universal nature, both the positive and the negative influences can be used in technical practice, one for the improvement of mechanical properties and the other for the modification of the surface layer. However, this universality is strongly dependent on the absorbed radiation dose, especially the ratio between both coexisting processes, i.e., cross-linking and degradation (oxidation). Up to now, many important studies have been written on the topic of the influence of high-energy radiation on the useful properties of polymer materials, e.g., studies of Holik, Danek, Manas et al. [33,34] and other authors [5,18–29]. Nevertheless, the aforementioned correlation between the absorbed radiation dose and the required surface and mechanical properties has not yet been comprehensively investigated.

Hence, the main goal of this study is to comprehensively describe the influence of radiation cross-linking (by suitably designed regression models) on the useful properties of injection-molded parts manufactured from commodity and technical plastics (PE and PA). Furthermore, it is necessary to investigate the universality of this method, especially regarding its ability to modify both the surface (free surface energy and adhesive properties) and the mechanical (tensile and bending strength) properties in a wide spectrum of working temperatures. The designed regression models could make finding the optimal value of applied radiation dose with regard to the required surface and mechanical properties of injection-molded parts easier.

2. Materials and Methods

2.1. Materials and Specimen Preparation

Verification of high-energy radiation's influence on the useful properties of injection-molded parts was performed on one representative from group of commodity thermoplastics, specifically high-density polyethylene (HDPE) with trade name DOW HDPE 25055E provided by DOW (Midland, MI, USA). The other representative was chosen out of the technical plastics group, specifically, polyamide 66 (PA66) filled by 30 wt. % of glass fibers with trade name V-PTS-CREAMID-A3H7.2G6*M0129A provided by PTS (Adelshofen, Germany). The PA66 was also filled with polyfunctional monomer TAIC, which was performed to increase the recombination frequency of polymer radicals [27–29,34].

Specimens were molded using the Arburg Allrounder 420C and 470H (Loßburg, Germany). The cavities of the tools (injection molds) were machined with specific dimensions, which are given by the individual standards for testing mechanical and surface properties [35–38] and the strength of bonded joints in shear stressing [39]. The injection molding process parameters for each material were chosen according to the manufacturer's recommendations and can be seen in Table 1.

Table 1. The injection molding process parameters.

Processing Conditions	HDPE	PA66
Injection Rate (mm/s)	60	60
Injection Pressure (MPa)	80	88
Holding Pressure (MPa)	60	70
Holding Time (s)	30	25
Cooling Time (s)	20	30
Mold Temperature (°C)	40	75
Plastic Unit Temperature Bands		
Zone 1 (°C)	200	265
Zone 2 (°C)	205	280
Zone 3 (°C)	210	285
Zone 4 (°C)	225	290

2.2. Modification of Specimens by High-Energy Radiation

After the manufacturing, the specimens were exposed to high-energy radiation, specifically an electron beam produced by an electron accelerator (β radiation). The irradiation was performed in standard atmospheric conditions at ambient temperature in BGS Beta-Gamma Service, Saal an der Donau, Germany. Source of radiation was high-voltage electron accelerator, type Rhodotron 10 MeV–200 kW (Tongeren, Belgium). Magnitude and range of doses were set according to the industrial practice to 33, 66, 99, 132, 165, and 198 kGy. Since the thickness of modified samples was, at maximum, 10 mm, the one-sided irradiation method could be used (the penetration depth of β radiation was determined based on the density of irradiated material and energy of the electron beam) [3,36]. In order to prevent the induction of thermal stresses in the irradiated specimens, the exposure was performed in cycles (the specimens were exposed to 33 kGy each cycle). The magnitude of the radiation dose was measured by Nylon FTN 60-00 dosimeter (Goleta, CA, USA) [40]. The subsequent analysis of absorbed dose was performed by Genesys 5 spectrometer in accordance with ASTM 51261 [41].

2.3. Determination of Surface and Adhesive Properties

The influence of high-energy radiation on the wettability of injection-molded surfaces was determined and quantified by its free surface energy and the subsequent spectroscopic analysis that focused on the degree of oxidation. The influence of radiation on the adhesive properties of injection-molded surfaces was determined by change in the strength of bonded joints in shear and the type of failure.

2.3.1. Free Surface Energy

The determination of free surface energy of the tested specimens was performed by regression model Owens–Wendt–Rabel–Kaelble (OWRK) [5,40,42–46], which uses the values of the wetting contact angles. The individual values of wetting contact angles were measured by sessile drop method on SeeSystem device made by Advex Instruments (Brno, Czech Republic). Each wetting contact angle measurement was performed according to CSN EN 15802 [38]. Three reference liquids with varying surface tension (distilled water: 72.8 mJ/m^2, glycerine: 64 mJ/m^2, ethylene glycol: 48 mJ/m^2) were used [46]. Each specimen was measured 15 times for each reference liquid. Reference liquids' drops were applied on surface layer by micropipette (volume of each applied drop was 4 μL).

The analysis of drops profile (Figure 1) was used in following equations, which led to determination of wetting contact angles [5,40,47]:

$$h = R(1 - \cos\theta), \tag{1}$$

$$r_b = R\sin\theta, \tag{2}$$

and

$$\frac{h}{r_b} = \frac{1 - \cos\theta}{\sin\theta} = \tan\left(\frac{\theta}{2}\right). \tag{3}$$

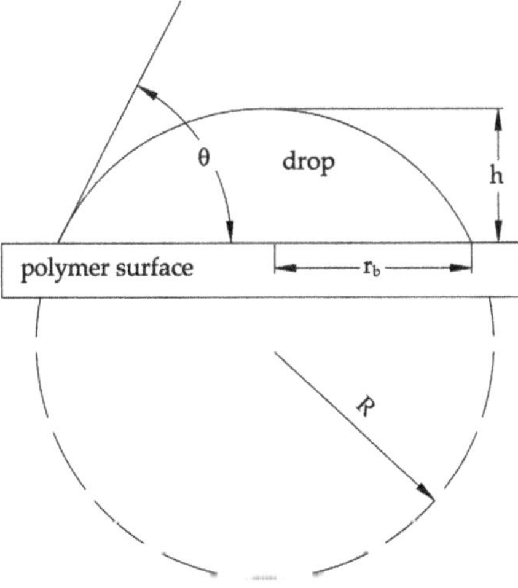

Figure 1. Droplet profile analysis: θ—wetting contact angle; h—droplet height, r_b—droplet radius at contact point; and R—entire droplet radius [5,40].

2.3.2. Fourier-Transform Infrared (FTIR) Spectroscopy

The degree of specimens' surface layer changes was investigated by ATR-FTIR spectroscopy. This was performed on Nicolet iS50 FTIR device (Thermo Scientific™, Waltham, MA, USA) [40] equipped with diamond ATR crystal. The spectra were recorded with $4\ cm^{-1}$ resolution and 60 scans, while the evaluation was performed by Omnic® software (version 9.3.32, Thermo Scientific™, Waltham, MA, USA). Specifically, the spectra from three different places of the sample were acquired, and their baseline was corrected by means of automatic algorithm of the Omnics software. From these spectra, an average spectrum of the sample was calculated.

2.3.3. Strength of Bonded Joints in Shear

The creation of bonded joints, including shape and dimensions of specimens (Figure 2), was performed in accordance with CSN EN 1456 [39]. The specimens were bonded by commercially available adhesive with cyanoacrylate basis PR100 manufactured by 3M (Saint Paul, MN, USA). A consistent thickness of the adhesive layer was ensured by spacers placed between the bonded samples. The shear strength of joint's bond was measured by testing on Zwick 1456 (ZwickRoell, Ulm, Germany) with crossbar velocity of 50 mm/min. The bonded joint was symmetrically placed in grips (distance between grips was (50 ± 1) mm). The spacers were used in order to ensure the point of force application was in bonded joint's plane. The measured data were evaluated by TestXpert® II software (version 2.1, ZwickRoell, Ulm, Germany).

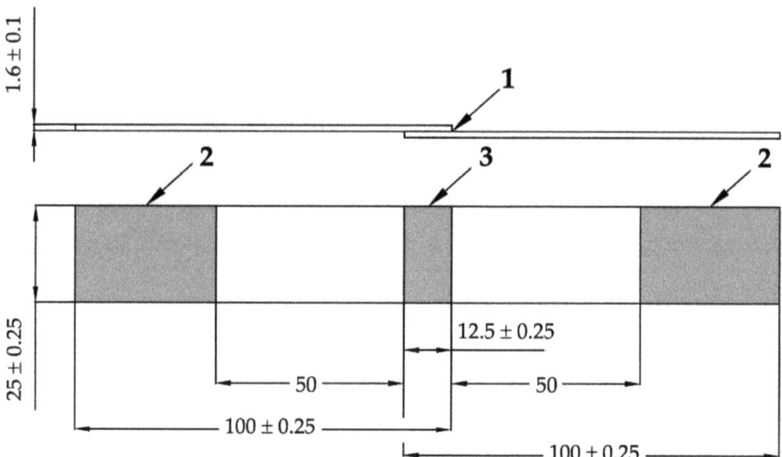

Figure 2. Bonded joint (dimensions in mm): (1) adhesive layer; (2) area of test machine grips; (3) shear area [39,40].

2.3.4. Analysis of Bonded Surfaces

The analysis of failure type of bonded joints was performed by evaluation of bonded joints' images after strength testing by optical profilometer NewView 8000 (Zygo, Middlefield, OH, USA). The profilometer used interferometric scanning method and its vertical variation. The measurement is based on phase shift (use of white and monochromatic light). Speed of vertical scanning was 96 μm/s, and Streamlined Mx™ software (version 8.0.0.33, Zygo, Middlefield, OH, USA) was used for processing of measured data.

2.4. Measurement of Mechanical Properties and Gel Content

The evaluation and quantification of the influence of high-energy radiation on the mechanical properties of tested specimens was performed by the investigation of tensile and bending strength, while the determination of cross-linked phase was performed by gel test. As injection-molded parts modified by high-energy radiation can be used in higher working temperatures, these properties were also tested under wide spectrum of working temperatures.

2.4.1. Mechanical Properties

The testing of mechanical properties was performed by tensile strength and three-point bending test on universal testing machine Zwick 1456 (ZwickRoell, Ulm, Germany) in accordance with CSN EN ISO 527-1 [35], CSN EN 527-2 [36], and CSN EN ISO 178 [37] standards. The working temperatures for tensile strength and three-point bending tests can be seen in Table 2. A temperature chamber W91255 (ZwickRoell, Ulm, Germany) was used to heat the test specimens (in the case of tests at elevated temperatures). The test speed was set to 50 mm/min (tensile strength) and 5 mm/min (bending test). Distance between the supports for three-point bending test was (64 ± 1) mm. As with the testing of bonded joint strength, the measured data were evaluated by TestXpert® II software (version 2.1, ZwickRoell, Ulm, Germany).

Table 2. The mechanical properties–working temperatures.

Material	Mechanical Properties (MPa)	Working Temperatures (°C)
HDPE	Tensile and Bending Strength	23, 30, 40, 50, 60, 70, 80
PA66		23, 50, 80, 110, 140, 170, 200

2.4.2. Gel Content (Degree of Cross-Linking)

The determination of degree of cross-linking was performed by gel test in accordance with ASTM D2765 standard–Test Method C [48]. In this case, a 0.5 g of solid sample-cut from the whole modified specimen (measured to five decimal numbers) was mixed with 100 mL of solvent, i.e., xylene. The solvent dissolved amorphous part of tested polymers, while the cross-linked part remained undissolved. The mixture was boiled for 24 h. After that, the gel and the dissolved phase were separated. After separation, the gel (undissolved cross-linked part) was dried for 8 h in vacuum at 100 °C. Dried remnant of sample was once again weighted to five decimal numbers and compared with original weight. Degree of cross-linking was then determined from following equation [48,49]:

$$G_i = \frac{m_3 - m_1}{m_2 - m_1} \cdot 100, \tag{4}$$

in which G_i represents the degree of cross-linking of tested sample (in percent); m_1 is the weight of equipment (in milligrams); m_2 is the overall weight of original sample and equipment (in milligrams), and m_3 is the overall weight of sample's remnant and equipment.

3. Results

The individual measurements were performed 15 times at atmospheric conditions and room temperature (23 °C). Based on measured data, a suitable regression model together with parameters was prepared to describe the influence of radiation dose on observed characteristics. The models were designed with the help of the following software: Minitab®17 (Minitab Inc., State College, PA, USA), QC-Expert 3.3 (TriloByte, Staré Hradiště, Czech Republic). The designed model has the following form:

$$y = b_0 + b_1 x + b_2 x^2 \tag{5}$$

where y is the observed characteristic, x is the radiation dose (kGy), and b_0, b_1, b_2 are the estimates of regression parameters.

During the search for a regression model, a test of statistical significance was performed. The output of regression model testing was the "rejection of hypothesis of insignificance". Furthermore, a calculation of "predicted correlation coefficient" and "median quadratic error of prediction" was performed with the intent to estimate the regression parameters and find the type of regression function on confidential level $1 - \alpha = 0.9$, i.e., $\alpha = 0.1$. This step, i.e., choice of $\alpha = 0.1$, was necessary to correctly process the results, as due to the difficult preparation of specimens, a certain noise in the data must be assumed. At the end of data processing, a regression triplet was tested [50]. The subsequent spatial display of designed regression models was performed by OriginPro® 2023 software (version 10.0, OriginLab, Northhampton, MA, USA).

3.1. Surface and Adhesive Properties

The effect of β radiation on the surface properties of tested materials was evaluated by the free surface energy since previously submitted studies [5,7,40] indicate that a high value of the free surface energy is the key factor for the good wettability of surfaces and the subsequent creation of quality adhesive bonds, which significantly affect the load capacity of bonded joints. For this reason, the load capacity of bonded joints was chosen as demonstrative test of practical application of β radiation effect on the adhesive properties.

The models' designed parameters for the description of the change in free surface energy and strength of the bonded join in dependence on absorbed radiation dose can be seen in Table 3, while individual models are displayed in Figure 3. The characteristics of designed regression models are shown in Table 4.

Table 3. The estimations of regression parameters–surface and adhesive properties.

Material	Tested Parameter	Estimations of Regression Parameters		
		b_0	b_1	b_2
HDPE	Free Surface Energy (mJ/m^2)	2.457×10^1	1.675×10^{-1}	-5.010×10^{-4}
	Load-Bearing of Adhered Joints (MPa)	5.324×10^{-1}	8.312×10^{-3}	-2.700×10^{-5}
PA66	Free Surface Energy (mJ/m^2)	3.170×10^1	7.673×10^{-2}	-3.553×10^{-4}
	Load-Bearing of Adhered Joints (MPa)	9.508×10^0	4.459×10^{-3}	-2.300×10^{-5}

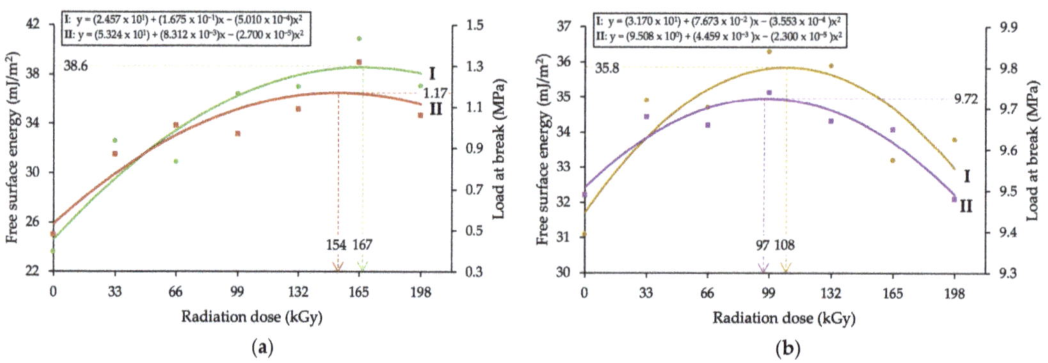

Figure 3. Proposed regression models–the effect of radiation dose on surface and adhesion properties: (**aI**) HDPE, free surface energy; (**aII**) HDPE, load-bearing of adhered joints; (**bI**) PA66, free surface energy; (**bII**) PA66, load-bearing of adhered joints.

The results shown in Table 3 and Figure 3 demonstrate that β radiation influences the value of free surface energy. Out of both tested polymers, a higher growth of free surface energy was found in HDPE, in which the value rose from 24.6 mJ/m^2 to 38.6 mJ/m^2, which was a total increase of 57%. In the case of PA66, the growth was 13%, which was not as impressive. The increase in free surface energy influenced adhesive properties of tested polymers, which resulted in an increase in the shear strength in bonded joints.

Table 4. The characteristics of designed regression models–surface and adhesive properties.

Parameters	HDPE		PA66	
	Free Surface Energy	Load-Bearing of Adhered Joints	Free Surface Energy	Load-Bearing of Adhered Joints
Coefficient of Multiple Correlation	9.352×10^{-1}	9.223×10^{-1}	8.579×10^{-1}	9.424×10^{-1}
Coefficient of Determination	8.745×10^{-1}	8.507×10^{-1}	7.360×10^{-1}	8.881×10^{-1}
Predicted Correlation Coefficient	3.503×10^{-1}	2.188×10^{-1}	1.626×10^{-1}	4.422×10^{-1}
Mean Squared Error of Prediction	1.117×10^{1}	3.016×10^{-2}	3.763×10^{0}	2.842×10^{-3}
Testing of Regression Triplet				
Fisher–Snedecor Test of Model Significance	model is significant			
Scott's Criteria of Multicollinearity	model is correct			
Cook–Weisberg Score Test for Heteroskedasticity	residue demonstrating homoskedasticity			
Jarque–Berra Test of Normality	residue has normal distribution			
Wald Test of Auto Correlation	autocorrelation is insignificant			
Durbin–Watson Test of Auto Correlation	negative autocorrelation of residues not demonstrated			

The shear strength of HDPE modified by radiation rose from 0.53 MPa to 1.17 MPa, which was an increase of 120%. For PA66, the growth of shear strength was significantly lower, approximately 3%. The highest growth for both polymers and each observed characteristic was found in materials irradiated by higher doses of radiation, specifically the doses higher than 132 kGy for HDPE and the doses in the range of 99 to 132 kGy for PA66.

Table 4 shows the characteristics of designed regression models designated for the description of changes in the surface and adhesive properties of tested polymers modified by the β radiation. As can be interpreted from the results, the designed models for the description of β radiation effect on the magnitude of free surface energy and the strength of bonded joints are significant and correct. The residues demonstrate homoskedasticity, a normal distribution, and insignificant autocorrelation.

Besides the growth of the strength of bonded joints and adhesive properties, the effect of the radiation modification can also be observed in the change in failure type in the bonded joint. Figures 4 and 5 show images of bonded surfaces after the strength test. For HDPE, the non-modified surfaces generally experienced failure on the phase interface of adherent/adhesive (Figure 4a). Due to irradiation, adhesive properties were improved, which also impacted the change in failure type. After the modification by β radiation, a mixed failure type was observed in HDPE, i.e., a combination of adhesive and cohesive failure (Figure 4b). Figure 4b displays the bonded surface (after shear strength testing), which was modified by β radiation. The blue and red areas indicate assumed adhesive failure, while green and yellow areas indicate cohesive failure in a layer of adhesive (height of adhesive was 80 μm). In the case of PA66, it is difficult to unequivocally determine the type of failure (Figure 5a,b). However, due to irradiation, the change in the topography of the bonded surface occurred in both HDPE (Figure 4c,d) and PA66 (Figure 5c,d).

As can be seen in Figure 4a,c, the non-modified surface has a distribution of Z coordinates quite close to the standard normal distribution with parameters $\mu = 0$ and $\sigma = 1$. This is given not only by the normality test, according to Anderson–Darling, which has not refused normality, but also from the shape of the histogram for the Z coordinate. On the other hand, Figure 4b,d shows the surface of the specimen modified by the β radiation, which shows a significant breach of normality of the Z coordinate. The Anderson–Darling test showed that the normality of the Z coordinate was refused. This can also be interpreted from the asymmetrical shape of a given histogram.

As can be seen in Figure 5a,c, the non-modified surface demonstrates the Z coordinate distribution that is very close to the standard normal distribution with parameters $\mu = 0$ and $\sigma = 1$. The coordinate was once again tested by the Anderson–Darling test, which has not refused normality. However, it is also possible to say that the Z coordinate was slightly sloped. This observation does not appear significant. The surface in Figure 5b,d, which was modified by β radiation, does not demonstrate a breach of normality of the Z coordinate.

The repeated application of the Anderson–Darling test did not lead to the refusal of the Z coordinate.

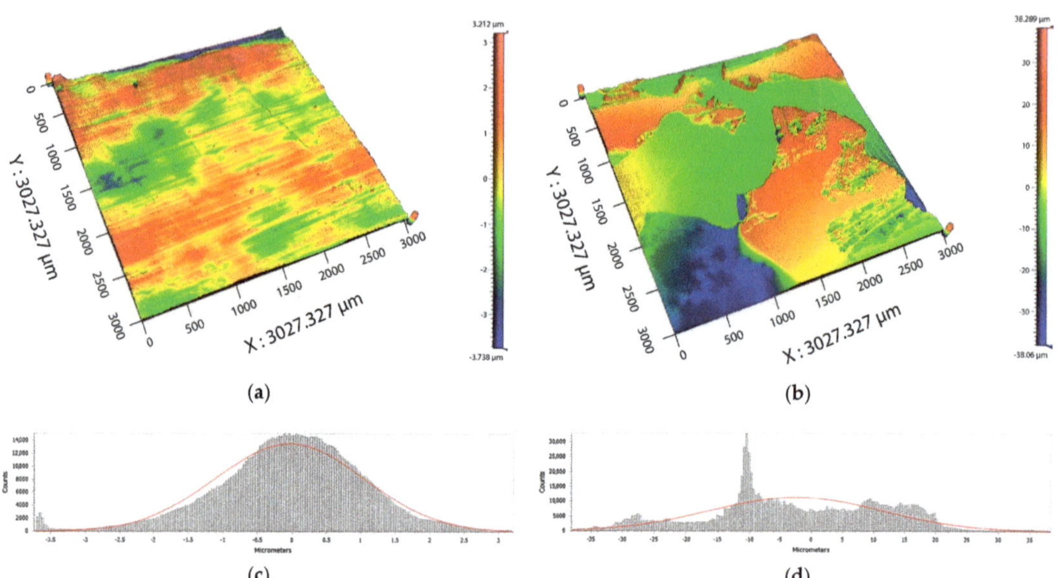

Figure 4. The bonded surface for HDPE after shear strength test: (**a**) Image of original surface; (**b**) image of surface modified by β radiation; (**c**) topography of non-modified surface; (**d**) topography of surface modified by β radiation.

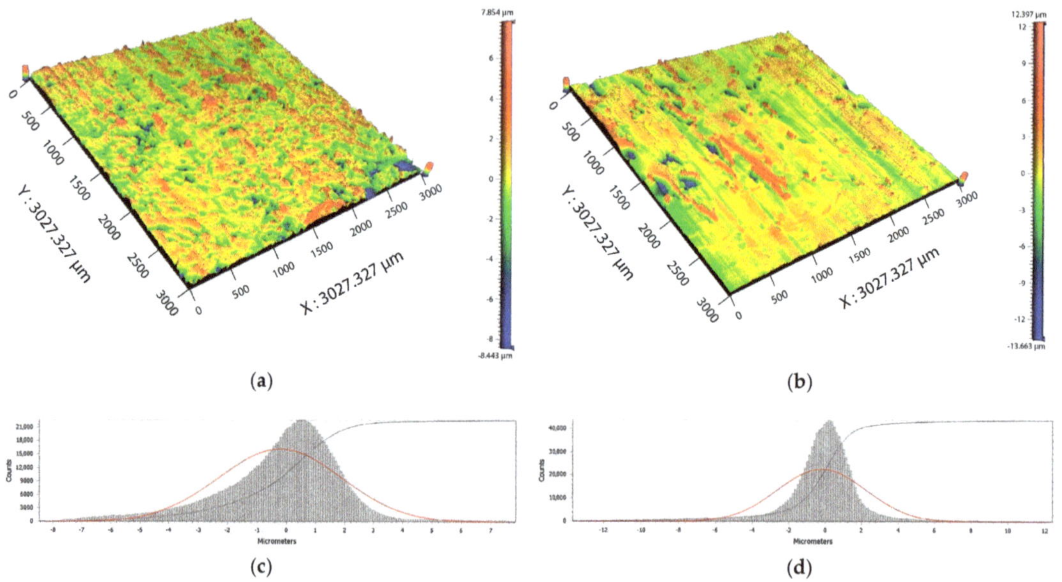

Figure 5. The bonded surface for PA66 after shear strength test: (**a**) Image of original surface; (**b**) image of surface modified by β radiation; (**c**) topography of non-modified surface; (**d**) topography of surface modified by β radiation.

Figures 6 and 7 show the results from infrared spectroscopy. The evaluation of spectra of the non-modified HDPE (Figure 6a) and the material modified by the β radiation demonstrated the characteristic absorption bands in the range of 1680 cm^{-1} to 1740 cm^{-1}, which confirms the expected creation of functional carbonyl groups in tested material (Figure 6b). These results correspond with an earlier study [5], which focused on the change in surface layer properties, including the oxidation and relative representation of carbonyl and hydroxyl functional groups, dependent upon the absorbed dose of high-energy radiation.

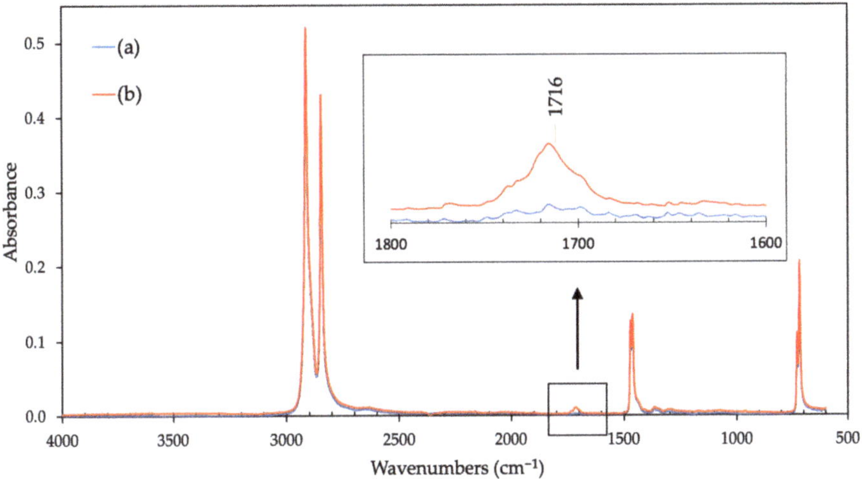

Figure 6. The infrared spectra: (a) HDPE, untreated; (b) HDPE, after β radiation treatment.

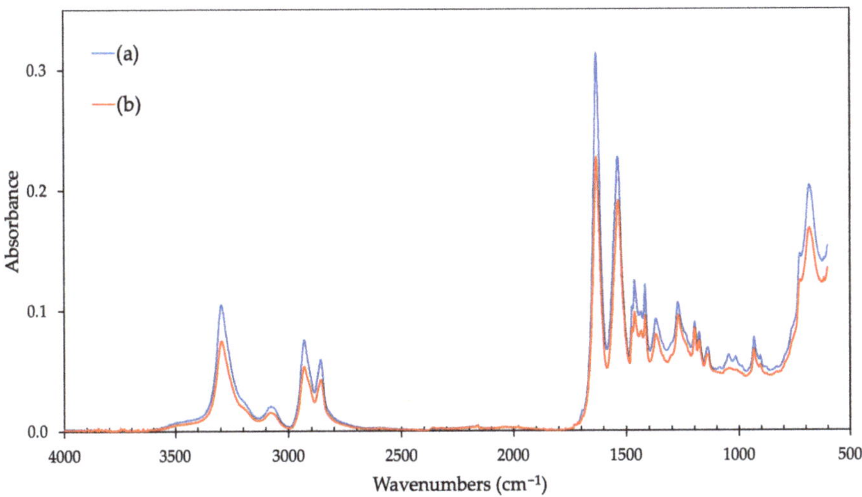

Figure 7. The infrared spectra: (a) PA66, untreated; (b) PA66, after β radiation treatment.

In the case of PA66 modified by the β radiation, there are some notable changes between the spectra of the untreated and irradiated samples (Figure 7). Specifically, a decreased intensity of the peak area at 1684 cm^{-1}, lower ratio of Amide I (1632 cm^{-1}) to Amide II (1537 cm^{-1}) respective bands, and changes in the absorption bands between 1060 and 1000 cm^{-1} can be observed in the spectrum of the irradiated sample (Figure 7b).

This indicates the reduction of C=O groups in the polymer chain and possible conformational changes.

3.2. Mechanical Properties

The changes in mechanical properties induced by the β radiation were determined by tensile and bending strength tests. Both tested characteristics were measured in a wide spectrum of working temperatures (see Table 2). The designed parameters of the regression model designated for the description of changes in tensile and bending strength dependent upon the absorbed dose can be seen in Tables 5 and 6.

Table 5. The estimation of regression parameters–mechanical properties (HDPE).

Tested Property (MPa)	Working Temperature (°C)	Estimations of Regression Parameters		
		b_0	b_1	b_2
Tensile Strength	23	2.189×10^1	4.892×10^{-2}	-1.727×10^{-4}
	30	1.965×10^1	3.950×10^{-2}	-1.520×10^{-4}
	40	1.674×10^1	5.195×10^{-2}	-1.968×10^{-4}
	50	1.406×10^1	5.952×10^{-2}	-2.230×10^{-4}
	60	1.267×10^1	3.409×10^{-2}	-1.432×10^{-4}
	70	1.082×10^1	2.835×10^{-2}	-1.006×10^{-4}
	80	9.150×10^0	2.165×10^{-2}	-8.527×10^{-5}
Bending Strength	23	2.642×10^1	8.323×10^{-2}	-3.290×10^{-4}
	30	2.183×10^1	5.530×10^{-2}	-2.285×10^{-4}
	40	1.769×10^1	5.519×10^{-2}	-2.033×10^{-4}
	50	1.410×10^1	2.781×10^{-2}	-9.073×10^{-5}
	60	1.071×10^1	4.004×10^{-2}	-1.749×10^{-4}
	70	8.898×10^0	2.597×10^{-2}	-1.028×10^{-4}
	80	7.386×10^0	3.517×10^{-2}	-1.541×10^{-4}

Table 6. The estimation of regression parameters–mechanical properties (PA66).

Tested Property (MPa)	Working Temperature (°C)	Estimations of Regression Parameters		
		b_0	b_1	b_2
Tensile strength	23	1.669×10^2	2.908×10^{-1}	-9.806×10^{-4}
	50	1.262×10^2	1.735×10^{-1}	-5.302×10^{-4}
	80	1.056×10^2	8.377×10^{-2}	-3.280×10^{-4}
	110	8.970×10^1	9.665×10^{-2}	-3.903×10^{-4}
	140	7.853×10^1	2.489×10^{-2}	-2.274×10^{-4}
	170	6.714×10^1	2.771×10^{-2}	-7.871×10^{-5}
	200	5.525×10^1	2.435×10^{-2}	-7.980×10^{-5}
Bending strength	23	1.889×10^2	9.491×10^{-1}	-3.594×10^{-3}
	50	1.651×10^2	4.370×10^{-1}	-1.675×10^{-3}
	80	1.385×10^2	2.314×10^{-1}	-9.030×10^{-4}
	110	1.192×10^2	6.753×10^{-2}	-1.968×10^{-4}
	140	1.032×10^2	7.727×10^{-2}	-1.793×10^{-4}
	170	8.870×10^1	5.942×10^{-2}	-2.241×10^{-4}
	200	7.257×10^1	1.054×10^{-1}	-3.170×10^{-4}

A spatial portrayal of designed regression models with their respective areas, which characterize the degree of influence of radiation on given characteristics (high, medium, and low influence), can be seen in Figures 8 and 9, while examples of bending curves are presented in Figure 10. The characteristics of designed regression models are shown in Tables 7 and 8.

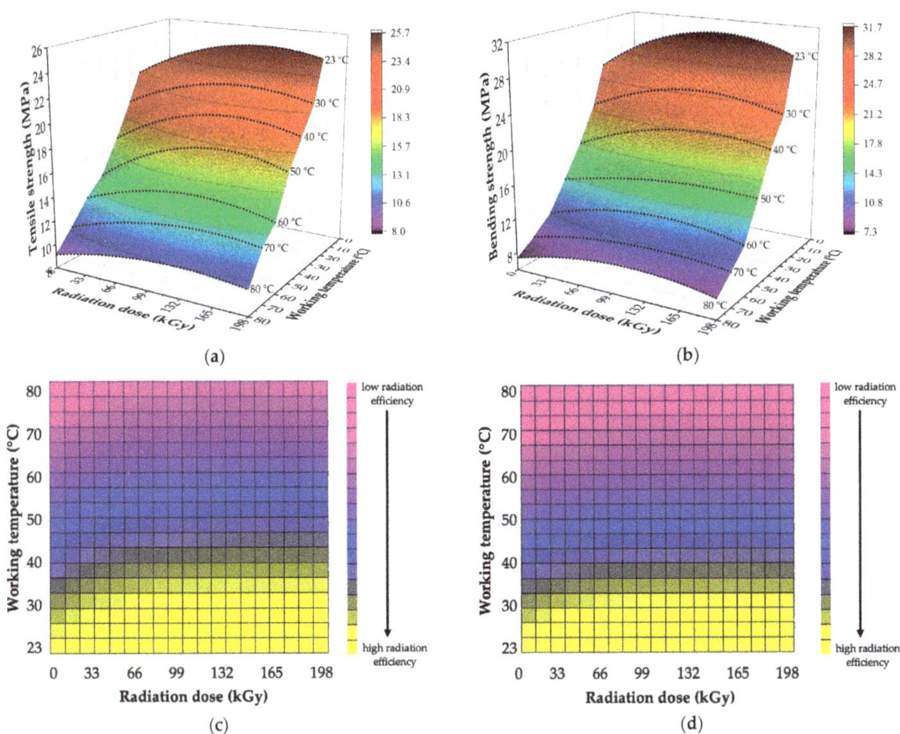

Figure 8. Mechanical properties of HDPE: (**a**) spatial portrayal of regression models (tensile strength); (**b**) spatial description of regression models (bending strength); (**c**) responsive area (tensile strength); (**d**) responsive area (bending strength).

Table 7. The characteristics of designed regression models (HDPE).

Parameters		Working Temperature (°C)						
		23	30	40	50	60	70	80
Coefficient of Multiple Correlation	Tensile strength	9.099×10^{-1}	9.187×10^{-1}	8.646×10^{-1}	8.810×10^{-1}	8.590×10^{-1}	9.297×10^{-1}	8.524×10^{-1}
	Bending strength	9.121×10^{-1}	9.266×10^{-1}	8.771×10^{-1}	9.445×10^{-1}	8.315×10^{-1}	8.631×10^{-1}	8.756×10^{-1}
Coefficient of Determination	Tensile strength	8.279×10^{-1}	8.439×10^{-1}	7.475×10^{-1}	7.762×10^{-1}	7.378×10^{-1}	8.643×10^{-1}	7.265×10^{-1}
	Bending strength	8.318×10^{-1}	8.586×10^{-1}	7.693×10^{-1}	8.922×10^{-1}	6.913×10^{-1}	7.449×10^{-1}	7.666×10^{-1}
Predicted Correlation Coefficient	Tensile strength	2.090×10^{-2}	4.271×10^{-3}	3.806×10^{-1}	7.883×10^{-2}	2.528×10^{-1}	1.113×10^{-1}	7.581×10^{-1}
	Bending strength	3.031×10^{-2}	1.986×10^{-2}	2.075×10^{-2}	4.968×10^{-2}	6.614×10^{-1}	3.936×10^{-1}	2.980×10^{-2}
Mean Squared Error of Prediction	Tensile strength	1.892×10^{0}	9.023×10^{-1}	2.781×10^{0}	2.865×10^{0}	8.863×10^{-1}	3.503×10^{-1}	5.207×10^{-1}
	Bending strength	4.183×10^{0}	1.185×10^{0}	2.320×10^{0}	4.701×10^{-1}	1.476×10^{0}	6.297×10^{-1}	4.667×10^{-1}

Testing of Regression Triplet

Fisher–Snedecor Test of Model Significance	model is significant
Scott's Criteria of Multicollinearity	model is correct
Cook–Weisberg Score Test for Heteroskedasticity	residue demonstrating homoskedasticity
Jarque–Berra Test of Normality	residue has normal distribution
Wald Test of Auto Correlation	autocorrelation is insignificant
Durbin–Watson Test of Auto Correlation	negative autocorrelation of residues not demonstrated

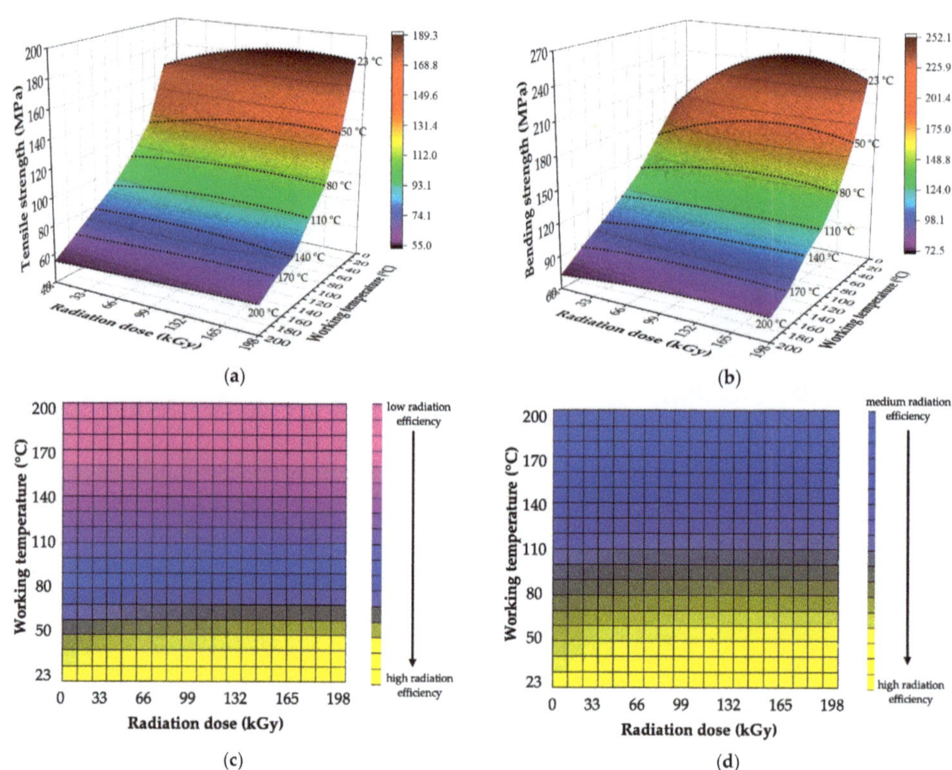

Figure 9. Mechanical properties of PA66: (**a**) spatial portrayal of regression models (tensile strength); (**b**) spatial description of regression models (bending strength); (**c**) responsive area (tensile strength); (**d**) responsive area (bending strength).

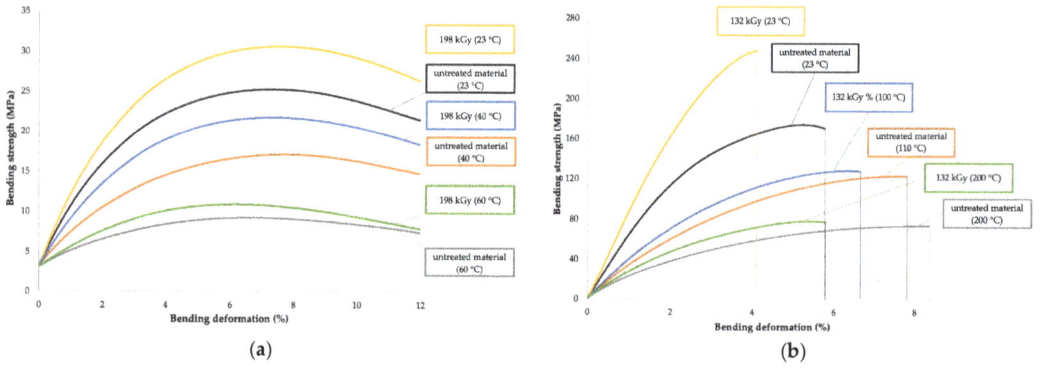

Figure 10. The dependence of bending stress on deformation: (**a**) HDPE; (**b**) PA66.

The results displayed in Table 5, Figures 8 and 10 show that β radiation increased the tensile and bending strength in HDPE for a wide spectrum of working temperatures. In the case of the tensile strength, the increase was in the range of 13 to 28%, depending on the applied radiation dose and working temperature (Table 5 and Figure 8a). For example, at ambient temperature, the tensile strength grew by 3.5 MPa, while only a 1.4 MPa increase was measured for tests conducted at 80 °C. Regarding the bending strength, the growth was in the range of 15 to 27%, dependent on the radiation dose and the working temperature (Table 5, Figures 8b and 10a). For example, at ambient temperature, the bending strength grew by 5.2 MPa, while at the highest working temperature, the growth was only 2.1 MPa. On responsive areas (Figure 8c,d), the effect of radiation on the tensile and bending strength can be observed for concrete working temperatures. Both tested characteristics revealed a medium-to-high effect of radiation for working temperatures up to 60 °C (yellow and blue area). The effect of radiation continually decreased for working temperatures over 60 °C (pink area).

Table 8. The characteristics of designed regression models (PA66).

Parameters		Working Temperature (°C)						
		23	50	80	110	140	170	200
Coefficient of Multiple Correlation	Tensile strength	9.053×10^{-1}	9.478×10^{-1}	9.311×10^{-1}	8.935×10^{-1}	9.772×10^{-1}	9.721×10^{-1}	9.468×10^{-1}
	Bending strength	8.888×10^{-1}	8.509×10^{-1}	8.679×10^{-1}	9.759×10^{-1}	9.801×10^{-1}	8.549×10^{-1}	9.559×10^{-1}
Coefficient of Determination	Tensile strength	8.196×10^{-1}	8.983×10^{-1}	8.670×10^{-1}	7.984×10^{-1}	9.548×10^{-1}	9.449×10^{-1}	8.964×10^{-1}
	Bending strength	7.900×10^{-1}	7.240×10^{-1}	7.532×10^{-1}	9.523×10^{-1}	9.605×10^{-1}	7.308×10^{-1}	9.138×10^{-1}
Predicted Correlation Coefficient	Tensile strength	2.192×10^{-2}	3.816×10^{-1}	1.299×10^{-2}	3.082×10^{-2}	6.202×10^{-1}	5.790×10^{-1}	3.429×10^{-1}
	Bending strength	7.791×10^{-2}	9.397×10^{-2}	2.320×10^{-1}	7.983×10^{-1}	7.307×10^{-1}	1.908×10^{-2}	1.739×10^{-1}
Mean Squared Error of Prediction	Tensile strength	7.616×10^{1}	1.040×10^{1}	3.143×10^{0}	5.599×10^{0}	5.565×10^{-1}	1.842×10^{-1}	1.891×10^{-1}
	Bending strength	6.953×10^{2}	1.594×10^{2}	4.653×10^{1}	4.594×10^{-1}	1.218×10^{0}	2.006×10^{0}	5.965×10^{0}
Testing of regression triplet								
Fisher–Snedecor Test of Model Significance		model is significant						
Scott's Criteria of Multicollinearity		model is correct						
Cook–Weisberg Score Test for Heteroskedasticity		residue demonstrating homoskedasticity						
Jarque–Berra Test of Normality		residue has normal distribution						
Wald Test of Auto Correlation		autocorrelation is insignificant						
Durbin–Watson Test of Auto Correlation		negative autocorrelation of residues not demonstrated						

The effect of β radiation on the tensile and bending strength of PA66 can be seen in Table 6 and Figures 9 and 10. In the case of tensile strength, the irradiation led to an increase of up to 13%, dependent on the radiation dose and working temperature (Table 6 and Figure 9a). At ambient temperature, the tensile strength grew by 21.6 MPa, while at the highest temperature, the same characteristic rose only by 4.8 MPa. The increase in bending strength was up to 33%, dependent on the radiation dose and working temperature (Table 6, Figures 9b and 10b). At the lowest tested temperature, the growth induced by radiation was 62.7 MPa, while at the highest tested temperature, it was 8.8 MPa. The responsive area (Figure 9c) displays that for tensile strength, the effect of radiation was medium or high for working temperatures up to 130 °C (yellow and blue area). For working temperatures over 130 °C, the effect of irradiation continually decreased (pink area). In the case of the bending strength (Figure 9d), the effect of radiation was medium or high up to the highest working temperature, i.e., 200 °C (yellow and blue area).

The characteristics of designed regression models for the description of changes in the mechanical properties of tested polymers induced by the β radiation can be seen in Tables 7 and 8. The interpretation of measured results shows that the designed models for the description of β radiation's influence on the magnitude of tensile and bending strength are significant and correct. The residues demonstrate homoskedasticity and normal distribution, and the autocorrelation is insignificant

4. Discussion

This study is primarily focused on a quantitative description of changes in useful properties in injection-molded parts due to irradiation by β radiation. Specimens were manufactured out of one representative for the commodity plastics (HDPE) and one representative for the technical plastics (PA66). First, the properties of the surface layer, such as free surface energy and adhesion, were tested. The measured results indicate that the β radiation influences the properties of the surface layer (Table 3 and Figure 3). In the case of free surface energy in HDPE, the highest growth was seen in specimens irradiated by higher doses of radiation (more than 132 kGy) (Table 3 and Figure 3a). On the other hand, for PA66, the highest growth was observed in specimens irradiated by the medium intensity of radiation, i.e., 99 to 132 kGy (Table 3 and Figure 3b). The increase in free surface energy (in comparison with the non-modified material) was up to 57% for HDPE and up to 13% for PA66. The change in free surface energy value had a significant effect on the adhesive properties of tested materials. The practical effect of β radiation on adhesive properties was tested by evaluation of the shear strength of bonded joints. Figure 3 demonstrates that the β radiation modification influenced the shear strength of bonded joints for both HDPE and PA66. In specific cases, the strength of the bonded joint rose by 120% (Table 3 and Figure 3).

The presented changes in terms of free surface energy and adhesive properties taken in the context of the strength of bonded joints were most likely caused by oxidation, which could have occurred during the irradiation process or after it. Oxidation is one of many secondary reactions which can occur when β radiation interacts with polymers. A significant factor of this study is that the process of specimen modification was the same as with common industrial applications performed at ambient temperature and in standard atmospheric conditions (with oxygen). In this case, the free radicals created due to irradiation could easily react with oxygen molecules, which results in the creation of peroxide radicals that could lead to oxidation. Moreover, it is the oxygen and humidity which help oxidation reactions (oxidation, oxidative scission). As stated by Rivaton et al. [51], most post-radiation effects are driven by the migration of radicals from crystalline areas to amorphous/crystalline mesophases, where the radicals remain more accessible for oxygen. As mentioned previously, oxidation can occur during the irradiation process itself (contemporaneously with cross-linking and scission) or after it if the polymer can react with oxygen. This conclusion is supported by the results of infrared spectra (Figure 6) for HDPE and corresponds with conclusions reached by Hama et al. [30], Carpentieri et al. [52], and Costa et al. [53].

The oxidation usually occurs together with oxidative degradation, and the combination of these effects can cause a different color (yellow) and fragility (or decrease in other mechanical properties) in the materials; thus, it is best to avoid these interactions [20,21]. However, new functional groups (carbonyl and others) can enhance the surface with new properties which can find their application in practice. Among these properties are adhesion, the increase in polarity, and others, which all positively affect the load capability of bonded joints (Figure 3). The change in adhesive properties is shown in Figures 4 and 5, which both display bonded surfaces after shear strength testing. The recorded changes were most significant in HDPE (Figure 4), in which the non-modified surface displayed mostly adhesive failure, i.e., the failure occurred at the adhesive/adherent interface (Figure 4a). After the modification by β radiation, the failure changed to combined, i.e., a combination of adhesive and cohesive failure (Figure 4b). Besides the change in the type of failure, the irradiation also led to a change in the topography of bonded surfaces (Figure 4c,d). The change in the type of bond failure corresponds with improved adhesive bonds. In the case of PA66 (Figure 5a,b), the type of failure was not unequivocally evident. On the other hand, irradiation definitely led to changes in the topography of the bonded surface (Figure 5c,d).

The second tested group focused on mechanical properties, such as tensile and bending strength. Measured results indicate that β radiation influences even the mechanical properties in a wide spectrum of working temperatures (Tables 5 and 6, Figures 8–10). In both cases, the characteristics increased due to irradiation for both tested materials.

For tensile strength, the growth was almost 28% (Figures 8a and 9a) in comparison with non-altered material. For bending strength, this increase was even greater, up to 33% (Figures 8b, 9b and 10). The changes in mechanical properties induced by the β radiation correspond with the content of the cross-linked phase (gel), which also rose with increasing radiation dose (Figure 11). Furthermore, these results correspond with the findings of other authors [18,33,34,54,55], who focused on the effects of high-energy radiation on the mechanical properties of polymers. The increasing content of gel induced by radiation and its influence on mechanical properties was also detected by Lee et al. [22], who recorded an increase in tensile strength together with the rising content of gel, which was caused by radiation. The aforementioned changes (improvements) of mechanical properties were caused above all by cross-linking, which is one of the secondary processes that occur in polymer materials (prevalently in amorphous areas [56]) due to β irradiation [33,34,54,55,57].

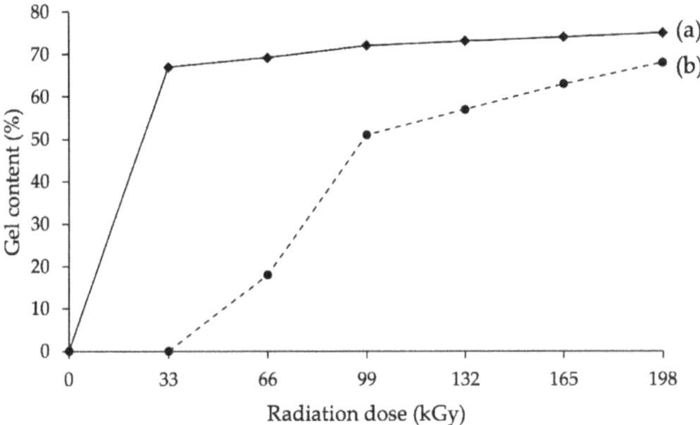

Figure 11. The effect of radiation on gel content. (a) PA66, (b) HDPE.

When the tested polymers were exposed to radiation, C-H bonds started to scission (release of the hydrogen atom), which led to the creation of free radicals. Afterward, gradual bonding (creation of C-C bond) of two free radicals of neighboring chains commenced. In the end, a 3D spatial network was created in which the polymer chains were interconnected. This spatial network was the bearer of improvements in the mechanical properties of tested polymers. The creation of a spatial network is proved by the aforementioned results of gel content (insoluble phase) increase due to radiation (Figure 11). In the case of PA66 with 30 wt. % of glass fibers, the irradiation incurred improvement of adhesion between individual fibers, which resulted in the increase in mechanical properties [18,34,55].

The description of changes in properties of tested materials due to irradiation was performed by suitably designed regression models. All regression models were tested with regression triplet. The regression models were significant and correct; residues demonstrated homoskedasticity and normal distribution, and autocorrelation was insignificant. The designed regression models showed that the highest growth of the given parameter was reached in specimens exposed to radiation dose lower than 198 kGy, i.e., after the maximum was reached, the observed parameter decreased with increasing radiation dose (Figures 3, 8a,b and 9a,b). This course was also noted in studies of other authors [58], who used the second-degree polynomial equation (quadratic polynomial) to describe the changes in properties of specimens exposed to high-energy radiation. This effect can be explained by parameter G, which is commonly used in practice to determine the reactions ongoing in the material during the irradiation. As presented by Makuuchi, Cheng [20], and Drobny [21], parameter G can be defined as the chemical gain of radiation in dependence on the number of reacting molecules per 100 eV of absorbed energy. At a certain point, the regression curve (extreme of function) experiences a breakpoint, following which the degradation reactions start to prevail over cross-linking, which results in a decline of useful properties and an overall decline in the regression curve. In the case of free surface energy and adhesive properties (Figure 3), a gradual increase in radiation dose led to a point where chain scission started to prevail over cross-linking. At this point, even the chains that cross-linked started to undergo scission, and it resulted in a decrease in free surface energy and an increase in the hydrophobic nature of the surface. This corresponds with findings presented by Egghe et al. [59]. This is also true for mechanical properties which started to decrease with the increasing radiation dose after reaching a certain breakpoint (specific radiation dose) (Figures 8 and 9) despite the continuous growth of gel content. This effect was recorded in studies of Gheysari and Behjat [23,60]. Although the exposure to higher radiation doses led to a gradual increase in the cross-linking phase, the degradation processes started to prevail at a certain point. This likely caused the decrease in the quality of the created spatial network and, thus, the decline of useful properties.

The designed regression models can be used to find a suitable (optimal) dose of radiation, which, when applied, leads to the best results in both the surface properties and the mechanical properties (Figures 12 and 13). The optimal dose was determined by standard parameters (by transformed data), which were calculated from specific values subtracted by the average value and then divided by standard deviation. Figures 12 and 13 show optimal values which were close to the extremes of functions. In the case of HDPE, the optimal dose was in the range of 145 to 150 kGy, depending on the combination of surface and mechanical properties (Figure 12). The optimal dose for PA66 was found in the range of 128 to 135 kGy, depending on the combination of surface and mechanical properties (Figure 13).

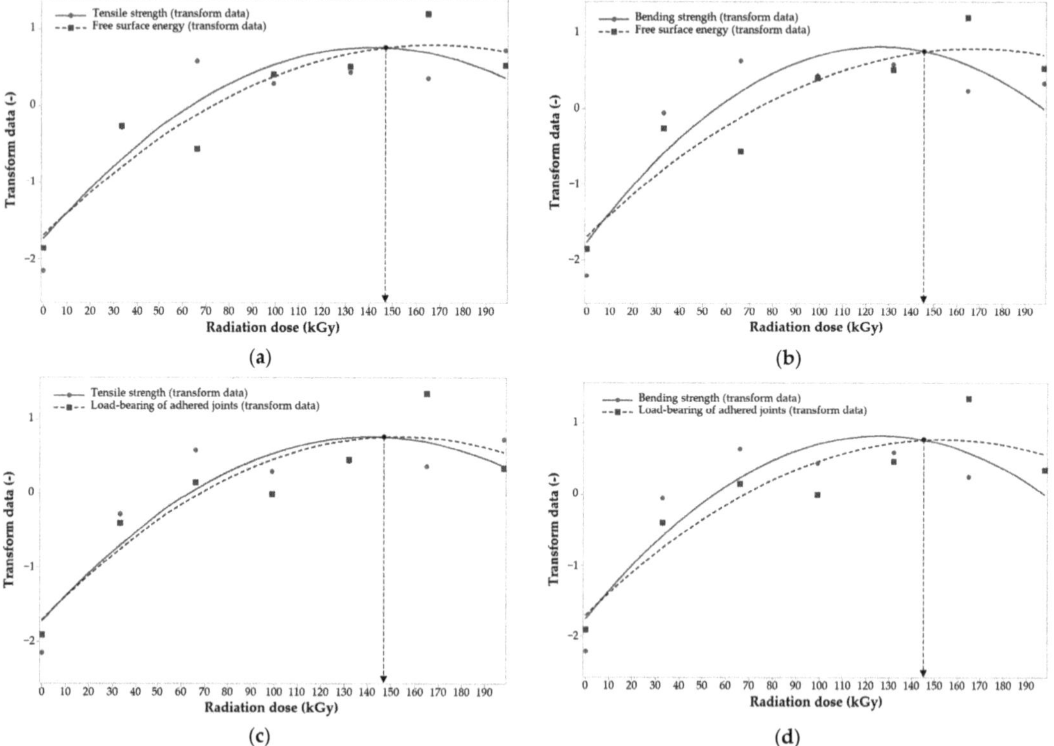

Figure 12. The optimal dose for HDPE: (**a**) tensile strength/free surface energy; (**b**) bending strength/free surface energy; (**c**) tensile strength/load-bearing of adhered joints; (**d**) bending strength/load-bearing of adhered joints.

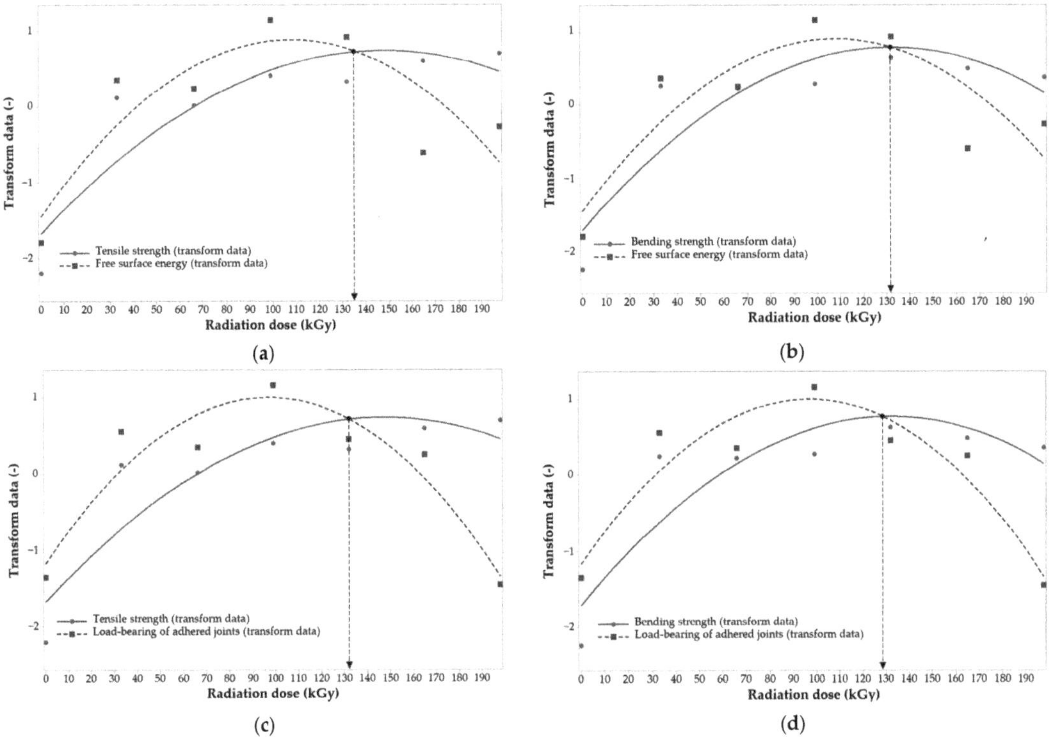

Figure 13. The optimal dose for PA66: (**a**) tensile strength/free surface energy; (**b**) bending strength/free surface energy; (**c**) tensile strength/load-bearing of adhered joints; (**d**) bending strength/load-bearing of adhered joints.

5. Conclusions

Tests and measurements performed in this study lead to the following conclusions:

- The modification of polymer materials by β radiation leads to the improvement of useful properties of injection-molded parts.
- Well-chosen doses of radiation can lead to improvement of both the mechanical and the surface properties.
- The designed regression models can be used as a suitable tool for choosing the optimal dose of radiation in terms of the required properties of the given part and its application in a specific working environment.

Future research in this area should focus on the effects of very low doses of radiation (in the range of 0 to 20 kGy) on the useful properties of injection-molded parts and the stability of gained properties, which is specifically true for surface layer properties.

Author Contributions: Conceptualization, M.B., V.P. and M.M.; methodology, A.M.; validation, M.O., J.H. (Jakub Husar) and J.H. (Jan Hanzlik); formal analysis, M.K.; investigation, M.O.; resources, J.H. (Jan Hanzlik); data curation, V.P.; writing—original draft preparation, M.B.; writing—review and editing, M.B., A.M. and M.M.; visualization, M.O.; supervision, M.B. and M.M.; project administration, A.M. All authors have read and agreed to the published version of the manuscript.

Funding: This research was funded by the Internal Grant Agency of Tomas Bata University in Zlin, supported under project No. IGA/FT/2024/003 and IGA/CebiaTech/2024/002.

Institutional Review Board Statement: Not applicable.

Data Availability Statement: The data presented in this study are available upon request from the corresponding author.

Acknowledgments: We would like to thank Michal Danek from BGS GmbH & Co. KG for professional consultation and technical support in modifying test specimens by ionizing radiation.

Conflicts of Interest: The authors declare no conflicts of interest.

References

1. Dua, R.; Rashad, Z.; Spears, J.; Dunn, G.; Maxwell, M. Applications of 3D-printed PEEK via fused filament fabrication: A Systematic review. *Polymers* **2021**, *13*, 4046. [CrossRef]
2. Friedman, M.; Walsh, G. High performance films: Review of new materials and trends. *Polym. Eng. Sci.* **2002**, *42*, 1756–1788. [CrossRef]
3. BGS: Radiation Crosslinking and Radiation Sterilization. Available online: https://en.bgs.eu (accessed on 13 November 2023).
4. Peng, Y.Y.; Srinivas, S.; Narain, R. Modification of polymers. In *Polymer Science and Nanotechnology*; Elsevier: Amsterdam, The Netherlands, 2020; pp. 95–104. [CrossRef]
5. Manas, D.; Bednarik, M.; Mizera, A.; Manas, M.; Ovsik, M.; Stoklasek, P. Effect of beta radiation on the quality of the bonded joint for difficult to bond polyolefins. *Polymers* **2019**, *11*, 1863. [CrossRef]
6. Sanchis, M.; Blanes, V.; Blanes, M.; Garcia, D.; Balart, R. Surface modification of low density polyethylene (LDPE) film by low pressure O_2 plasma treatment. *Eur. Polym. J.* **2006**, *42*, 1558–1568. [CrossRef]
7. Ebnesajjad, S. *Surface Treatment of Materials for Adhesion Bonding*; William Andrew Publishing: Norwich, NY, USA, 2006; p. 260. ISBN 0-8155-1523-5.
8. Pappas, D.; Bujanda, A.; Demaree, J.D.; Hirvonen, J.K.; Kosik, W.; Jensen, R.; McKnigh, S. Surface modification of polyamide fibers and films using atmospheric plasmas. *Surf. Coat. Technol.* **2006**, *201*, 4384–4388. [CrossRef]
9. Tang, J.; Tang, W.; Yuan, H.; Jin, R. Mechanical behaviors of ethylene/styrene interpolymer compatibilized polystyrene /polyethylene blends. *J. Appl. Polym. Sci.* **2007**, *104*, 4001–4007. [CrossRef]
10. Veilleux, J.; Rodrigue, D. Properties of recycled PS/SBR blends: Effect of SBR pretreatment. *Prog. Rubber Plast. Recycl. Technol.* **2016**, *32*, 111–128. [CrossRef]
11. Alves, A.C.L.; Grande, R.; Carvalho, A.J.F. Thermal and mechanical properties of thermoplastic starch and poly(vinyl alcohol-co-ethylene) blends. *J. Renew. Mater.* **2019**, *7*, 245–252. [CrossRef]
12. Nemani, S.K.; Annavarapu, R.K.; Mohammadian, B.; Raiyan, A.; Heil, J.; Haque, M.A.; Abdelaal, A.; Sojoudi, H. Surface modification of polymers: Methods and applications. *Adv. Mater. Interfaces* **2018**, *5*, 1801247. [CrossRef]
13. Encinas, N.; Díaz-Benito, B.; Abenojar, J.; Martínez, M.A. Extreme durability of wettability changes on polyolefin surfaces by atmospheric pressure plasma torch. *Surf. Coat. Technol.* **2010**, *205*, 396–402. [CrossRef]
14. Dryakhlov, V.O.; Shaikhiev, I.G.; Galikhanov, M.F.; Sverguzova, S.V. Modification of polymeric membranes by corona discharge. *Membr. Membr. Technol.* **2020**, *2*, 195–202. [CrossRef]
15. Yasuda, H.; Gazicki, M. Biomedical applications of plasma polymerization and plasma treatment of polymer surfaces. *Biomaterials* **1982**, *3*, 68–77. [CrossRef]
16. Zhu, G.; Su, Y.; Bai, P.; Chen, J.; Jing, Q.; Yang, W.; Wang, Z.L. Harvesting water wave energy by asymmetric screening of electrostatic charges on a nanostructured hydrophobic thin-film surface. *ACS Nano* **2014**, *8*, 6031–6037. [CrossRef]
17. Van Deynse, A.; Cools, P.; Leys, C.; Morent, R.; De Geyter, N. Influence of ambient conditions on the aging behavior of plasma-treated polyethylene surfaces. *Surf. Coat. Technol.* **2014**, *258*, 359–367. [CrossRef]
18. Bradler, P.R.; Fischer, J.; Wallner, G.M.; Lang, R.W. Characterization of irradiation crosslinked polyamides for solar thermal applications—Basic thermo-analytical and mechanical properties. *Polymers* **2018**, *10*, 969. [CrossRef]
19. Manas, D.; Mizera, A.; Navratil, M.; Manas, M.; Ovsik, M.; Sehnalek, S.; Stoklasek, P. The electrical, mechanical and surface properties of thermoplastic polyester elastomer modified by electron beta radiation. *Polymers* **2018**, *10*, 1057. [CrossRef]
20. Makuuchi, K.; Cheng, S. *Radiation Processing of Polymer Materials and Its Industrial Applications*; Wiley: Hoboken, NJ, USA, 2012; p. 415. ISBN 978-0-470-58769-0.
21. Drobny, J.G. *Ionizing Radiation and Polymers: Principles, Technology and Applications*; Elsevier/William Andrew: Oxford, UK, 2013; p. 298. ISBN 978-1-4557-7881-2.
22. Lee, J.-G.; Jeong, J.-O.; Jeong, S.-I.; Park, J.-S. Radiation-based crosslinking technique for enhanced thermal and mechanical properties of HDPE/EVA/PU blends. *Polymers* **2021**, *13*, 2832. [CrossRef]
23. Gheysari, D.; Behjat, A.; Haji-Saeid, M. The effect of high-energy electron beam on mechanical and thermal properties of LDPE and HDPE. *Eur. Polym. J.* **2001**, *37*, 295–302. [CrossRef]
24. Jeong, J.-O.; Oh, Y.-H.; Jeong, S.-I.; Park, J.-S. Optimization of the Physical Properties of HDPE/PU Blends through Improved Compatibility and Electron Beam Crosslinking. *Polymers* **2022**, *14*, 3607. [CrossRef] [PubMed]
25. Ovsik, M.; Stanek, M.; Dockal, A.; Vanek, J.; Hylova, L. Influence of cross-linking agent concentration/beta radiation surface modification on the micro-mechanical properties of polyamide 6. *Materials* **2021**, *14*, 6407. [CrossRef] [PubMed]

26. Kopal, I.; Vršková, J.; Labaj, I.; Ondrušová, D.; Hybler, P.; Harničárová, M.; Valíček, J.; Kušnerová, M. The effect of high-energy ionizing radiation on the mechanical properties of a melamine resin, phenol-formaldehyde resin, and nitrile rubber blend. *Materials* **2018**, *11*, 2405. [CrossRef] [PubMed]
27. Ovsik, M.; Stanek, M.; Bednarik, M. Evaluation of cross-linked polyamide 6 micro-indentation properties: TAIC concentration and electron radiation intensity. *Materials* **2023**, *16*, 2391. [CrossRef]
28. Malinowski, R. Application of the electron radiation and triallyl isocyanurate for production of aliphatic-aromatic co-polyester of modified properties. *Int. J. Adv. Manuf. Technol.* **2016**, *87*, 3307–3314. [CrossRef]
29. Ovsik, M.; Manas, M.; Stanek, M.; Dockal, A.; Vanek, J.; Mizera, A.; Adamek, M.; Stoklasek, P. Polyamide surface layer nano-indentation and thermal properties modified by irradiation. *Materials* **2020**, *13*, 2915. [CrossRef] [PubMed]
30. Hama, Y.; Oka, T.; Uchiyama, J.; Kanbe, H.; Nebeta, K.; Yatagai, F. Long-term oxidative degradation in polyethylene irradiated with ion beams. *Radiat. Phys. Chem.* **2001**, *62*, 133–139. [CrossRef]
31. Buttafava, A.; Tavares, A.; Arimondi, M.; Zaopo, A.; Nesti, S.; Dondi, D.; Mariani, M.; Faucitano, A. Dose rate effects on the radiation induced oxidation of polyethylene. *Nucl. Instrum. Methods Phys. Res. Sect. B Beam Interact. Mater. At.* **2007**, *265*, 221–226. [CrossRef]
32. Mouaci, S.; Saidi, M.; Saidi-Amroun, N. Oxidative degradation and morphological properties of gamma-irradiated isotactic polypropylene films. *Micro Nano Lett.* **2017**, *12*, 478–481. [CrossRef]
33. Holik, Z.; Danek, M.; Manas, M.; Cerny, J.; Malochova, M. The influence of ionizing radiation on chemical resistance of polymers. *Int. J. Mech.* **2011**, *5*, 210–217.
34. Holik, Z.; Danek, M.; Manas, M.; Cerny, J. Influence of the amount of cross-linking agent on properties of irradiated polyamide 6. *Int. J. Mech.* **2011**, *5*, 218–225.
35. *CSN EN ISO 527-1*; Plasticss—Determination of Tensile Properties—Part 1: General Principles. CEN: Brussels, Belgium, 2019.
36. *CSN EN ISO 527-2*; Plasticss—Determination of Tensile Propertiess—Part 2: Test Conditions for Moulding and Extrusion Plastic. CEN: Brussels, Belgium, 2012.
37. *CSN EN ISO 178*; Plastics—Determination of Flexural Properties. CEN: Brussels, Belgium, 2019.
38. *EN 15802*; Conservation of Cultural Property—Test Methods—Determination of Static Contact Angle. CEN: Brussels, Belgium, 2009.
39. *CSN EN 1465*; Adhesives—Determination of Tensile Lap-Shear Strength of Bonded Assemblies. CEN: Brussels, Belgium, 2009.
40. Bednarik, M.; Mizera, A.; Manas, M.; Navratil, M.; Huba, J.; Achbergerova, E.; Stoklasek, P. Influence of the β^- radiation/cold atmospheric-pressure plasma surface modification on the adhesive bonding of polyolefins. *Materials* **2021**, *14*, 76. [CrossRef]
41. *ASTM 51261*; Practice for Calibration of Routine Dosimetry Systems for Radiation Processing, 2nd ed. ASTM International: West Conshohocken, PA, USA, 2013.
42. Kaelble, D.H. Dispersion-polar surface tension properties of organic solids. *J. Adhes.* **1970**, *2*, 66–81. [CrossRef]
43. Rabel, W. Aspekte der benetzungstheorie und ihre anwendung auf die untersuchung und veränderung der oberflächeneigenschaften von polymeren. *Farbe Lacke* **1971**, *77*, 997–1005.
44. Owens, D.K.; Wendt, R.C. Estimation of the surface free energy of polymers. *J. Appl. Polym. Sci.* **1969**, *13*, 1741–1747. [CrossRef]
45. Schrader, M.E.; Loeb, G.I. *Modern Approaches to Wettability*; Online; Springer: Boston, MA, USA, 1992; ISBN 978-1-4899-1178-0. [CrossRef]
46. Kwok, D.Y.; Neumann, A.W. Contact angle measurement and contact angle interpretation. *Adv. Colloid Interface Sci.* **1999**, *81*, 167–249. [CrossRef]
47. Erbil, H. *Surface Chemistry of Solid and Liquid Interfaces*; Online; Blackwell: Oxford, UK, 2006; p. 352. ISBN 978-1-444-30540-1. [CrossRef]
48. *ASTM D2765*; Standard Test Methods for Determination of Gel Content and Swell Ratio of Crosslinked Ethylene Plastics. ASTM International: West Conshohocken, PA, USA, 2016.
49. Manas, D.; Ovsik, M.; Mizera, A.; Manas, M.; Hylova, L.; Bednarik, M.; Stanek, M. The Effect of irradiation on mechanical and thermal properties of selected types of polymers. *Polymers* **2018**, *10*, 158. [CrossRef]
50. Meloun, M.; Militky, J. *A Compendium of Statistical Data Processing*; Karolinum: Prague, Czech Republic, 2013; p. 984. ISBN 978-80-246-2196-8.
51. Rivaton, A.; Lalande, D.; Gardette, J.L. Influence of the structure on the γ- irradiation of polypropylene and on the post-irradiation effects. *Nucl. Instrum. Methods Phys. Res. Sect. B Beam Interact. Mater. At.* **2004**, *222*, 187–200. [CrossRef]
52. Carpentieri, I.; Brunella, V.; Bracco, P.; Paganini, M.C.; Del Prever, E.M.B.; Luda, M.P.; Bonomi, S.; Costa, L. Post-irradiation oxidation of different polyethylenes. *Polym. Degrad. Stab.* **2011**, *96*, 624–629. [CrossRef]
53. Costa, L.; Carpentieri, I.; Bracco, P. Post electron-beam irradiation oxidation of orthopaedic UHMWPE. *Polym. Degrad. Stab.* **2008**, *93*, 1695–1703. [CrossRef]
54. Holik, Z.; Manas, M.; Stanek, M.; Danek, M.; Kocourek, J. Improvement of mechanical and termomechanical properties of polyethylene by irradiation crosslinking. *Chem. Listy* **2009**, *103*, 60–63.
55. Manas, M.; Stanek, M.; Manas, D.; Danek, M.; Holik, Z. Modification of polyamides properties by irradiation. *Chem. Listy* **2009**, *103*, 24–28.
56. Khonakdar, H.A.; Morshedian, J.; Mehrabzadeh, M.; Wagenknecht, U.; Jafari, S.H. Thermal and shrinkage behaviour of stretched peroxide-crosslinked high-density polyethylene. *Eur. Polym. J.* **2003**, *39*, 1729–1734. [CrossRef]
57. Piątek-Hnat, M.; Bomba, K.; Pęksiński, J.; Kozłowska, A.; Sośnicki, J.G.; Idzik, T.J. Effect of E-beam irradiation on thermal and mechanical properties of ester elastomers containing multifunctional alcohols. *Polymers* **2020**, *12*, 1043. [CrossRef] [PubMed]

58. Motaleb, K.Z.M.A.; Abakeviciene, B.; Milasius, R. Development and characterization of bio-composites from the plant wastes of water hyacinth and sugarcane bagasse: Effect of water repellent and gamma radiation. *Polymers* **2023**, *15*, 1609. [CrossRef] [PubMed]
59. Egghe, T.; Van Guyse, F.R.; Ghobeira, R.; Morent, R.; Hoogenboom, R.; De Geyter, N. Evaluation of cross-linking and degradation processes occurring at polymer surfaces upon plasma activation via size-exclusion chromatography. *Polym. Degrad. Stab.* **2021**, *187*, 109543. [CrossRef]
60. Gheysari, D.; Behjat, A. Radiation crosslinking of LDPE and HDPE with 5 and 10 MeV electron beams. *Eur. Polym. J.* **2001**, *37*, 2011–2016. [CrossRef]

Article

Enhancing Amplification in Compliant Mechanisms: Optimization of Plastic Types and Injection Conditions

Pham Son Minh [1], Van-Thuc Nguyen [1], Tran Minh The Uyen [1], Vu Quang Huy [1], Hai Nguyen Le Dang [1] and Van Thanh Tien Nguyen [2,*]

[1] Faculty of Mechanical Engineering, Ho Chi Minh City University of Technology and Education, Ho Chi Minh City 71307, Vietnam; minhps@hcmute.edu.vn (P.S.M.); nvthuc@hcmute.edu.vn (V.-T.N.)
[2] Faculty of Mechanical Engineering, Industrial University of Ho Chi Minh City, Nguyen Van Bao Street, Ward 4, Go Vap District, Ho Chi Minh City 70000, Vietnam
* Correspondence: thanhtienck@naver.com

Abstract: This study surveys the impacts of injection parameters on the deformation rate of the injected flexure hinge made from ABS, PP, and HDPE. The flexure hinges are generated with different filling time, filling pressure, filling speed, packing time, packing pressure, cooling time, and melt temperature. The amplification ratio of the samples between different injection parameters and different plastic types is measured and compared to figure out the optimal one with a high amplification ratio. The results show that the relationship between the input and output data of the ABS, PP, and HDPE flexure hinges at different injection molding parameters is a linear relation. Changing the material or many injection molding parameters of the hinge could lead to a great impact on the hinge's performance. However, changing each parameter does not lead to a sudden change in the input and output values. Each plastic material has different optimal injection parameters and displacement behaviors. With the ABS flexure hinge, the filling pressure case has the greatest amplification ratio of 8.81, while the filling speed case has the lowest value of 4.81. With the optimal injection parameter and the input value of 105 μm, the ABS flexure hinge could create a maximum average output value of 736.6 μm. With the PP flexure hinge, the melt temperature case achieves the greatest amplification ratio of 6.73, while the filling speed case has the lowest value of 4.1. With the optimal injection parameter and the input value of 128 μm, the PP flexure hinge could create a maximum average output value of 964.8 μm. The average amplification ratio values of all injection molding parameters are 6.85, 5.41, and 4.01, corresponding to ABS, PP, and HDPE flexure hinges. Generally, the ABS flexure hinge has the highest amplification ratios, followed by the PP flexure hinge. The HDPE flexure hinge has the lowest amplification ratios among these plastic types. With the optimal injection parameter and the input value of 218 μm, the HDPE flexure hinge could create a maximum average output value of 699.8 μm. The results provide more insight into plastic flexure hinges and broaden their applications by finding the optimal injection parameters and plastic types.

Keywords: ABS; HDPE; PP; amplification ratio; filling time; packing time

1. Introduction

Recently, flexure hinges have been intensively investigated for precision actuation due to their complexity and low cost [1–4]. Flexure hinges work as a displacement amplification mechanism with no friction force and backlash; therefore, they can replace the conventional system such as motors in super precision machines in lithography, space telescopes, micro-electromechanical systems, and optical devices [5–8]. Some metals and alloys are usually stainless steel, titanium-based, copper-based, and aluminum alloys to create a flexure hinge [9–12]. The metal-based alloy flexure hinges provide good elastic properties and high corrosion resistance. For example, Wei et al. [13] created a SUS 316L stainless steel hinge from 3D printing. The report showed that the porous layer thickness of the hinge is like the

Citation: Minh, P.S.; Nguyen, V.-T.; Uyen, T.M.T.; Huy, V.Q.; Le Dang, H.N.; Nguyen, V.T.T. Enhancing Amplification in Compliant Mechanisms: Optimization of Plastic Types and Injection Conditions. *Polymers* **2024**, *16*, 394. https://doi.org/10.3390/polym16030394

Academic Editors: Andrew N. Hrymak and Shengtai Zhou

Received: 4 November 2023
Revised: 19 January 2024
Accepted: 25 January 2024
Published: 31 January 2024

average diameter of the powder particle. Coemert et al. [14] generated Ti-6Al-4V flexure hinges via laser cutting. The results indicated that the payload value is proportional to the hinge width but inversely proportional to the length. Interestingly, Roopa et al. [15] studied the impacts of geometry and material on the displacement of stainless steel, beryllium copper, and brass flexure hinges. They pointed out that the deflection values of the flexure hinge are 6.75 μm, 10.89 μm, and 12.10 μm, corresponding to stainless steel, beryllium copper, and brass materials. Schlick et al. [16] reported an aluminum gripper based on a flexure hinge mechanism. The gripper was applied as a micro-assembly station, and the momentum can be limited by controlling the oscillation via the actuator, lowering the gripping jaw speed, and moving coil actuators.

Flexure hinges with compliant mechanisms could have many designs, depending on the purposes, such as displacement amplifiers, smart structures, robotics, and quasi-static mechanisms. Typically, compliant flexure hinges for displacement amplifier applications have many types such as bridge type, Rhombus type, differential amplifier, lever mechanism, five bar structure, and tensural types. Compared to other designs, a bridge-type flexure hinge has the advantages of a simple and symmetrical shape, high load capacity, and high amplification ratio. Generally, flexure hinges are usually created using metals. However, fabricating a flexure hinge from these material types often requires a lot of time and cost because of the complicated structure and the high precision of the hinge. Using alternative materials with better productivity in fabricating flexure hinges is a rigorous demand. For example, Mutlu et al. [17] used TPE to produce a 3D printing flexure hinge and pinpointed that the elliptic and non-symmetric shapes have the best quality. Rosa et al. [18] created a flexure-based nanopositioner using cyclic olefin copolymer (COC) via the mesoscale injection molding process. The travel range of the COC nanopositioner is 15 μm with the highest standard deviation being 52.3 nm. Notably, Sahota et al. [19] simulated a fiber Bragg grating pressure sensor by comparing ABS polymer and stainless steel flexural hinges. The simulation results showed that the sensitivity of the ABS flexural hinge is 16 times higher than that of the stainless steel one. Shen et al. [20] combined bridge type and lever type mechanisms to generate a hybrid one, achieving a high amplification ratio and low overstress. Abedi et al. [21] designed a bridge-type compliant mechanism with an S-shape, achieving an amplification ratio of 4.5–5.5. Shi et al. [22] created a microgripper with two bridge-type parallelogram amplification mechanisms, obtaining a high amplification ratio of 30.3. Chen et al. [23] used a hybrid-compliant mechanism with a bridge-type and a lever-type design to harvest the vibration energy. Wan et al. [24] optimized the design process of the flexure-based bridge-type amplification using a reliability-based design optimization method. This method provided good accuracy compared to the other traditional methods. Overall, plastic flexural hinge investigations are not popular and need more insight.

Injection molding is a popular technique that could produce plastic parts at a mass production level due to its advantages of industrial availability, high accuracy, short time, and affordable cost [25–28]. Injection molding, therefore, is widely applied in generating household products and industrial parts. Moreover, advanced injection molding techniques such as microinjection molding, gas-assisted molding systems, conformal cooling channels, metal injection molding, foam injection molding, and water-assisted injection molding have received much interest from authors [29–31]. Injection molding can be applied with many thermoplastic polymers such as HDPE, LDPE, PP, ABS, and PA [32–35]. However, the reports about plastic flexure hinges generated using the injection molding process are rarely discussed despite the advantages of this fabrication technique.

This investigation focuses on creating plastic bridge-type mechanism flexure hinges with a leaf hinge from different plastic types including ABS, PP, and HDPE. The injection parameters including filling time, filling pressure, filling speed, packing time, packing pressure, cooling time, and melt temperature are also surveyed to optimize the hinge performance. The amplification ratio of the samples is evaluated and compared across several injection parameters and plastic types to determine the optimal one with a high amplification ratio. The results provide more insight into plastic flexure hinges and broaden their applications by finding the optimal injection parameters and plastic types.

2. Experimental Methods

Figure 1 shows the flexure hinge design and the injected hinge. The injection gate dimension is 5 mm in diameter. The compliant mechanism of the flexure hinge with input and output position, the fixture of the hinge, and the measurement device with chronographs for measuring the displacement of the hinge are presented in Figure 2. The measurement procedure consists of three steps. Step 1: Attach the sample to the bracket in the proper position and then attach the chronographs. Step 2: Align the pusher with the measuring model and adjust it. All chronographs had to be reset to zero. Step 3: Turn the pusher handle, then watch the chronographs and record the results. For each sample number, the study injected three plastic hinges for an individual measurement. The average number of three samples is calculated and presented in this report. The standard deviation varies by around 1–7%. Figure 3 presents the flexure hinge at different plastics and injection conditions. Besides holding time and holding pressure, the study also applied packing time and packing pressure. They are the extra steps of the injection molding cycle in which pressure is applied to the polymer melt to compress the polymer and drive additional material into the mold. This compensates for shrinkage as the polymer cools from the melt temperature to the room temperature. The gate is closed during the packing operation to prevent material from entering the mold. Moreover, the molding part weight is about 12–13.7 g, which is suitable for the holding and cooling time.

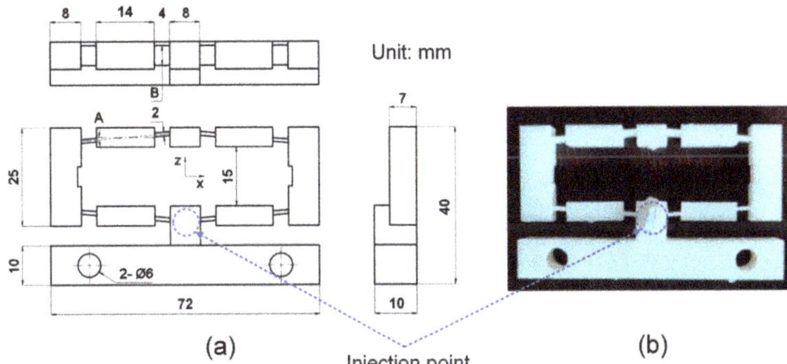

Figure 1. Plastic bridge-type flexure hinge: (**a**) design and (**b**) injected hinge [36].

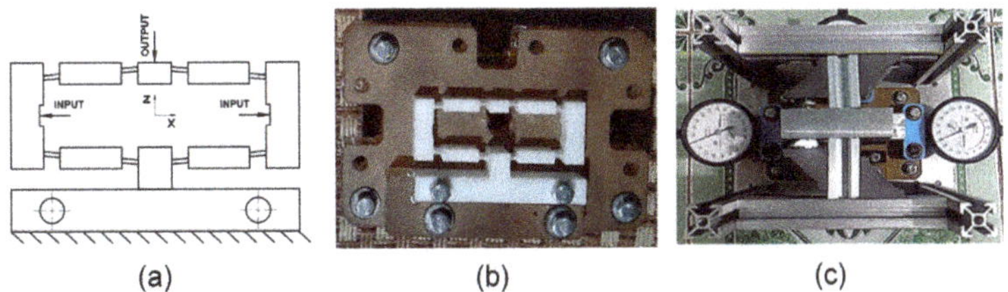

Figure 2. Compliant mechanism of the flexure hinge, fixture, and measurement device: (**a**) compliant mechanism, (**b**) fixture, and (**c**) measurement device setup [36].

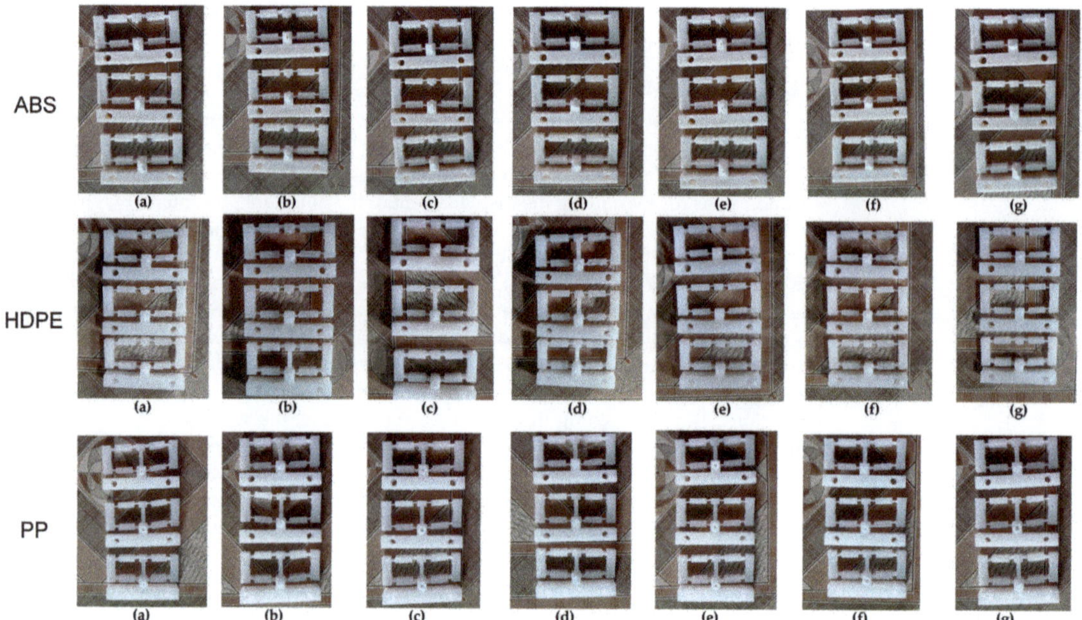

Figure 3. The flexure hinges at different plastics and injection conditions: (**a**) cooling time, (**b**) filling pressure, (**c**) filling speed, (**d**) filling time, (**e**) melt temperature, (**f**) packing pressure, and (**g**) packing time.

After the setup of the experiment machines, the plastics are dried at 85 °C for 12 h to remove the humidity. Then, they are injected into the mold using an injection molding. Finally, the injection flexure hinge is tested using a fixture and measurement device. This machine has a screw diameter of 36 mm, an injection capacity of 157 g, a screw speed of 122 mm/s, a 3 mm nozzle with a temperature limit switch control, and a clamping force of 1200 kN. Injection molding parameters depend on the geometric features of the moldings (especially thickness), the flow path of the polymer melt, the number of injection points, the type of polymeric material, the injection machine, etc. During the initial step, which is the experimental setup, this study tried many injection parameters for different plastic types to ensure that the samples were successfully injected. Firstly, this study attempted with relatively low parameters, in which the samples were partially injected. An attempt to inject the molded part was made by gradually increasing the values of the parameters. After that, the injection parameters are gradually adjusted to improve the injection quality. Finally, after having enough information, the injection parameters are set up as shown in Tables 1 and 2. The injection machine in this report is MA 1200III (Haitian, China). The weights of the moldings are 12 g, 12.7 g, and 13.7 g, corresponding to the PP, HDPE, and ABS hinges.

The injection rate of the screw is 154 cm^3/s. The filling speed is presented based on the number "154 cm^3/s". The percentage presents the ratio between the selected filling speed and the number "154 cm^3/s". Regarding the mold temperature, this study does not apply a mold temperature assistance system. Therefore, the mold temperature could be 30 °C. These hinges are injected with different injection molding parameters and different plastics. Table 1 shows the injection mold parameters of the ABS hinge, while Table 2 illustrates the injection molding parameters of the PP and HDPE hinges. The filling time is set at 1.5–2.8 s because the study did not apply the maximum filling pressure and filling speed of the injection molding machine MA 1200III. Therefore, it requires more time to fill the mold cavity than the maximum condition. The injection pressure is quite low; however, it is enough to fill the mold cavity with a suitable filling time (1.5–2.8 s). Because the study did not use the injection molding machine MA 1200III's maximum filling pressure and filling speed, the filling speed is set from 50% to 70%. As a result, a higher filling speed is required to fill the mold cavity than under the maximum condition. The study applied only 75–83% of this number, and according to the moderate pressure and volume of the runner system, the filling time must be greater than 2.0 s to ensure a good filling stage. Moreover, the subsequent packing stages (or holding stages) will ensure the product is filled with melt during the cooling phase, as shown in Table 1.

Table 1. Injection molding parameters of ABS hinges.

Group	Filling Time (s)	Filling Pressure (bar)	Packing Time (s)	Packing Pressure (bar)	Cooling Time (s)	Melt Temperature (°C)	Filling Speed (%)
1	2 2.2 2.4 2.6 2.8	41	0.6	40	24	210	79
2	2.4	39 40 41 42 43	0.6	40	24	210	79
3	2.4	41	0 0.3 0.6 0.9 1.2	40	24	210	79
4	2.4	41	0.6	38 39 40 41 42	24	210	79
5	2.4	41	0.6	40	20 22 24 26 28	210	79
6	2.4	41	0.6	40	24	206 208 210 212 214	79
7	2.4	41	0.6	40	24	210	75 77 79 81 83

ABS 750 SW polymer is supplied by Kumho Petrochemical, Seoul, Republic of Korea. PP polymer named Advanced-PP 1100 N is manufactured by Advanced Petrochemical Company, Al Jubail, Saudi Arabia. HDPE polymer HTA 108 is manufactured by Exxon Mobile Petroleum and Chemical Company, Riyadh, Saudi Arabia. The Young's modulus of ABS, PP, and HDPE are 3110 MPa, 1550 MPa, and 1300 MPa, respectively.

Table 2. Injection molding parameters of HDPE and PP hinges.

Group	Filling Time (s)	Filling Pressure (bar)	Packing Time (s)	Packing Pressure (bar)	Cooling Time (s)	Melt Temperature (°C)	Filling Speed (%)
1	1.5 1.7 1.9 2.1 2.3	27	0.6	25	54	214	54
2	1.9	25 26 27 28 29	0.6	25	54	214	54
3	1.9	41	0 0.3 0.6 0.9 1.2	25	54	214	54
4	1.9	41	0.6	23 24 25 26 27	54	214	54
5	1.9	41	0.6	25	50 52 54 56 58	214	54
6	1.9	41	0.6	25	54	210 212 214 216 218	54
7	1.9	41	0.6	25	54	214	50 52 54 56 58

3. Results and Discussion

3.1. ABS Flexure Hinge

In this section, the effects of injection parameters on the displacement of the ABS flexure hinge are examined. Firstly, the impacts of the filling stage, including filling time, filling pressure, and filling speed, are surveyed.

Figures 4–6 present the displacement diagrams of the ABS flexure hinge at different filling times, filling pressures, and filling speeds. In the filling time case, with the maximum input value of 128 µm, the average output value is 760.4 µm, indicating the magnification effect of the hinge, as shown in Figure 4. Generally, changing the filling time, filling pressure, and filling speed does not strongly affect the displacement rate of the ABS flexure hinge due to the similar values of the curves at different filling times and filling pressures. In addition, improving the input value mostly leads to a linear increase in the output value. The regression equations between the average values of the input and the output displacements are

$$y_1 = 5.21x_1, \tag{1}$$

$$y_2 = 8.81x_2, \tag{2}$$

$$y_3 = 4.81x_3, \tag{3}$$

where y_1, y_2, and y_3 are the output data and x_1, x_2, and x_3 are the input data of the ABS flexure hinge at different filling times, filling pressures, and filling speeds.

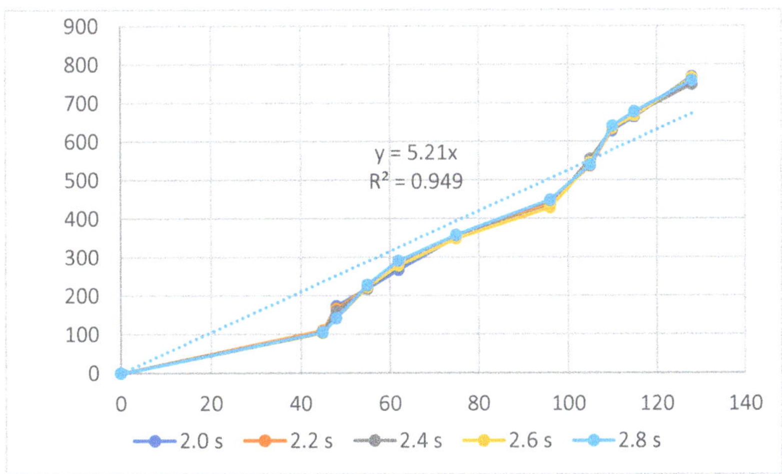

Figure 4. Displacement of the ABS flexure hinge at different filling times.

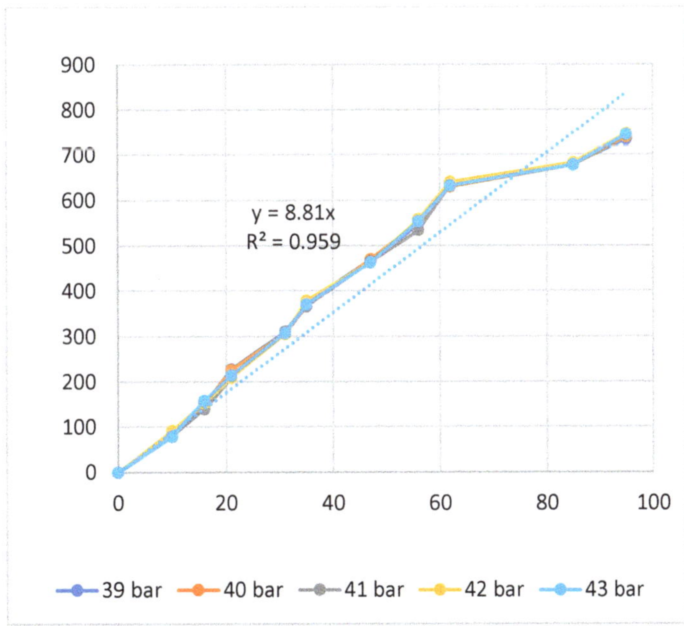

Figure 5. Displacement of the ABS flexure hinges at different filling pressures.

The trendline diagrams are set as a linear relationship $y = a.x + b$, where "b" is zero because the intercept of the trendline is set at zero. Therefore, there is only the "a" regression coefficient in the equation $y = a.x$. Moreover, the R-squared value of these trendlines is very high, which is higher than 0.9, indicating that the equation has a good prediction value.

This equation indicates a linear relationship between the input and the output data, as mentioned above. The amplification ratio of the filling time case is 5.21, the filling pressure case is 8.81, and the filling speed is 4.81. These values mean that the filling pressure cases have the highest amplification ratio, indicating the strong effect of the filling pressure on the amplification ratio. On the contrary, the filling speed has the lowest amplification ratio, meaning a weaker effect rate than the filling pressure and the filling time.

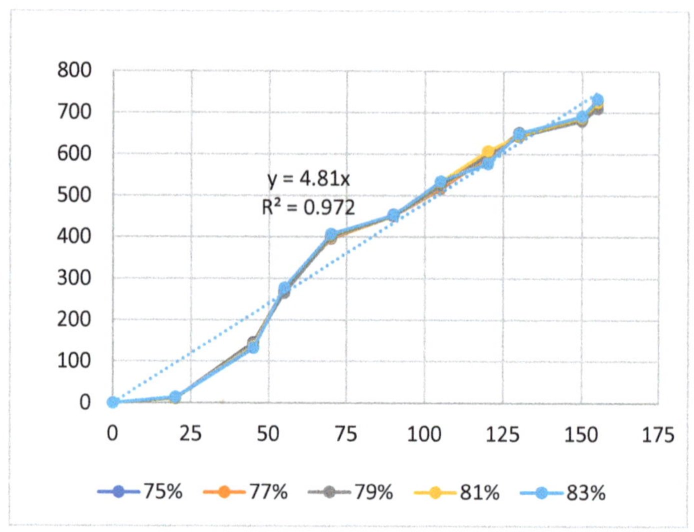

Figure 6. Displacement of the ABS flexure hinges at different filling speeds.

Figures 7 and 8 present the displacement of the ABS flexure hinge at different packing times and packing pressures. Consistent with the filling stage results, the similar values at different packing times and packing pressures reveal that altering these values does not cause much change in the displacement rate. Moreover, the regression equations between the average values of the input and the output displacements are

$$y_4 = 7.68x_4, \tag{4}$$

$$y_5 = 8.08x_5, \tag{5}$$

where y_4 and y_5 are the output data and x_4 and x_5 are the input data of the ABS flexure hinge at different packing times and packing pressures.

These data have a linear relationship, which is similar to the filling stage. The amplification ratio values for the packing time and packing pressure cases are 7.68 and 8.08, respectively. These ratios are comparable, unlike the filling stage, where the filling pressure has a larger amplification ratio than the filling time.

Figures 9 and 10 display the displacement of the ABS flexure hinge at different cooling times and melt temperatures. The curve diagrams for different parameters in every figure are almost equivalent, demonstrating that changing the cooling time and melt temperature has little effect on the displacement rate. Furthermore, the regression equations between the average values of the input and output displacements are as follows:

$$y_6 = 6.92x_6, \tag{6}$$

$$y_7 = 6.47x_7, \tag{7}$$

where y_6 and y_7 are the output data and x_6 and x_7 are the input data of the ABS flexure hinge at different cooling times and melt temperatures.

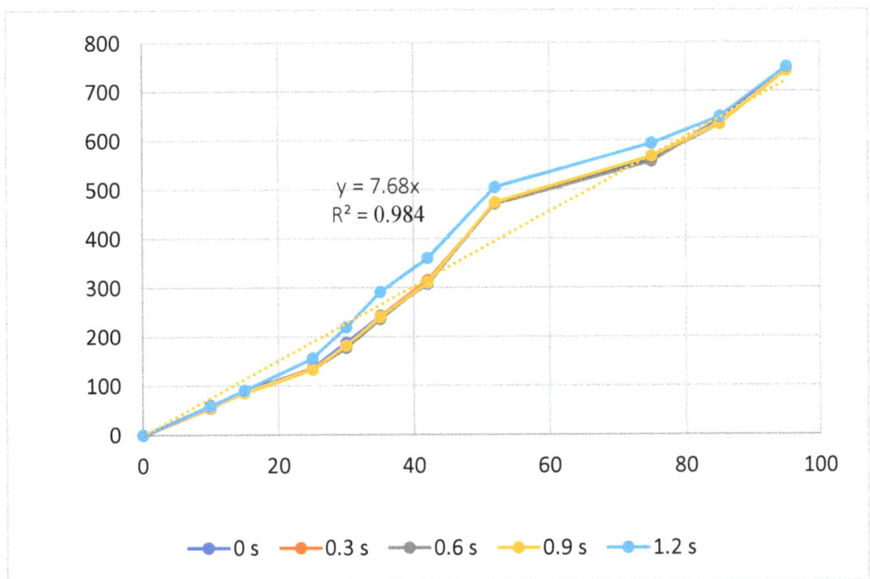

Figure 7. Displacement of the ABS flexure hinges at different packing times.

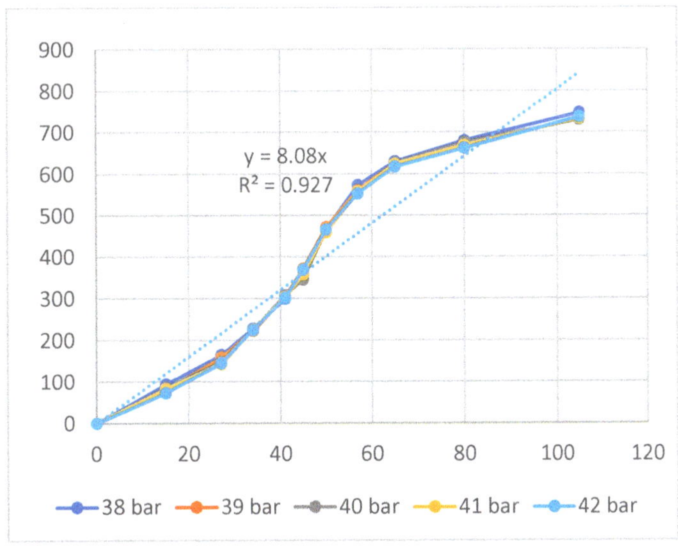

Figure 8. Displacement of the ABS flexure hinges at different packing pressures.

The linear relationship between the input and output data is revealed by Equations (6) and (7), which are similar to the filling and packing stages. The amplification ratio values for the cooling time and melt temperature cases are 6.92 and 6.47, respectively. The results reveal that the cooling time case has a higher amplification ratio, resulting in a higher effect rate.

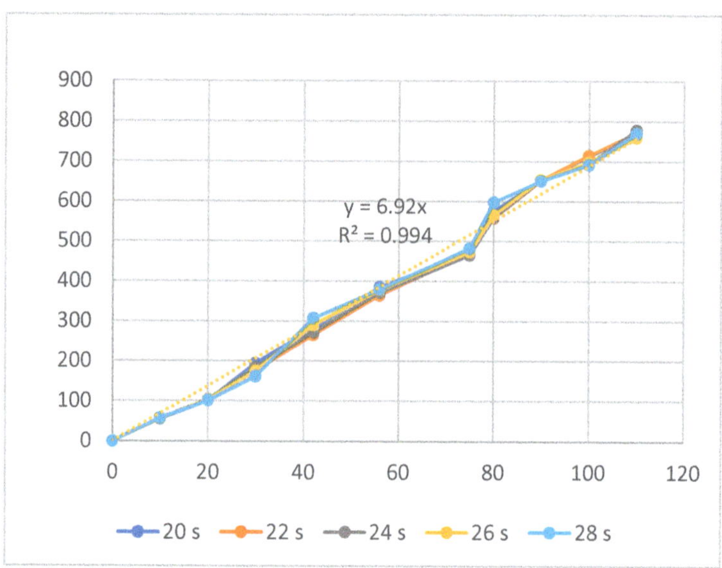

Figure 9. Displacement of the ABS flexure hinges at different cooling times.

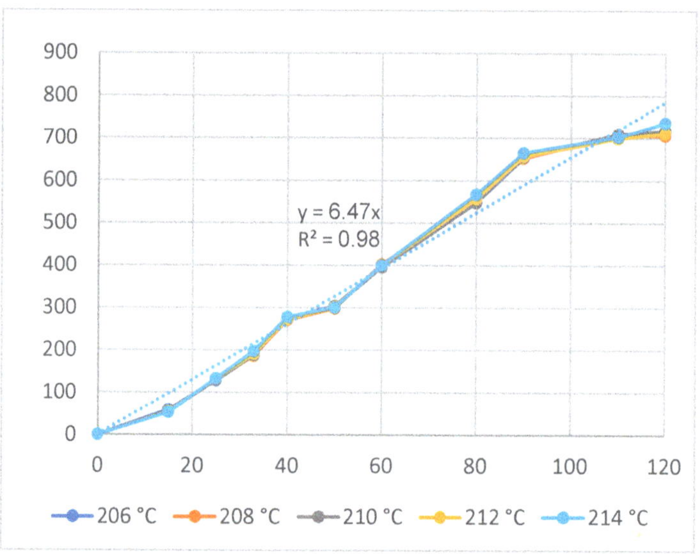

Figure 10. Displacement of the ABS flexure hinges at different melt temperatures.

Overall, the relationship between the input and output data of the ABS flexure hinges at different injection molding parameters is linear. The amplification ratio values for filling time, filling pressure, filling speed, packing time, packing pressure, cooling time, and melt temperature are 5.21, 8.81, 4.81, 7.68, 8.08, 6.92, and 6.47. The amplification ratio of filling pressure is the highest, while that of filling speed is the lowest. The range of the filling pressure examined is 39–43 bar. In this range, the amplification ratio is highest due to the good quality of the sample following the parameters of filling time 2.4 s, filling pressure 39–43 bar, packing time 0.6 s, packing pressure 40 bar, cooling time 24 s, melt temperature 210 °C, and filling speed 79%, while the filling speed range is 75–83%. In this range, the amplification ratio is lowest due to the lower quality of the sample following the parameters of filling time 2.4 s, filling pressure 41 bar, packing time 0.6 s, packing pressure 40 bar, cooling time 24 s, melt temperature 210 °C, and filling speed 75–83%.

Interestingly, the packing stage also shows high amplification ratio values of 7.68 and 8.08, pinpointing the important role of this stage in optimizing the ABS flexure hinge. Most importantly, the results reveal that the curve diagrams of each figure are mostly similar because changing each parameter does not lead to a sudden change in the input and output value. In general, the injection parameters strongly impact the amplification ratio of the ABS hinge. In other words, changing a set of injection parameters could lead to a greater impact on the hinge's performance. The amplification ratio could range from 4.81 to 8.81, which is a relatively large deviation. The reason for this phenomenon is the sensitivity of the ABS flexure hinge performance when changing the injection parameters. The plastic injection hinge design with thin hinge corners especially requires suitable injection parameters to successfully fill the mold cavity because the plastic melt flow is hindered in these thin areas. The presented injection parameters are ideal for forming the ABS flexure hinge.

Overall, the amplification ratio of the ABS flexure hinge varies in the range of 4.81–8.81. In Kim et al.'s report [37], a similar aluminum alloy flexure hinge has an amplification of 5–25. The amplification ratio of the ABS hinge is lower than that of the aluminum alloy one. The reason is the higher strength and stiffness of the aluminum alloys compared to the ABS plastic.

3.2. PP Flexure Hinge

This section examines the influences of injection parameters on the displacement of the PP flexure hinge. In the previous section, the results reveal that the curve diagrams of each figure are mostly similar because changing each parameter does not lead to a sudden change in the input and output value. Furthermore, linear relationships are popular in all cases of the ABS flexure hinge. As a result, this section concentrates on the average value of each parameter. First, the impact of the filling stage is addressed, which includes filling time, filling pressure, and filling speed.

Figures 11–13 display the displacement diagrams of the PP flexure hinge at different filling times, filling pressures, and filling speeds. Following calculations, the regression equations demonstrating the linear relationship between the average values of the input and output displacements are

$$y_8 = 6.73x_8, \tag{8}$$

$$y_9 = 4.58x_9, \tag{9}$$

$$y_{10} = 4.89x_{10}, \tag{10}$$

where y_8, y_9, and y_{10} are the output data and x_8, x_9, and x_{10} are the input data of the PP flexure hinge at different filling times, filling pressures, and filling speeds.

The amplification ratio of the filling time case is 6.73, the filling pressure case is 4.58, and the filling speed case is 4.89. Different from the ABS flexure hinge, in which the filling pressure has the highest amplification ratio, the filling time case achieves the highest amplification ratio of 6.83. On the other hand, the filling pressure case has the lowest amplification ratio of 4.58. Compared to the steel alloy flexure hinge with an amplification ratio of 16.2 in Na et al.'s report [38], the PP flexure hinge has a lower value. The reason is that steel alloy has a much higher elastic modulus and strength than PP plastic.

Figures 14 and 15 show the displacement of the PP flexure hinge at different packing times and packing pressures. The regression equations between the average values of the input and the output displacements of the PP flexure hinge are

$$y_{11} = 5.08x_{11}, \tag{11}$$

$$y_{12} = 6.39x_{12},\tag{12}$$

where y_{11} and y_{12} are the output data and x_{11} and x_{12} are the input data of the PP flexure hinge at different packing times and packing pressures.

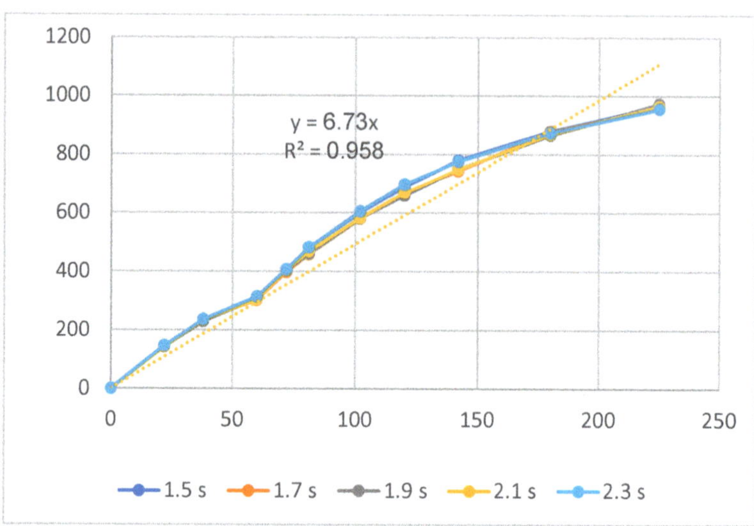

Figure 11. Displacement of PP flexure hinges at different filling times.

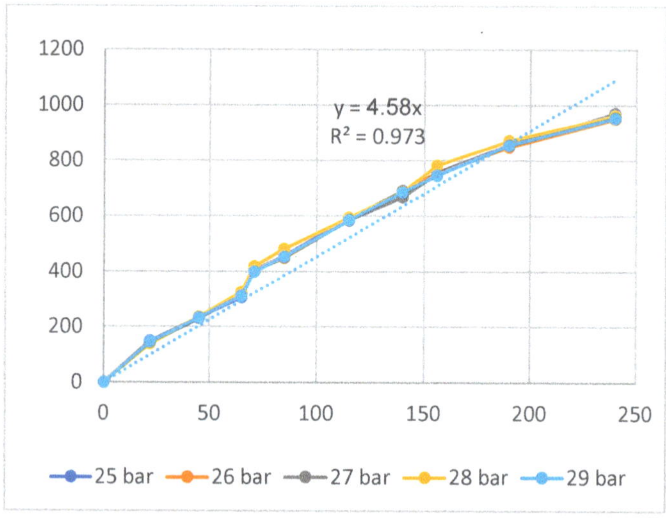

Figure 12. Displacement of PP flexure hinges at different filling pressures.

The correlations between these factors are linear. The amplification ratio values for the packing time and packing pressure situations are 5.08 and 6.39, respectively. Because of the lower elastic modulus and tensile strength, these ratios are significantly lower than those of the ABS flexure hinge, which are 7.68 and 8.08.

Figures 16 and 17 display the displacement of the PP flexure hinge at different cooling times and melt temperatures. The curve diagrams of the various parameters in each figure are most comparable, demonstrating that changing the cooling time and melt temperature has little effect on the displacement rate. Furthermore, the regression equations between the average values of the input and output displacements are as follows:

$$y_{13} = 6.13x_{13},\qquad(13)$$

$$y_{14} = 4.1x_{14},\qquad(14)$$

where y_{13} and y_{14} are the output data and x_{13} and x_{14} are the input data of the PP flexure hinge at different cooling times and melt temperatures.

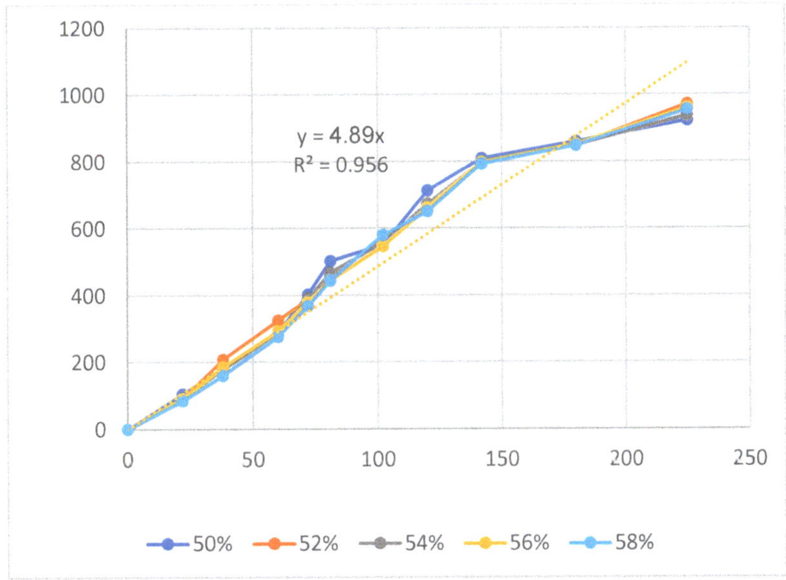

Figure 13. Displacement of PP flexure hinges at different filling speeds.

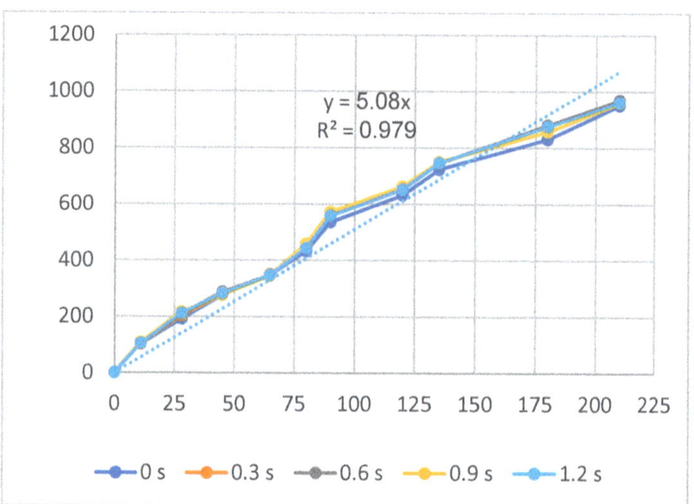

Figure 14. Displacement of PP flexure hinges at different packing times.

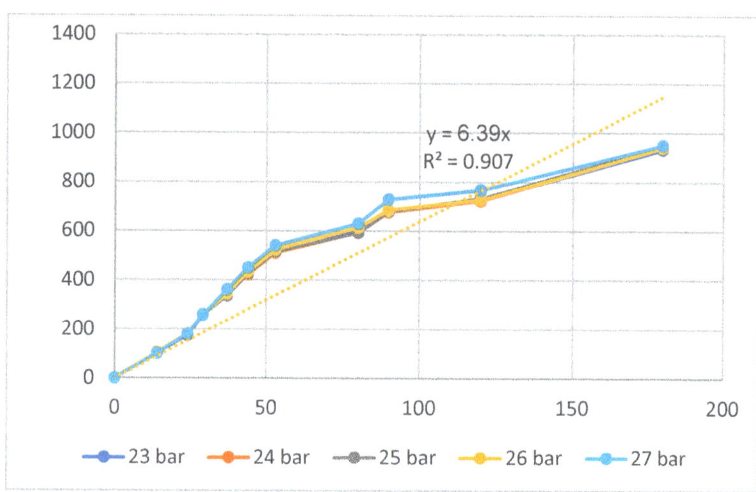

Figure 15. Displacement of PP flexure hinges at different packing pressures.

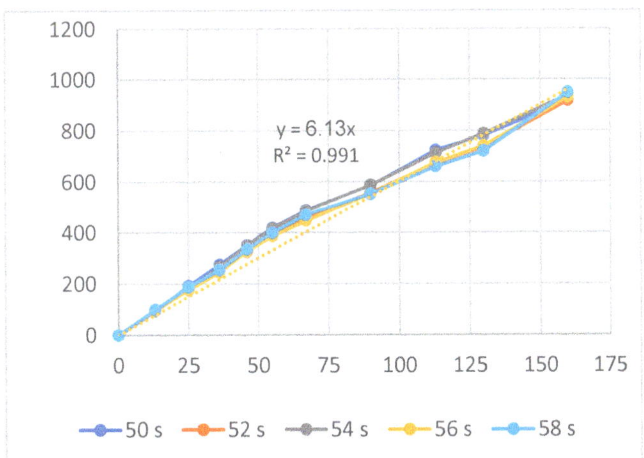

Figure 16. Displacement of PP flexure hinges at different cooling times.

The amplification ratio values for the cooling time and melt temperature cases are 6.13 and 4.1, respectively. The PP flexure hinge has lower amplification ratios than the ABS flexure hinge, which has values of 6.92 and 6.47 due to its lower elastic modulus and tensile strength.

Overall, the amplification ratio values are 6.73, 4.58, 4.89, 5.08, 6.39, 6.13, and 4.1, corresponding to the filling time, filling pressure, filling speed, packing time, packing pressure, cooling time, and melt temperature. The filling time achieves the greatest amplification ratio, while the melt temperature has the lowest one. Notably, these values are mainly lower than the ABS flexure hinge because of the weaker strength. The amplification ratio of the ABS flexure hinge could range from 4.81 to 8.81, while the amplification ratio of the PP flexure hinge could range from 4.1 to 6.73. The suitable amplification ratio could be achieved using two methods: changing the plastic-type or changing the injection parameters. In the next section, the displacement of the HDPE flexure hinge is investigated and then compared to the PP and ABS flexure hinge. Furthermore, the PP flexure hinge's linear relationship between input and output data is similar to the ABS one.

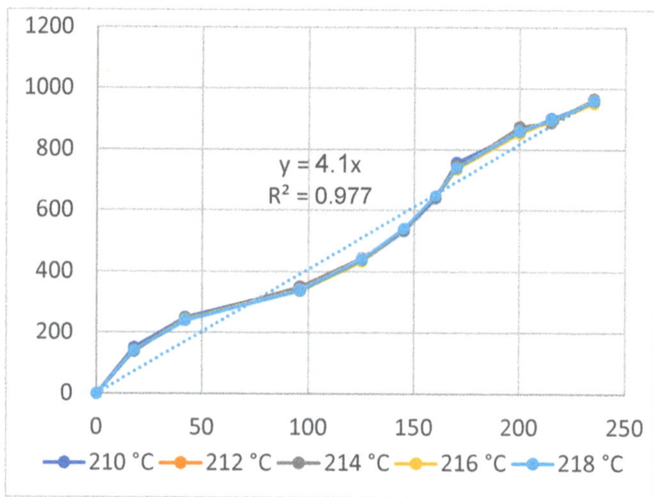

Figure 17. Displacement of PP flexure hinges at different melt temperatures.

In general, the amplification ratio of the PP flexure hinge varies in the range of 4.1–6.73. Compared to the amplification ratio of the ABS hinge, the amplification ratio of the PP flexure hinge is slightly lower. Moreover, compared to the Na et al. [38] report, the similar steel flexure hinge has an amplification of 16.2. The amplification ratio of the ABS hinge is lower due to the lower strength and stiffness of the PP plastic compared to the steel material.

3.3. HDPE Flexure Hinge and Comparison

In this section, displacements of the HDPE flexure hinge are surveyed. After that, these data are compared with the previous results of the ABS and PP flexure hinges. The input and output equations at the different filling times, filling pressures, filling speeds, packing times, packing pressures, cooling times, and melt temperatures are presented in Table 3. Figure 18 shows a more visual representation of these results.

Table 3. Displacement equations of the HDPE flexure hinge.

Injection Parameters	Equations
Filling time	$y_{15} = 5.0x_{15}$; $R^2 = 0.938$
Filling pressure	$y_{16} = 4.33x_{16}$; $R^2 = 0.967$
Filling speed	$y_{17} = 4.23x_{17}$; $R^2 = 0.988$
Packing time	$y_{18} = 5.4x_{18}$; $R^2 = 0.965$
Packing pressure	$y_{19} = 2.89x_{19}$; $R^2 = 0.952$
Cooling time	$y_{20} = 2.83x_{20}$; $R^2 = 0.942$
Melt temperature	$y_{21} = 3.4x_{21}$; $R^2 = 0.993$

Figure 18 shows the comparison of the amplification ratio among ABS, PP, and HDPE flexure hinges. The average amplification ratio values of all injection molding parameters are 6.85, 5.41, and 4.01, corresponding to ABS, PP, and HDPE flexure hinges. Generally, the ABS flexure hinge has the highest amplification ratios, followed by the PP flexure hinge. HDPE flexure hinge has the lowest amplification ratios among these plastic types. The reason is this order's gradual reduction in the tensile strength and elastic modulus [39]. Moreover, the highest amplification ratio is 8.81, gained by the ABS flexure hinge in the packing pressure case. In reverse, the HDPE flexure hinge in the cooling time case has the lowest amplification ratio of 2.83. The mechanical properties strongly affect the performance of the plastic flexure hinge. The flexure hinge shape, in this case, is not suitable for finding

Young's modulus, as it is not a tensile test sample. The shape of the flexure hinge and the process parameters could impact Young's modulus value via the degree of crystallinity. The Young's modulus of ABS, PP, and HDPE are 3110 MPa, 1550 Mpa, and 1300 Mpa, respectively. These values are very useful to explain the average amplification ratio values of these plastic hinges. For ABS, PP, and HDPE flexure hinges, the average amplification ratio values of all injection molding parameters are 6.85, 5.41, and 4.01. Therefore, the higher the Young's modulus value, the higher the amplification ratio value of the plastic hinges.

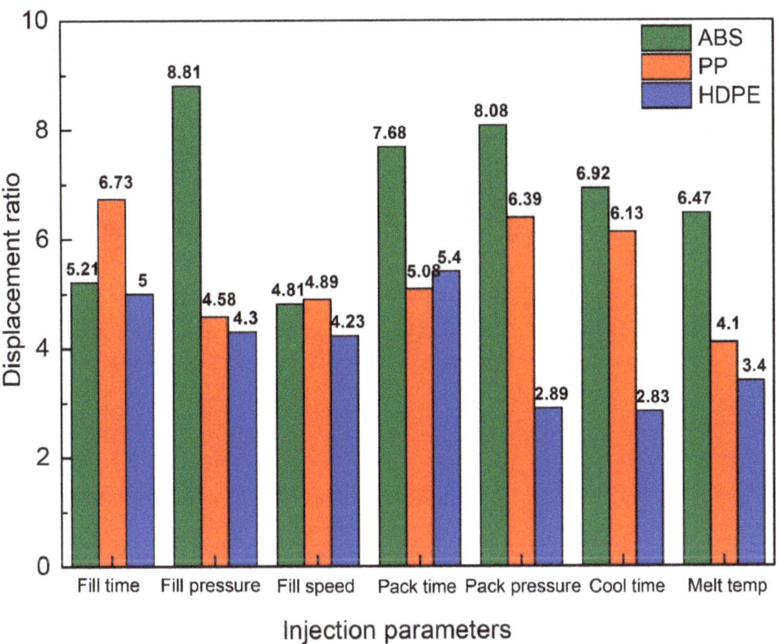

Figure 18. Comparison of displacement ratio among ABS, PP, and HDPE flexure hinges.

The optimal function here is the maximum amplification ratio, which is the "a" factor" in the equation $y = a.x$. This study tries to find the injection parameters that generate the maximum amplification ratio. In addition, the optimal injection parameter for the ABS flexure hinge with the highest amplification ratio is a filling time of 2.4 s, a filling pressure of 39–43 bar, a filling speed of 79%, a packing time of 0.6 s, a packing pressure of 40 bar, a cooling time of 24 s, and a melt temperature of 210 °C. With these optimal injection parameters and the input value of 105 μm, the ABS flexure hinge could create a maximum average output value of 736.6 μm. The optimal injection parameter for the PP flexure hinge is a filling time of 1.5–2.3 s, a filling pressure of 27 bar, a filling speed of 54%, a packing time of 0.6 s, a packing pressure of 25 bar, a cooling time of 24 s, and a melt temperature of 214 °C. With the optimal injection parameter, and the input value of 128 μm, the PP flexure hinge could create a maximum average output value of 964.8 μm. The optimal injection parameter for the HDPE flexure hinge is like the PP hinge, which is a filling time of 1.5–2.3 s, a filling pressure of 27 bar, a filling speed of 54%, a packing time of 0.6 s, a packing pressure of 25 bar, a cooling time of 24 s, and a melt temperature of 214 °C. With the optimal injection parameter, and the input value of 218 μm, the HDPE flexure hinge could create a maximum average output value of 699.8 μm.

Table 4 presents the amplification ratios of different flexure hinge materials. The aluminum alloys and steel alloys could produce a flexure hinge with a greater amplification ratio compared to the ABS, PP, and HDPE flexure hinges [37,38]. However, the amplification ratios of the ABS, PP, and HDPE flexure hinges are compatible with smart memory alloys and titanium alloys [40,41]. Most importantly, the injection plastic hinges may be mass-produced at a low cost and high productivity level. On the other hand, the cost and difficulty of producing the metallic flexure hinges are significantly higher than those of injection plastic ones. Therefore, the injection plastic hinges could facilitate the application of flexural hinges. In the future, we could examine how different hinge forms and corner shapes affect the functionalities of injection flexure plastic hinges. The fatigue strength of the injection plastic hinge could also be investigated.

Table 4. Amplification ratios of different flexure hinge materials.

Materials	Amplification Ratio	References
ABS	5.01–8.15	This study
PP	3.99–7.9	This study
HDPE	2.17–6.24	This study
Aluminum alloys	5–25	Kim et al. [37]
Steel alloys	16.2	Na et al. [38]
Smart memory alloys	2.2	Maffiodo et al. [40]
Titanium alloys	6.0	Fiaz et al. [41]

4. Conclusions

This study investigates the amplification of the injected plastic flexure hinge. The effects of injection plastics and injection parameters on the amplification ratio of the compliant mechanism flexural hinges are surveyed. The flexural hinge is injected with different plastic types: filling time, filling pressure, filling speed, packing time, packing pressure, cooling time, and melt temperature. Some noteworthy remarks that may be noted include the following:

The injection plastic hinges could be produced at the mass production level with high productivity and low cost. The relationship between the input and output data of the ABS, PP, and HDPE flexure hinges at different injection molding parameters is a linear relation. Changing the hinge's material or numerous injection molding settings could have a significant impact on its performance.

With the ABS flexure hinge, the packing pressure case has the greatest amplification ratio of 8.81, while the filling speed case has the lowest value of 4.81. The optimal injection parameter for an ABS flexure hinge is a filling time of 2.4 s, a filling pressure of 41 bar, a filling speed of 79%, a packing time of 0.6 s, a packing pressure of 38–42 bar, a cooling time of 24 s, and a melt temperature of 210 °C. The ABS flexure hinge could produce a maximum average output value of 736.6 μm with this parameter and an input value of 105 μm.

The filling time case has the highest amplification ratio of 6.73 with the PP flexure hinge, while the melt temperature case has the lowest value of 4.1. The optimal injection parameter for a PP flexure hinge is a filling time of 1.5–2.3 s, a filling pressure of 27 bar, a filling speed of 54%, a packing time of 0.6 s, a packing pressure of 25 bar, a cooling time of 24 s, and a melt temperature of 214 °C. With this parameter, and the input value of 128 μm, the PP flexure hinge could create a maximum average output value of 964.8 μm.

With the HDPE flexure hinge, the packing time case achieves the greatest amplification ratio of 5.4, while the cooling case has the lowest value of 2.83. The optimal injection parameter for the HDPE flexure hinge is like the PP hinge, which is a filling time of 1.5–2.3 s, a filling pressure of 27 bar, a filling speed of 54%, a packing time of 0.6 s, a packing pressure of 25 bar, a cooling time of 24 s, and a melt temperature of 214 °C. The largest average output value that the HDPE flexure hinge could produce with the ideal injection parameter and an input value of 218 μm is 699.8 μm.

For the ABS, PP, and HDPE flexure hinges, the amplification ratio values of all injection molding parameters are 4.81–8.81, 4.1–6.73, and 2.83–5.4. In general, the ABS flexure hinge could be superior to the PP flexure hinge in terms of amplification ratios. The lowest amplification ratio among these plastic kinds is found in HDPE flexure hinges. The results provide more insight into plastic flexure hinges and broaden their applications by finding the optimal injection parameters and plastic types. In the future, we could analyze the effects of corner shapes and try other hinge shapes on the performance of the flexure plastic hinges.

Author Contributions: P.S.M. and V.-T.N.: conceptualization, funding acquisition; V.-T.N., V.Q.H. and H.N.L.D.: writing—original draft, investigation; T.M.T.U., P.S.M. and V.-T.N.: analyzing, visualization, project administration; V.T.T.N., P.S.M., H.N.L.D. and V.-T.N.: writing—review and editing. All authors have read and agreed to the published version of the manuscript.

Funding: This research was funded by HCMC University of Technology and Education.

Institutional Review Board Statement: Not applicable.

Informed Consent Statement: Not applicable.

Data Availability Statement: The data used to support the findings of this study are available from the corresponding author upon request.

Acknowledgments: The authors acknowledge the support of HCMC University of Technology and Education.

Conflicts of Interest: The authors declare no conflicts of interest.

References

1. Tian, Y.; Zhang, D.; Shirinzadeh, B. Dynamic modelling of a flexure-based mechanism for ultra-precision grinding operation. *Precis. Eng.* **2011**, *35*, 554–565. [CrossRef]
2. Gao, X.; Yang, J.; Wu, J.; Xin, X.; Li, Z.; Yuan, X.; Shen, X.; Dong, S. Piezoelectric actuators and motors: Materials, designs, and applications. *Adv. Mater. Technol.* **2020**, *5*, 1900716. [CrossRef]
3. Chen, F.; Zhang, Q.; Gao, Y.; Dong, W. A review on the flexure-based displacement amplification mechanisms. *IEEE Access* **2020**, *8*, 205919–205937. [CrossRef]
4. Mutlu, R.; Tawk, C.; Alici, G.; Sariyildiz, E. A 3D printed monolithic soft gripper with adjustable stiffness. In Proceedings of the IECON 2017-43rd Annual Conference of the IEEE Industrial Electronics Society, Beijing, China, 29 October–1 November 2017; pp. 6235–6240.
5. Judy, J.W. Microelectromechanical systems (MEMS): Fabrication, design and applications. *Smart Mater. Struct.* **2001**, *10*, 1115. [CrossRef]
6. Shaar, N.S.; Barbastathis, G.; Livermore, C. Integrated folding, alignment, and latching for reconfigurable origami microelectromechanical systems. *J. Microelectromech. Syst.* **2014**, *24*, 1043–1051. [CrossRef]
7. Bagolini, A.; Ronchin, S.; Bellutti, P.; Chistè, M.; Verotti, M.; Belfiore, N.P. Fabrication of novel MEMS microgrippers by deep reactive ion etching with metal hard mask. *J. Microelectromech. Syst.* **2017**, *26*, 926–934. [CrossRef]
8. Solgaard, O.; Godil, A.A.; Howe, R.T.; Lee, L.P.; Peter, Y.A.; Zappe, H. Optical MEMS: From micromirrors to complex systems. *J. Microelectromech. Syst.* **2014**, *23*, 517–538. [CrossRef]
9. Noveanu, S.; Lates, D.; Fusaru, L.; Rusu, C. A new compliant microgripper and study for flexure hinges shapes. *Procedia Manuf.* **2020**, *46*, 517–524. [CrossRef]
10. Hinkley, D.; Simburger, E. A multifunctional flexure hinge for deploying omnidirectional solar arrays. In Proceedings of the 19th AIAA Applied Aerodynamics Conference, Anaheim, CA, USA, 11–14 June 2001; p. 1260.
11. Marsh, D.M. The construction and performance of various flexure hinges. *J. Sci. Instrum.* **1962**, *39*, 493. [CrossRef]
12. Smith, S.T.; Badami, V.G.; Dale, J.S.; Xu, Y. Elliptical flexure hinges. *Rev. Sci. Instrum.* **1997**, *68*, 1474–1483. [CrossRef]
13. Wei, H.; Shirinzadeh, B.; Niu, X.; Zhang, J.; Li, W.; Simeone, A. Study of the hinge thickness deviation for a 316L parallelogram flexure mechanism fabricated via selective laser melting. *J. Intell. Manuf.* **2021**, *32*, 1411–1420. [CrossRef]
14. Coemert, S.; Wegener, L.G.; Yalvac, B.; Fuckner, J.; Lueth, T.C. Experimental and FEM-Based Payload Analysis of Ti-6Al-4V Flexure Hinges. In Proceedings of the ASME International Mechanical Engineering Congress and Exposition, Salt Lake City, UT, USA, 11–14 November 2019; Volume 83518, p. V014T14A001.
15. Roopa, R.; Karanth, P.N.; Kulkarni, S.M. Effect of flexure beam geometry and material on the displacement of piezo actuated diaphragm for micropump. In *IOP Conference Series: Materials Science and Engineering*; IOP Publishing: Bristol, UK, 2018; Volume 310, p. 012111.
16. Schlick, J.; Zuehlke, D. Design and application of a gripper for microparts using flexure hinges and pneumatic actuation. In *Microrobotics and Microassembly III*; SPIE: Bellingham, DC, USA, 2001; Volume 4568, pp. 1–11.
17. Mutlu, R.; Alici, G.; in het Panhuis, M.; Spinks, G. Effect of flexure hinge type on a 3D printed fully compliant prosthetic finger. In Proceedings of the 2015 IEEE International Conference on Advanced Intelligent Mechatronics (AIM), Busan, Republic of Korea, 7–11 July 2015; pp. 790–795.

18. Rosa, F.; Scaccabarozzi, D.; Cinquemani, S.; Bizzozero, F. Sensor Embedding in a 3D Printed Flexure Hinge. In *Design Tools and Methods in Industrial Engineering, Proceedings of the International Conference on Design Tools and Methods in Industrial Engineering, ADM 2019, Modena, Italy, 9–10 September 2019*; Springer International Publishing: Berlin/Heidelberg, Germany, 2020; pp. 848–859.

19. Sahota, J.K.; Dhawan, D.; Gupta, N. Modeling and Analysis of Temperature-Compensated Fiber Bragg Grating Sensor Based on Flexure Hinge Beam and Diaphragm for Low-Pressure Detection. *Arab. J. Sci. Eng.* **2023**, *49*, 1095–1115. [CrossRef]

20. Shen, X.; Zhang, L.; Qiu, D. A lever-bridge combined compliant mechanism for translation amplification. *Precis. Eng.* **2021**, *67*, 383–392. [CrossRef]

21. Abedi, K.; Shakhesi, E.; Seraj, H.; Mahnama, M.; Shirazi, F.A. Design and analysis of a 2-DOF compliant serial micropositioner based on "S-shaped" flexure hinge. *Precis. Eng.* **2023**, *83*, 228–236. [CrossRef]

22. Shi, C.; Dong, X.; Yang, Z. A microgripper with a large magnification ratio and high structural stiffness based on a flexure-enabled mechanism. *IEEE/ASME Trans. Mechatron.* **2021**, *26*, 3076–3086. [CrossRef]

23. Chen, X.; Li, Y. Design, modeling and testing of a vibration absorption device with energy harvesting based on force amplifier and piezoelectric stack. *Energy Convers. Manag.* **2022**, *255*, 115305. [CrossRef]

24. Wan, L.; Chen, H.; Ouyang, L.; Chen, Y. A new ensemble modeling approach for reliability-based design optimization of flexure-based bridge-type amplification mechanisms. *Int. J. Adv. Manuf. Technol.* **2020**, *106*, 47–63. [CrossRef]

25. McCormick, R.M.; Nelson, R.J.; Alonso-Amigo, M.G.; Benvegnu, D.J.; Hooper, H.H. Microchannel electrophoretic separations of DNA in injection-molded plastic substrates. *Anal. Chem.* **1997**, *69*, 2626–2630. [CrossRef] [PubMed]

26. Yu, L.; Koh, C.G.; Lee, L.J.; Koelling, K.W.; Madou, M.J. Experimental investigation and numerical simulation of injection molding with micro-features. *Polym. Eng. Sci.* **2002**, *42*, 871–888. [CrossRef]

27. Agrawal, A.R.; Pandelidis, I.O.; Pecht, M. Injection-molding process control—A review. *Polym. Eng. Sci.* **1987**, *27*, 1345–1357. [CrossRef]

28. Chen, Z.; Turng, L.S. A review of current developments in process and quality control for injection molding. *Adv. Polym. Technol. J. Polym. Process. Inst.* **2005**, *24*, 165–182. [CrossRef]

29. Czepiel, M.; Bańkosz, M.; Sobczak-Kupiec, A. Advanced Injection Molding Methods. *Materials* **2023**, *16*, 5802. [CrossRef] [PubMed]

30. Yang, W.; Jian, R. Polymer 3D Printing and 3D Copying Processes. In *Polymer 3D Printing and 3D Copying Technology*; Springer Nature: Singapore, 2023; pp. 17–82.

31. Raju, A.; Samanta, D.; Rajendrakumar, K. A Review of Recent Advances in the Development of Superhydrophobicity over Various Substrate Surfaces Using Polymers. *ChemistrySelect* **2023**, *8*, e202204262. [CrossRef]

32. Begum, S.A.; Rane, A.V.; Kanny, K. Applications of compatibilized polymer blends in automobile industry. In *Compatibilization of Polymer Blends*; Elsevier: Amsterdam, The Netherlands, 2020; pp. 563–593.

33. Goodship, V. Injection molding of thermoplastics. In *Design and Manufacture of Plastic Components for Multifunctionality: Structural Composites, Injection Molding, and 3D Printing*; Elsevier: Amsterdam, The Netherlands, 2015; p. 103.

34. Strapasson, R.; Amico, S.C.; Pereira, M.F.R.; Sydenstricker, T.H.D. Tensile and impact behavior of polypropylene/low density polyethylene blends. *Polym. Test.* **2005**, *24*, 468–473. [CrossRef]

35. Ozcelik, B.; Ozbay, A.; Demirbas, E. Influence of injection parameters and mold materials on mechanical properties of ABS in plastic injection molding. *Int. Commun. Heat Mass Transf.* **2010**, *37*, 1359–1365. [CrossRef]

36. Uyen, T.M.T.; Minh, P.S.; Nguyen, V.-T.; Do, T.T.; Nguyen Le Dang, H.; Nguyen, V.T.T. Amplification Ratio of a Recycled Plastics-Compliant Mechanism Flexure Hinge. *Appl. Sci.* **2023**, *13*, 12825. [CrossRef]

37. Kim, J.H.; Kim, S.H.; Kwak, Y.K. Development and optimization of 3-D bridge-type hinge mechanisms. *Sens. Actuators A Phys.* **2004**, *116*, 530–538. [CrossRef]

38. Na, T.W.; Choi, J.H.; Jung, J.Y.; Kim, H.G.; Han, J.H.; Park, K.C.; Oh, I.K. Compact piezoelectric tripod manipulator based on a reverse bridge-type amplification mechanism. *Smart Mater. Struct.* **2016**, *25*, 095028. [CrossRef]

39. Ravisankar, H.; Hariprasadarao, P.; Das, V.C.; Bhanukiran, G. Effect of nanoclay inclusion on mechanical and fatigue behavior of ABS/PP/HDPE thermoplastics. *Mater. Today Proc.* **2020**, *24*, 209–217. [CrossRef]

40. Maffiodo, D.; Raparelli, T. Three-fingered gripper with flexure hinges actuated by shape memory alloy wires. *Int. J. Autom. Technol.* **2017**, *11*, 355–360. [CrossRef]

41. Fiaz, H.S.; Settle, C.R.; Hoshino, K. Metal additive manufacturing for microelectromechanical systems: Titanium alloy (Ti-6Al-4V)-based nanopositioning flexure fabricated by electron beam melting. *Sens. Actuators A Phys.* **2016**, *249*, 284–293. [CrossRef]

Article

Flexible Polyolefin Elastomer/Paraffin Wax/Alumina/Graphene Nanoplatelets Phase Change Materials with Enhanced Thermal Conductivity and Mechanical Performance for Solar Conversion and Thermal Energy Storage Applications

Jie Tian [1], Chouxuan Wang [2], Kaiyuan Wang [2], Rong Xue [2], Xinyue Liu [2] and Qi Yang [3,*]

[1] School of Civil Engineering and Architecture, Shaanxi University of Technology, Hanzhong 723099, China; 13366086793@163.com

[2] National and Local Engineering Laboratory for Slag Comprehensive Utilization and Environment Technology, School of Materials Science and Engineering, Shaanxi University of Technology, Hanzhong 723099, China; wangchouxuanxuan@163.com (C.W.); wangkaiyuan6@126.com (K.W.); 18582481950@163.com (R.X.); 1809179387@163.com (X.L.)

[3] College of Polymer Science and Engineering, Sichuan University, Chengdu 610065, China

* Correspondence: yangqi@scu.edu.cn; Tel.:+86-28-85405401

Abstract: In this study, electrically insulating polyolefin elastomer (POE)-based phase change materials (PCMs) comprising alumina (Al_2O_3) and graphene nanoplatelets (GNPs) are prepared using a conventional injection moulding technique, which exhibits promising applications for solar energy storage due to the reduced interfacial thermal resistance, excellent stability, and proficient photo-thermal conversion efficiency. A synergistic interplay between Al_2O_3 and GNPs is observed, which facilitates the establishment of thermally conductive pathways within the POE/paraffin wax (POE/PW) matrix. The in-plane thermal conductivity of POE/PW/GNPs 5 wt%/Al_2O_3 40 wt% composite reaches as high as 1.82 W m^{-1}K^{-1}, marking a remarkable increase of $\approx$269.5% when compared with that of its unfilled POE/PW counterpart. The composite exhibits exceptional heat dissipation capabilities, which is critical for thermal management applications in electronics. Moreover, POE/PW/GNPs/Al_2O_3 composites demonstrate outstanding electrical insulation, enhanced mechanical performance, and efficient solar energy conversion and transportation. Under 80 mW cm^{-2} NIR light irradiation, the temperature of the POE/PW/GNPs 5 wt%/Al_2O_3 40 wt% composite reaches approximately 65 °C, a notable 20 °C improvement when compared with the POE/PW blend. The pragmatic and uncomplicated preparation method, coupled with the stellar performance of the composites, opens a promising avenue and broader possibility for developing flexible PCMs for solar conversion and thermal storage applications.

Keywords: phase change materials; injection moulding; solar energy storage; photo-thermal conversion efficiency

Citation: Tian, J.; Wang, C.; Wang, K.; Xue, R.; Liu, X.; Yang, Q. Flexible Polyolefin Elastomer/Paraffin Wax/Alumina/Graphene Nanoplatelets Phase Change Materials with Enhanced Thermal Conductivity and Mechanical Performance for Solar Conversion and Thermal Energy Storage Applications. *Polymers* **2024**, *16*, 362. https://doi.org/10.3390/polym16030362

Academic Editors: Andrew N. Hrymak and Shengtai Zhou

Received: 30 December 2023
Revised: 23 January 2024
Accepted: 25 January 2024
Published: 29 January 2024

1. Introduction

As electronics and energy storage devices become more tightly packed, high-performance thermal interface materials (TIMs) are gaining attention [1–3]. Overheating leads to the decline of performance, reduced lifespan, and safety risks due to a sudden increase in internal heat flux. To ensure safety within a temperature range, it is vital to advance the development of effective materials and technologies for thermal management to ensure the prompt dissipation of accumulated heat. Moreover, given the pressing issues and concerns relating to the population growth, the environment, and energy challenges, the demand for eco-friendly energy has intensified [4–6].

Phase change materials (PCMs) are used to store energy and release excess heat during phase changes, making them a key component of thermal management and storage [4,7–9].

In general, PCMs mainly include organic PCMs and inorganic PCMs [10]. Compared with inorganic PCMs, organic PCMs have a low supercooling degree, high storage density, and no corrosivity, making them more widely used in industrial sectors [4]. Paraffin wax (PW), which is recognized for its versatility as an organic material, is highly valued for its capacity to adjust its phase change temperature, substantial latent heat capacity, and cost effectiveness [11–13]. These attributes make it applicable across various thermal management domains, such as solar energy, electronics, and power batteries. However, during practical use, PW-based phase change composites (PWPCMs) face substantial challenges, exhibiting an inherent low thermal conductivity, potential leakage, inefficient light-to-heat conversion, and lack of flexibility, which decreases the efficiency of thermal storage devices [14]. To overcome these challenges, flexible PCMs are developed through methods like encapsulation with elastic shells, incorporating flexible porous structures, and forming polymer networks [15]. Various polymers like unsaturated polyester resin, polymethyl methacrylate, polyvinyl chloride, and thermoplastic elastomer have been used [16–20]. Yang et al. [21] found that using a melamine sponge with excellent elasticity enhanced stability and thermal conductivity. Bing et al. [22] created a UPR/EG/PEG composite that could be used for solar energy applications, and it was found effective in absorbing sunlight, converting photo-thermal energy, and energy storage. However, flexible PCMs that include polymer-supporting structures experience sluggish thermal response due to the low thermal conductivity of organic solid-liquid PCMs.

Researchers have adopted diverse approaches, including utilizing carbon-based materials, metallic fillers, and inorganic ceramic fillers such as MXene, graphite, graphene nanoplatelets (GNPs), carbon nanotubes (CNTs), silver particles, silver nanowires, and spherical alumina (Al_2O_3) to enhance the photo-thermal conversion and the thermal conductivity of PCMs [23,24]. For example, Qi et al. [25] constructed a PCM that consisted of graphene foam and PW (i.e., GF/PW), which had a three-dimensional (3D) network, giving it excellent shape stability and a high thermal energy storage density. Notably, it demonstrated a remarkable 87% enhancement in thermal conductivity and an 89% improvement in solar-thermal conversion efficiency. Wei et al. [3] employed a simple physical blending method to integrate GNPs as thermal conductive fillers into a cross-linked polyolefin elastomer (POE), resulting in flexible PCMs with a thermal conductivity of about $\sim$5.11 W m^{-1}K^{-1} and a notable capacity for efficient solar-thermal conversion. Ishida and Rimdusit [26] achieved a thermal conductivity as high as 32.5 W m^{-1}K^{-1} in a polybenzoxazine composite containing boron nitride (BN) at a filler concentration of 78.5 vol% (88 wt%). Despite the improved thermal conductivity, the mechanical properties were compromised due to the formation of defects at the polymer–filler interface, which negatively affected the phonon transport and mechanical properties [27]. The observed decline in mechanical properties after filler addition is consistent across various flexible PCMs, posing a challenge to effectively improve thermal conductivity while preserving favorable mechanical properties.

Numerous studies showed that incorporating hybrid fillers improved the performance of functional fillers in polymer composites, enhancing properties like electromagnetic interference shielding [28,29], wave absorption [30,31], electrical conductivity [32], and thermal conductivity [33]. Ren et al. [34] observed enhancements in both mechanical properties and thermal conductivity of polypropylene (PP) through the addition of GNPs. Jin et al. [35] reported an increase in the thermal conductivity of polycarbonate (PC) from 0.19 to 1.42 W m^{-1}K^{-1} by adding 20 wt% BN, 1 wt% GNPs, and 1 wt% CNTs. Although significant research progress has been achieved, there remains a lack of comprehensive understanding regarding hybrid fillers in preparing functional polymer composites. Developing PCMs with latent heat and concurrently improving mechanical strength, light-to-heat conversion, and heat transport is necessary to overcome obstacles for experimental design. Further investigation is essential to uncovering the intricate relationship between the structure and performance of PCMs.

In this study, a series of POE/PW/GNPs/Al$_2$O$_3$ PCMs were prepared using an injection moulding technique. The mechanical strength, flexibility, thermal conductivity, and efficiency in converting light to heat were systematically studied. Additionally, the investigation delved into detailed analyses of morphology, structure, crystallization, thermal stability, and simulated thermal management scenarios. The objective is to develop flexible PCMs with noteworthy thermal conductivity, large latent heat storage, effective light-heat conversion, and high tensile strength. This work aimed to enable integrated functionalities such as photothermal conversion, efficient heat storage, and versatile utilization in various thermal management applications. The proposed method provides a facile method to prepare PCMs with enhanced performance that show promising applications in the energy storage, solar utilization, and advanced thermal management sectors.

2. Materials and Methods

2.1. Materials

Paraffin wax (PW), with an apparent density of 0.90 g cm^{-3}, was obtained from China Sinopharm Group, Beijing, China. Polyolefin elastomer (POE), commercially known as 3980, was provided by ExxonMobil Chemical Co., Ltd., Spring, TX, USA, and it has a density of 0.89 g cm^{-3}. Graphene nanoplatelets (GNPs), with a thickness of 4~20 nm and filler size ranging from 5~10 μm, were supplied by Chengdu Institute of Organic Chemistry, Chengdu, China. Spherical alumina (Al$_2$O$_3$), with an average particle size of 20 μm, was purchased from Tianxing New Materials Technology Co., Ltd., Xiaoyi, China.

2.2. Sample Preparation

Prior to blending, POE pellets, Al$_2$O$_3$, and GNPs were dried at 60 °C for at least 10 h. After this, a sequence of POE/PW/GNPs/Al$_2$O$_3$ composites were meticulously formulated through melt blending, employing an internal mixer (XSS-300) at 90 °C and 50 rpm for 5 min. The specific formulations for the prepared samples are tabulated in Table 1. Afterwards, the resulting mixtures were pulverized using a grinder, and they were used for injection moulding using a Thermo Scientific HAAKE Minijet apparatus (Waltham, MA, USA). The injection pressure was set at 500 bar, and the volumetric flow rate was about 3.0 cm^3 s^{-1}. The dimensions of the samples were 80 × 10 × 4 mm^3 (length × width × thickness), and the mold temperature was 40 °C.

Table 1. Formulations of flexible PCMs.

Designation	Composition			
	POE (wt%)	PW (wt%)	GNPs (wt%)	Al$_2$O$_3$ (wt%)
PPW	70	30	0	0%
PPWG$_5$	70	30	5	0
PPWG$_5$Al$_{10}$	70	30	5	10
PPWG$_5$Al$_{20}$	70	30	5	20
PPWG$_5$Al$_{30}$	70	30	5	30
PPWG$_5$Al$_{40}$	70	30	5	40
PPWG$_5$Al$_{40}$	70	30	0	40

2.3. Characterization

2.3.1. Differential Scanning Calorimetry (DSC)

To analyze the melting and crystallization behavior of the flexible phase change materials (FPCMs), a differential scanning calorimeter (DSC, TGA/DSC-1, Mettler Toledo, Zürich, Switzerland) was employed. Samples, weighing 5~8 mg, were heated to 200 °C at 10 °C min^{-1}, and they were held at 200 °C for 5 min to eliminate thermal history. Subsequently, samples were cooled down to 40 °C at 10 °C min^{-1}. All tests were carried out in a nitrogen (N$_2$) atmosphere.

2.3.2. Thermal Conductivity

The thermal diffusivity (α) of samples was determined by a laser flash method using a Discovery DXF-900 instrument (TA Instruments, New Castle, PA, USA). To determine the thermal conductivity (λ) of FPCMs, the following equation was employed: $\lambda = \alpha \times C_p \times \rho$, where ρ is density [36]. Here, α and C_p denote the thermal diffusivity and specific heat capacity, respectively.

The bulk density was measured using the water displacement method with the help of an MH-120E apparatus (Matsu Haku, Taiwan, China). The C_p was determined by DSC using sapphire as the reference.

2.3.3. Scanning Electron Microscope (SEM)

The morphology of FPCMs was observed using a scanning electron microscope (SEM, Nova Nano 450, FEI, Hillsboro, OR, USA). The samples were cryo-fractured in liquid nitrogen, followed by coating a thin layer of gold to enhance image resolution.

2.3.4. Electrical Conductivity

A high-resistance meter (model TH2684A, Changzhou Tonghui Electronics Co., Ltd., Changzhou, China) was used to evaluate the electrical conductivity (σ) of FPCMs. The testing range of the electrometer ranged from 10 kΩ to 100 TΩ. The values of σ for samples were determined using Equation (1) [37]:

$$\sigma = 1/\rho = L/RS \tag{1}$$

where, ρ is electrical resistivity, L is the distance between the copper electrodes, R is the volume resistance, and S is the cross-sectional area.

2.3.5. Viscoelastic Behavior

To delve into the viscoelastic properties, a dynamic rheometer (MCR302, Anton Paar, Graz, Austria) was employed at a temperature of 90 °C. Analyses were performed using a constant-strain mode with a strain rate of 1%. To investigate the storage modulus (G'), viscous modulus (G''), and complex viscosity (η^*) at various frequencies ranging from 100 to 0.01 rad s^{-1}, specimens with a diameter of 25 mm and thickness of 1 mm were employed.

2.3.6. X-ray Diffraction (XRD)

The crystalline structure and diffraction characteristics of FPCMs were studied using X-ray diffraction using an X-ray diffractometer (XRD, X'Pert Pro, PANalytical, Almelo, The Netherlands). The diffraction patterns were collected within a 2θ range of 5~60° at 5° min^{-1}.

2.3.7. Mechanical Properties

The mechanical properties of FPCMs were assessed using an Instron 4302 universal mechanical tester (Instron Co., Canton, MA, USA). The crosshead speed was 50 mm min^{-1}.

2.3.8. Light-to-Energy Conversion and Energy Storage

The energy transformation from light to electricity was accomplished with the assistance of a CELHXUV300 xenon lamp (CEAULIGHT, Beijing, China) and an AM 1.5 filter, along with a CEL-NP2000 optical power meter and a Seebeck thermoelectric device at 25 °C. A Keithley electrometer (2400, Cleveland, OH, USA) was utilized to record real-time voltage data.

2.3.9. Light-to-Temperature Energy Conversion

The energy conversion from light to temperature was conducted using a CELHXUV300 xenon lamp (CEAULIGHT, China) equipped with an AM 1.5 filter, a CEL-NP2000 optical power meter (CEAULIGHT, China), and a real-time temperature detector (TA612A, TASI,

Suzhou, China). The samples were positioned within a foam insulation system, and the lamp's simulated sunshine was radiated directly on the surface of samples. The light intensity was measured and adjusted using an optical power meter (CEL-NP2000, Beijing Zhongguo Jinyuan Science and Technology Co., Ltd., Beijing, China) to ensure accuracy. The calibration procedure allowed for exact control and measurement of light intensity during light-to-temperature energy conversions, providing vital insights into the performance of FPCMs under various lighting circumstances.

3. Results and Discussion

3.1. Rheological Properties

The flowability of composites plays a crucial role in the injection moulding process, which affects the filling properties and microstructural evolution of the products [24,38]. To investigate the effect of adding Al_2O_3 and GNPs on the flowability of POE/PW composites, the rheological properties of the composites were tested. Figure 1 presents the frequency-dependent viscoelastic properties of the POE/PW/Al_2O_3/GNPs composite, including storage modulus (G'), loss modulus (G''), and complex viscosity (η^*). Figure 1a,b indicate that introducing PW into the POE significantly reduces the values of both G' and G'', revealing that the elasticity of the POE/PW composites is enhanced. However, adding Al_2O_3 and GNPs into POE/PW composites displays a reverse phenomenon, in which the G' and G'' increases with the presence of inorganic fillers. The above revealed that the added fillers had a reinforcing effect on the POE/PW composites. Moreover, the η^* of all samples exhibits a gradual decrease with increasing shear frequency within the entire frequency range, known as the "shear thinning" phenomenon [39]. As shown in Figure 1c, all composites exhibit a plateau where the η^* of the composites does not change with an increase in a low-frequency region, indicating that the composites possess the characteristics of non-Newtonian fluids. Whereas the η^* of the composites decreases with further increasing angular frequency, exhibiting the phenomenon of Newtonian fluid. Nevertheless, the η^* for samples containing fillers demonstrates an increase with the incorporation of Al_2O_3 and GNPs. The results suggest that the addition of fillers impairs the mobility of polymer melts, causing an increase in η^*. It is noteworthy that the η^* of the composites is consistently lower than pure POE, indicating enhanced flowability and processability.

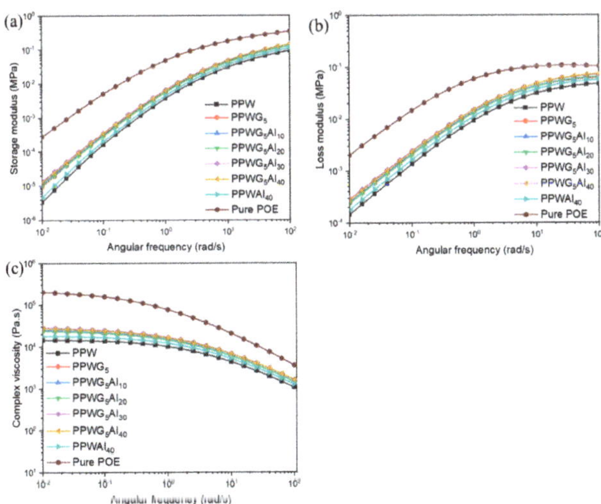

Figure 1. (**a**) Storage modulus (G'), (**b**) loss modulus (G''), and (**c**) complex viscosity (η^*) of POE/PW and POE/PW/GNPs/Al_2O_3 composites as a function of frequency at 90 °C.

3.2. Morphology

Figure 2 displays the SEM images illustrating the cross-sectional view of FPCMs. The PW exhibits a dense morphology, and the PW of spherical shapes that are irregular and vary in size are tightly integrated into a composite structure of POE. The irregular spherical shape of PW was attributed to the difference in the wettability between the POE and PW. In Figure 2b, it is evident that the addition of GNPs effectively improves the distribution of PW, resulting in smaller particle sizes. Interestingly, Figure 2c–g show that the presence of Al_2O_3 affects neither the morphology nor the structure of PW-based FPCMs. PW is well distributed among the inorganic particles, serving as an effective connector and contributing to enhancing the mechanical properties of the composites.

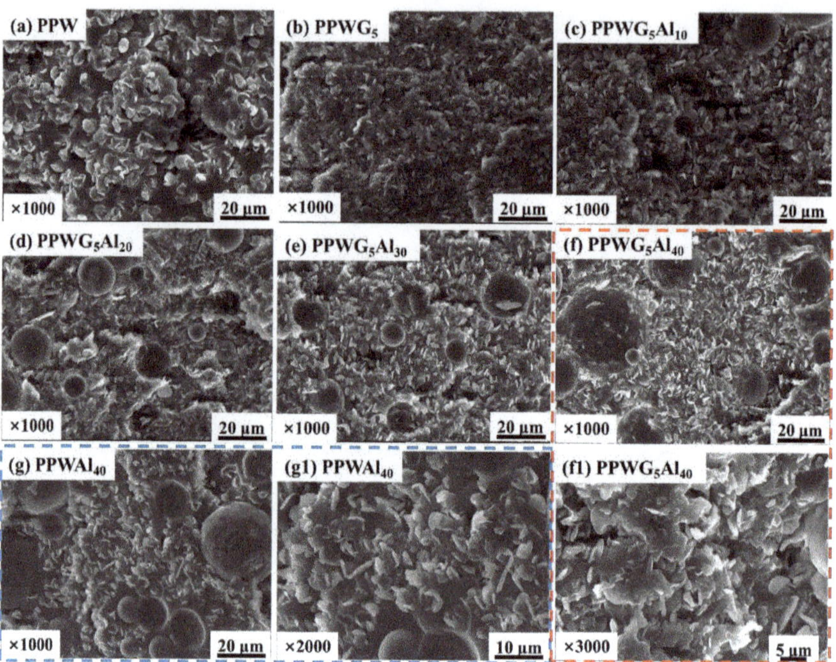

Figure 2. SEM images of FPCMs, in which (**f1**) and (**g1**) were the local magnification of (**f**) and (**g**), respectively.

3.3. Crystallization Behavior of FPCMs

The XRD patterns of FPCMs are shown in Figure 3a. The characteristic peaks of FPCMs are a combination of the peaks of several components (GNPs, Al_2O_3, and PW), which proves that the filling process is basically physical mixing. Moreover, compared with pure PPW, the position and intensity of the peaks at 21.4° and 23.8° of FPCMs do not change significantly. This suggests that PW retained its crystalline state within FPCMs, ensuring the effective release of latent heat.

The heat storage capacity, a critical parameter for FPCMs, was determined by DSC. Figure 3b illustrates the DSC curves for FPCMs with varying Al_2O_3 loadings. The corresponding results are shown in Table 2. During the melting process, there are two obvious phase transition peaks in the DSC curves of PW, among which, the peak of about 35 °C corresponds to the solid-solid phase transition of PW, and with the further increase in temperature, PW undergoes a solid-liquid phase change; a second endothermic peak appears at about 50 °C, which reflects the good heat storage ability of PW. The DSC curves of FPCMs with different loadings of Al_2O_3 and GNPs showed highly overlapping patterns, indicating a similar phase transition behavior, with two endothermal peaks appearing during the melting process. This suggested that the presence of Al_2O_3 and GNPs allowed for a normal phase transition process. However, a slight difference between FPCMs and pure PW was noted in the melting temperature. This is mainly because the cross-linking effect of FPCMs resulted in an expansion of PW particles during the melting process.

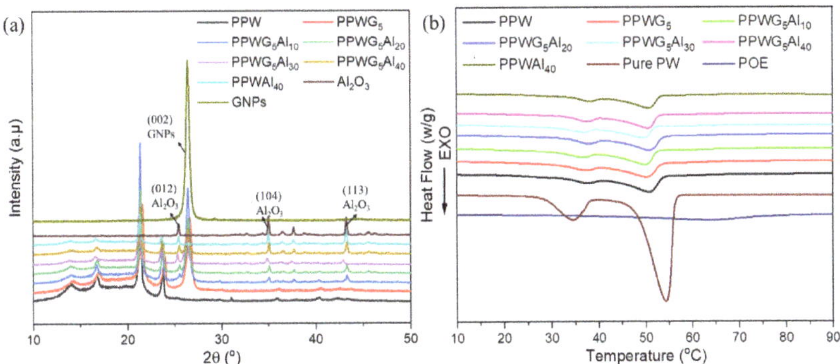

Figure 3. (a) The XRD curves and (b) DSC curves of FPCMs.

Table 2. The thermal analysis data of DSC.

Samples	Melting Temperature (°C)	Melting Enthalpy (J/g)	Theoretical Value (J/g)
PPW	50.7	64.9	66.4
PPWG$_5$	50.1	48.7	49.1
PPWG$_5$Al$_{10}$	49.9	53.4	57.7
PPWG$_5$Al$_{20}$	50.6	52.8	53.1
PPWG$_5$Al$_{30}$	49.9	64.6	66.4
PPWG$_5$Al$_{40}$	50.3	48.3	49.1
PPWAl$_{40}$	50.5	47.4	49.4
PW	53.7	221.2	/

3.4. Mechanical Properties

The application of FPCMs demands commendable mechanical properties, often presenting a challenge since PW has a high degree of rigidity in the solid state. Typically, the simultaneous improvement of mechanical strength and heat storage properties in FPCMs is considered contradictory. Figure 4a illustrates the stress–strain curves of FPCMs filled with varying amounts of Al_2O_3, and the results are shown in Figure 4b. Introducing PW into POE proves to be highly effective, significantly enhancing the elongation at break from approximately 650% to around 1100%, representing a twofold improvement. Notably, the PPW, with and without GNPs, demonstrated the ability to withstand tensile stresses of 15.3 and 14.8 MPa, respectively. Correspondingly, the tensile strain reached 930% and 1127%, respectively. The tensile strength and strain of PW-based PCMs demonstrated a regular decrease with increasing Al_2O_3 loadings. Even at an Al_2O_3 loading of 40 wt%, FPCMs exhibited a tensile stress of 12 MPa and a strain of 900%. FPCMs with mechanical strength serve as an effective thermal management solution. Despite the irregular distribution of Al_2O_3 and PW, the cross-linking structures contributed to improving the mechanical strength of FPCMs, enabling a simultaneous enhancement in strong mechanical strength, converting light into heat, and maintaining latent heat.

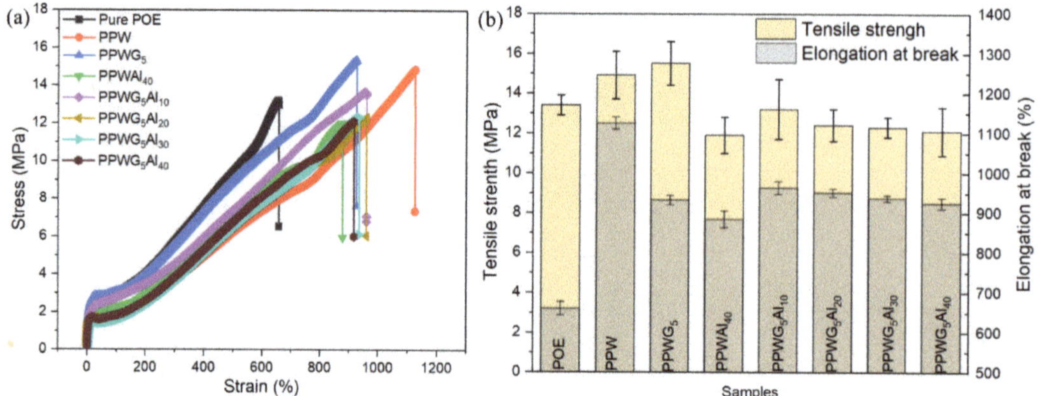

Figure 4. (**a**) Stress–strain curves of FPCMs and (**b**) the corresponding average data.

3.5. Thermal Conductivity

Figure 5a–c depicts the in-plane and through-plane λ as well as the enhancement rate of λ for POE/PW/Al$_2$O$_3$/GNPs composites. The λ of PPW without fillers is poor, with in-plane and through-plane λs of 0.49 and 0.18 W m^{-1}K^{-1}, respectively. The difference in λ between the in-plane and through-plane of PPW is due to the orientation of PPW chains under the action of a shear force field during injection moulding. The λ of PPW increased due to the addition of functional fillers. In particular, the PPWG$_5$ had an in-plane λ of 1.1 W m^{-1}K^{-1} and a through-plane λ of 0.22 W m^{-1}K^{-1}. This was related to the oriented arrangement of polymer chains that improved the λ by increasing the mean free path of phonons. Figure 5a,b show that the λ of POE/PW/GNPs/Al$_2$O$_3$ in different orientations increases linearly with the increase in Al$_2$O$_3$ content, which suggests that the increase in filler content facilitated the formation of 3D conductive pathways, thereby contributing to an increase in λ. The in-plane and through-plane λs of PPWG$_5$Al$_{40}$ are 1.42 and 0.38 W m^{-1}K^{-1}, which increased by 269% and 111%, respectively, when compared to pure PPW (Figure 5c). This is mainly due to the complementary shapes of Al$_2$O$_3$ and GNPs, which exert a synergistic effect inside the matrix and significantly improve the λ. Furthermore, the composites have commendable insulating properties, as confirmed by the insulation analysis in Figure 5d. Table 3 shows the λ using similar thermally conductive fillers or phase change materials. It can be noted that this work is the highest in both the values of λ and the degree of increase in λ. This further confirms that the spherical Al$_2$O$_3$ and lamellar GNPs are able to complement each other to exert a synergistic effect within the matrix, which greatly facilitates the construction of the thermal conductive network.

Table 3. Comparison of thermal conductivity and enhancement rate of thermal conductivity.

Filler Type	Phase Change Material	Thermal Conductivity (Wm^{-1}K^{-1})	Thermal Conductivity Enhancement (%)	Reference
PVA/Graphite	PW	0.29	20.8	[40]
Graphene	PW	0.32	18.5	[41]
OBC/GNPs	Palmitic acid	0.79	155.0	[4]
rGO	PW	0.46	84.2	[19]
Graphene	PW	0.41	17.1	[42]
PVA/MXene	PW	0.62	138.5	[43]
POE/PW/Al$_2$O$_3$/GNPs	PW	1.82	269.5	This work

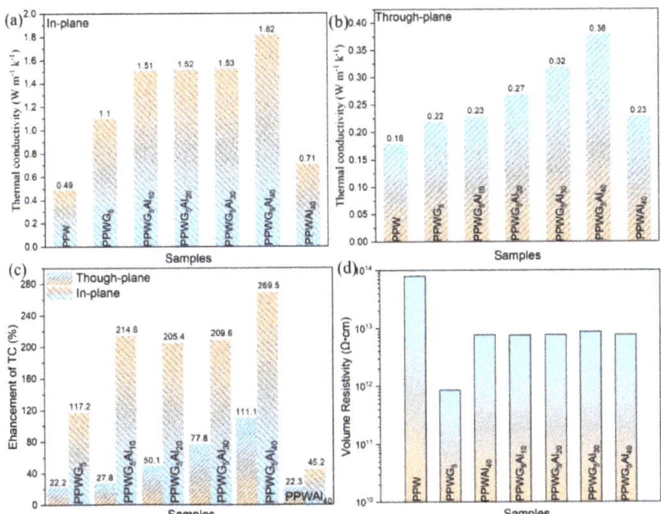

Figure 5. (**a**) In-plane and (**b**) through-plane thermal conductivity (λ) of FPCMs, (**c**) the enhancement of λ for FPCMs when compared with PPW, and (**d**) the volume resistivity of FPCMs.

The service life of LED lamps decreases exponentially with the increase in operating temperature, and the accumulation of heat causes problems such as wavelength shift and output power reduction. To verify the heat dissipation of thermal management materials, FPCMs with a thickness of 2 mm were placed between the LED chip and heat sink. A temperature recorder and a thermal infrared imager were also used to record the temperature of the LED lamp surface in real time, as shown in Figure 6a,b. The results showed that when using PPW as the thermal management material, the temperature of the LED surface increases gradually with the prolongation of working time, and when the time reaches 270 s, the temperature of the LED surface reaches about 75 °C, since PPW has a lower λ and it is not able to dissipate the heat generated by the LED chip. However, when using POE/PW/GNPs/Al$_2$O$_3$ composites, the temperature of the LED chip shows a slower increase and stays at a relatively low steady state of 77.3 °C (PPWAl$_{40}$) and 65.2 °C (PPWG$_5$), respectively. A comparison of the data reveals that the PPWG$_5$Al$_{40}$ has a higher thermal management capability with the concurrent addition of GNPs and Al$_2$O$_3$. The temperature of the LED chip sustained at 56.9 °C, which is about 13 °C higher than that of heat sink. This fully demonstrates that PPWG$_5$Al$_{40}$ has the best heat dissipation ability, which is consistent with its high λ.

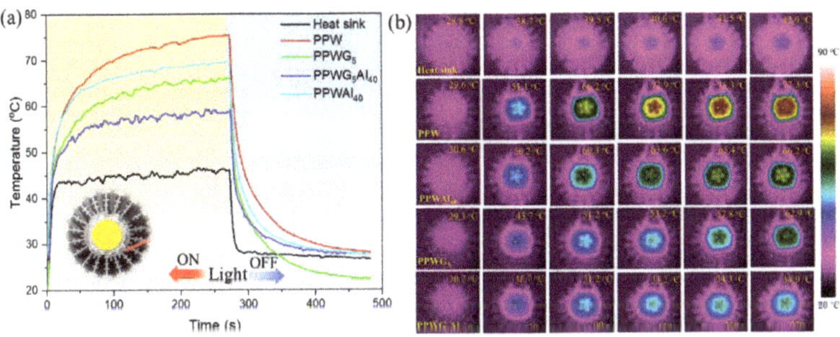

Figure 6. (**a**) The temperature variation of LED with time (the temperature of the center of LED chip was measured) and (**b**) the infrared images of working LED using FPCMs as a thermal management material after 5 min.

3.6. Light-to-Heat Conversion

As shown in Figure 7a, the ability of photo-thermal conversion was evaluated using a homemade device, in which a composite film was placed on a heat sink, and the change in surface temperature was recorded by an infrared thermographic camera under the simulated sunlight. As shown in Figure 7b, the surface temperature of POE/PW/GNPs/Al_2O_3 increases rapidly at a light density of 80 mW cm^{-2}, and a temperature plateau appears during the heating process. The maximum temperature of PPW as well as PPWAl$_{40}$ films reaches about 45 °C. However, with the addition of GNPs, the maximum temperature of PPWG$_5$ was substantially increased to 60 °C. Moreover, by introducing Al_2O_3 into PPWG$_5$, the maximum temperature of PPWG$_5$Al$_y$ can be further enhanced. For example, the PPWG$_5$Al$_{40}$ film shows the fastest heating rate with the highest temperature. Therefore, it can be concluded that GNPs and Al_2O_3 play a dominant role in photothermal conversion ability. As shown in Figure 7c, the surface temperatures of FPCMs were analyzed at the light intensities of 50, 80, and 120 mW cm^{-2}, respectively, and the surface temperatures of FPCMs increased with the increase in light power. PPWG$_5$Al$_{40}$ reaches a very high temperature of about 87 °C at a light intensity of 120 mW cm^{-2}.

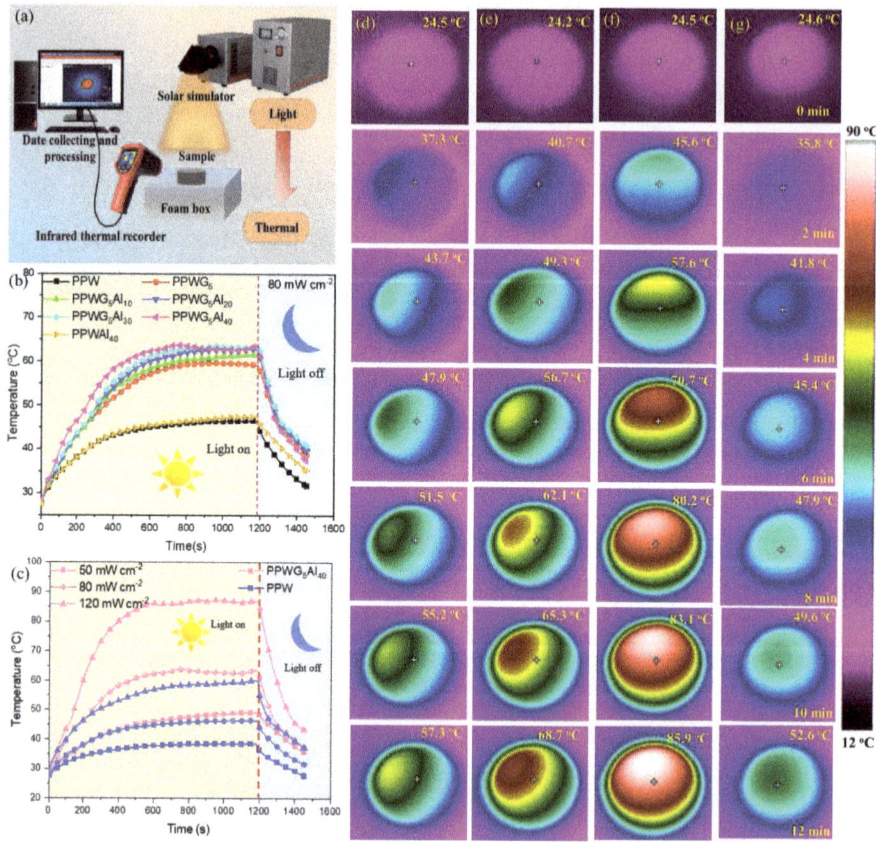

Figure 7. (a) Illustration of light-to-thermal energy conversion equipment; (b) surface temperatures of FPCMs under various NIR light irradiation; (c) surface temperature changes of FPCMs under 80 mW cm^{-2} NIR light irradiation and (d–f) the infrared thermal images of PPWG$_5$Al$_{40}$ composites under 50, 80, and 120 mW cm^{-2} NIR light irradiation, respectively; and (g) the infrared thermal images of FPCMs under 120 mW cm^{-2} NIR light irradiation.

To further validate the impact of GNPs and Al_2O_3 on light-to-thermal conversion performance, an analysis of the surface temperature distribution was conducted using an infrared thermal camera for FPCMs. As depicted in Figure 7d, with an increase in irradiation time, the surface temperature of FPCMs consistently increased. Notably, FPCMs with GNPs and Al_2O_3 exhibited higher surface temperatures when compared to those without fillers at equivalent irradiation times. Furthermore, a direct correlation was observed between higher sunlight irradiation and the increased surface temperature of FPCMs. For instance, after a 12 min irradiation period, the average surface temperature of $PPWG_5Al_{40}$ at 50, 80, and 120 mW cm^{-2} reached 57.3, 68.7, and 85.9 °C, respectively. In contrast, the average surface temperature of pure PPW at 120 mW cm^{-2} was only 52.6 °C. These findings further substantiated that the incorporation of GNPs and Al_2O_3 effectively enhances the light-to-thermal conversion performance of FPCMs.

The GNPs used in this study enhance the photon-absorbing capacity of FPCMs, allowing them to achieve the energy conversion of light property that graphene has demonstrated, which shows considerable promise for light-driven devices [39]. The "Seeback" thermoelectric device was utilized to construct a light-heat-electricity energy conversion that utilizes PCMs. This system utilizes a "Seeback" thermoelectric device, with a composite PCM and tap water serving as the heat source and cold source for the thermoelectric component, as depicted in Figure 8a. The intensity of the simulated sunshine was precisely regulated to 80 mW cm^{-2}, while a real-time electrostatic meter was employed to measure the voltage. As shown in Figure 8a, the voltage of POE/PW-based composites increases rapidly at a light intensity of 80 mW cm^{-2}, and a clear voltage plateau occurs during the heating process. The maximum voltage of PPW as well as $PPWAl_{40}$ reaches about 26 and 30 mV, respectively. However, with the addition of GNPs, the maximum voltages of $PPWG_5$ and $PPWG_5Al_y$ are substantially increased to about 40 mV. Therefore, it can be concluded that the addition of GNPs significantly improves the photo thermoelectric conversion efficiency. As shown in Figure 8b, the photo thermoelectric conversion efficiency of $PPWG_5Al_{40}$, which has the best photo thermoelectric conversion efficiency, is further investigated by analyzing the voltage of the films at light intensities of 50, 80, and 120 mW cm^{-2}, respectively, and the voltage of the films increases with the increase in light power. The $PPWG_5Al_{40}$ achieves a higher voltage of around 45 mV at a light intensity of 120 mW cm^{-2}. It further shows that the photothermal power conversion efficiency of the composite film is better. It is worth mentioning that this work uses tap water as a cold source, and it is close to the actual application scenario.

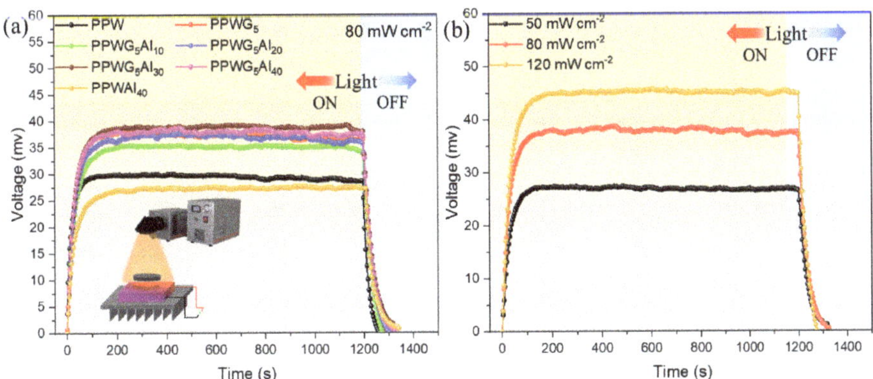

Figure 8. (a) Light-to-electric conversion and V-t curves of FPCMs under the sunlight irradiation of 80 mW cm^{-2} and (b) V-t curves of $PPWG_5Al_{40}$ under 50, 80, and 120 mW cm^{-2}, respectively.

3.7. Thermal Cycling Stability of FPCMs

The PPW and PPWG$_5$Al$_{40}$ were chosen to analyze the changes of crystalline structures undergoing 100 thermal cycles, as shown in Figure 9a. As can be seen from Figure 9a, no new diffraction peaks appeared in the XRD patterns of all samples before and after cycling, indicating that no change in the crystalline structure of the composites occurred. Moreover, the thermal cycling stability of PPWG$_5$Al$_{40}$ was characterized by DSC, as shown in Figure 9b. From the DSC curves in Figure 9b, it can be seen that the phase transition temperature and enthalpy of PPWG$_5$Al$_{40}$ did not change significantly after 100 DSC cycles, especially the enthalpy of the phase change basically did not have any loss, which indicates that PPWG$_5$Al$_{40}$ can still maintain stable thermophysical properties and chemical structure after 100 cycles of elevated/lowered temperatures, and it exhibits excellent cycling stability. To further validate its stability, multiple cyclic tests were performed on the FPCMs under 80 mW cm^{-2} NIR light irradiation, as displayed in Figure 9d. As can be seen from Figure 9d, the photothermal curves remained unchanged after multiple cycles, indicating that the PPWG$_5$Al$_{40}$ exhibits good stability.

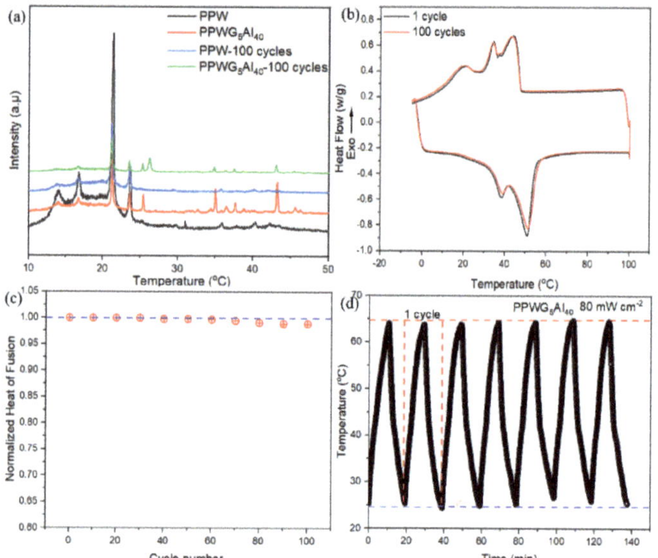

Figure 9. (**a**) The XRD patterns of FPCMs and (**b**) the DSC curves of PPWG$_5$Al$_{40}$ with 100 thermal cycles, (**c**) the mass specific heat of fusion normalized by the first heating cycle, and (**d**) the surface temperature changes of FPCMs under 80 mW cm^{-2} NIR light irradiation.

4. Conclusions

This work successfully prepared a series of thermal-induced flexible phase change composites (FPCMs) consisting of polyolefin elastomer (POE), paraffin wax (PW), graphene nanoplatelets (GNPs), and aluminum oxide (Al$_2$O$_3$) through blending and injection moulding. The composites exhibited good thermal management, coupled with superior performance in terms of photothermal conversion and heat transfer. A comprehensive analysis was conducted concerning the synergistic contribution of PW and GNPs to enhance the mechanical strength, the conversion of light to heat, and the heat storage capabilities of FPCMs. The underlying mechanism driving this enhancement was thoroughly elucidated. GNPs play a crucial role as photon-absorbing and molecular heaters that are effective under solar irradiation, inducing lattice vibrations in FPCMs, and facilitating efficient light-to-heat conversion. Furthermore, the combination of GNPs and Al$_2$O$_3$ established thermal conduction networks, thereby augmenting the heat transfer ability. The synergistic effect of hybrid fillers resulted in the fabrication of PPWG$_5$Al$_{40}$ with a tensile strength of 13 MPa, elongation at break of about 900%, in-plane thermal conductivity of 1.82 W m^{-1}K^{-1}, and efficient improvement in light-to-thermal conversion. In addition to exhibiting favorable crystallization properties, FPCMs displayed high light-driven shape recovery capabilities, acceptable thermal stability, and significant temperature control properties. Collectively, these attributes underscore the suitability of these composites for thermal management applications in electronics, among others.

Author Contributions: J.T.: conceptualization, methodology, formal analysis, writing—original draft preparation, project administration, and funding acquisition; K.W. and R.X.: resources and formal analysis; C.W. and X.L.: investigation, data curation, and resources; Q.Y.: conceptualization, writing—review and editing and supervision. All authors have read and agreed to the published version of the manuscript.

Funding: This research was funded by the Project of Basic Research Programme of Shaanxi Provincial Department of Education (No. 21JK0571).

Institutional Review Board Statement: Not applicable.

Informed Consent Statement: Not applicable.

Data Availability Statement: Data are contained within the article.

Conflicts of Interest: The authors declare no conflicts of interest. The funders had no role in the design of the study; in the collection, analyses, or interpretation of data; in the writing of the manuscript; or in the decision to publish the results.

References

1. Feng, C.P.; Bai, L.; Shao, Y.; Bao, R.Y.; Liu, Z.Y.Y.; Yang, M.B.; Chen, J.; Ni, H.Y.; Yang, W. A Facile Route to Fabricate Highly Anisotropic Thermally Conductive Elastomeric POE/NG Composites for Thermal Management. *Adv. Mater. Interfaces* **2018**, *5*, 1700946. [CrossRef]
2. Wang, J.; Li, W.; Zhang, X. An Innovative Graphene-based Phase Change Composite Constructed by Syneresis with High Thermal Conductivity for Efficient Solar-Thermal Conversion and Storage. *J. Mater. Sci. Technol.* **2024**, *178*, 179–187. [CrossRef]
3. Bai, Y.; Shi, Y.; Zhou, S.T.; Zou, H.W.; Liang, M. A Concurrent Enhancement of Both In-Plane and Through-Plane Thermal Conductivity of Injection Molded Polycarbonate/Boron Nitride/Alumina Composites by Constructing a Dense Filler Packing Structure. *Macromol. Mater. Eng.* **2021**, *306*, 2100267. [CrossRef]
4. Bing, N.; Wu, G.; Yang, J.; Chen, L.; Xie, H.; Yu, W. Thermally Induced Flexible Phase Change Composites with Enhanced Thermal Conductivity for Solar Thermal Conversion and Storage. *Sol. Energy Mater. Sol. Cells* **2022**, *240*, 111684. [CrossRef]
5. Zhang, Y.; Wang, J.; Qiu, J.; Jin, X.; Umair, M.M.; Lu, R.; Zhang, S.F.; Tang, B.T. Ag-Graphene/PEG Composite Phase Change Materials for Enhancing Solar-Thermal Energy Conversion and Storage Capacity. *Appl. Energy* **2019**, *237*, 83–90. [CrossRef]
6. Su, T.Y.; Liu, N.S.; Lei, D.D.; Wang, L.X.; Ren, Z.Q.; Zhang, Q.X.; Su, J.; Zhang, Z.; Gao, Y.H. Flexible MXene/Bacterial Cellulose Film Sound Detector Based on Piezoresistive Sensing Mechanism. *ACS. Nano* **2022**, *16*, 8461–8471. [CrossRef] [PubMed]
7. Zhang, C.; Zheng, X.; Fang, Y.; Zhang, Z.; Gao, X.; Xie, N. Flexible Composite Phase Change Materials with Enhanced Thermal Conductivity and Mechanical Performance for Thermal Management. *J. Energy Storage* **2023**, *11*, 18832–18842.
8. Lin, Y.; Kang, Q.; Liu, Y.; Zhu, Y.; Jiang, P.; Mai, Y.W.; Huang, X. Flexible, Highly Thermally Conductive and Electrically Insulating Phase Change Materials for Advanced Thermal Management of 5G Base Stations and Thermoelectric Generators. *Nano-Micro Lett.* **2023**, *15*, 31. [CrossRef] [PubMed]
9. Tang, X.; Deng, J.; Wu, Z.; Li, X.; Wang, C. Experimental Investigation on BN-based Flexible Composite Phase-Change Material for Battery Module. *Front. Energy Res.* **2022**, *10*, 801341. [CrossRef]
10. Tang, Z.D.; Gao, H.Y.; Chen, X.; Zhang, Y.F.; Li, A.; Wang, G. Advanced multifunctional composite phase change materials based on photo-responsive materials. *Nano Energy* **2021**, *80*, 105454. [CrossRef]
11. Li, X.; Chen, H.; Liu, L.; Lu, Z.; Sanjayan, J.G.; Duan, W.H. Development of Granular Expanded Perlite/Paraffin Phase Change Material Composites and Prevention of Leakage. *Sol. Energy* **2016**, *137*, 179–188. [CrossRef]
12. Gulfam, R.; Zhang, P.; Meng, Z. Advanced Thermal Systems Driven by Paraffin-Based Phase Change Materials—A Review. *Appl. Energy* **2019**, *238*, 582–611. [CrossRef]
13. Zhang, H.; Liu, Z.; Mai, J.; Wang, N.; Zhang, N. Super-Elastic Smart Phase Change Material (SPCM) for Thermal Energy Storage. *Chem. Eng. J.* **2021**, *411*, 128482. [CrossRef]
14. Kou, Y.; Sun, K.; Luo, J.; Zhou, F.; Huang, H.; Wu, Z.S.; Shi, Q. An Intrinsically Flexible Phase Change Film for Wearable Thermal Managements. *Energy Storage Mater.* **2021**, *34*, 508–514. [CrossRef]
15. Shi, J.; Qin, M.; Aftab, W.; Zou, R. Flexible Phase Change Materials for Thermal Energy Storage. *Energy Storage. Mater.* **2021**, *41*, 321–342. [CrossRef]

16. He, L.; Wang, H.; Yang, F.; Zhu, H. Preparation and Properties of Polyethylene Glycol/Unsaturated Polyester Resin/Graphene Nanoplates Composites as Form-Stable Phase Change Materials. *Thermochim. Acta* **2018**, *665*, 43–52. [CrossRef]

17. Zhou, S.T.; Yu, L.; Song, X.; Chang, J.; Zou, H.W.; Liang, M. Preparation of Highly Thermally Conducting Polyamide 6/Graphite Composites via Low-Temperature In Situ Expansion. *J. Appl. Polym. Sci.* **2013**, *131*, 131. [CrossRef]

18. Tang, Y.; Lin, Y.X.; Jia, Y.T.; Fang, G.Y. Improved thermal properties of stearyl alcohol/high density polyethylene/expanded graphite composite phase change materials for building thermal energy storage. *Energy Build.* **2017**, *153*, 41–49. [CrossRef]

19. Tao, Z.; Chen, X.; Yang, M.; Xu, X.; Wang, G. Three-Dimensional rGO@ Sponge Framework/Paraffin Wax Composite Shape-Stabilized Phase Change Materials for Solar-Thermal Energy Conversion and Storage. *Sol. Energy Mater. Sol. Cells* **2020**, *215*, 110600. [CrossRef]

20. Chen, Y.; Zhao, L.; Shi, Y. Preparation of Polyvinyl Chloride Capsules for Encapsulation of Paraffin by Coating Multiple Organic/Inorganic Layers. *J. Taiwan. Inst. Chem. Eng.* **2017**, *77*, 177–186. [CrossRef]

21. Yang, J.; Jia, Y.L.; Bing, N.C.; Wang, L.L.; Xie, H.Q.; Yu, W. Reduced Graphene Oxide and Zirconium Carbide Co-Modified Melamine Sponge/Paraffin Wax Composites as New Form-Stable Phase Change Materials for Photothermal Energy Conversion and Storage. *Appl. Therm. Eng.* **2019**, *163*, 114412. [CrossRef]

22. Bing, N.; Yang, J.; Gao, H.; Xie, H.; Yu, W. Unsaturated Polyester Resin Supported Form-Stable Phase Change Materials with Enhanced Thermal Conductivity for Solar Energy Storage and Conversion. *Renew. Energy* **2021**, *173*, 926–933. [CrossRef]

23. Li, X.; Sheng, X.; Fang, Y.; Hu, X.; Gong, S.; Sheng, M.; Qu, J. Wearable Janus-Type Film with Integrated All-Season Active/Passive Thermal Management, Thermal Camouflage, and Ultra-High Electromagnetic Shielding Efficiency Tunable by Origami Process. *Adv. Funct. Mater.* **2023**, *33*, 2212776. [CrossRef]

24. Bai, Y.; Shi, Y.; Zhou, S.; Zou, H.; Liang, M. Highly Thermally Conductive Yet Electrically Insulative Polycarbonate Composites with Oriented Hybrid Networks Assisted by High Shear Injection Molding. *Macromol. Mater. Eng.* **2022**, *307*, 2100632. [CrossRef]

25. Qi, G.; Yang, J.; Bao, R.; Xia, D.; Min, C.; Wei, Y.; Yang, M.; Wei, D. Hierarchical Graphene Foam-Based Phase Change Materials with Enhanced Thermal Conductivity and Shape Stability for Efficient Solar-To-Thermal Energy Conversion and Storage. *Nano Res.* **2017**, *10*, 802–813. [CrossRef]

26. Ishida, H.; Rimdusit, S. Very High Thermal Conductivity Obtained by Boron Nitride-Filled Polybenzoxazine. *Thermochim. Acta* **1998**, *320*, 177–186. [CrossRef]

27. Sharma, R.; Ganesan, P.; Tyagi, V.; Metselaar, H.; Sandaran, S. Developments in Organic Solid–Liquid Phase Change Materials and Their Applications in Thermal Energy Storage. *Energy Convers. Manag.* **2015**, *95*, 193–228. [CrossRef]

28. Kaushal, A.; Singh, V. Analysis of Mechanical, Thermal, Electrical and EMI Shielding Properties of Graphite/Carbon Fiber Reinforced Polypropylene Composites Prepared Via a Twin Screw Extruder. *J. Appl. Polym. Sci.* **2022**, *139*, 51444. [CrossRef]

29. Jia, Z.; Kou, K.; Yin, S.; Feng, A.; Zhang, C.; Liu, X.; Wu, G. Magnetic Fe Nanoparticle to Decorate N Dotted C as An Exceptionally Absorption-Dominate Electromagnetic Shielding Material. *Compos. Part B Eng.* **2020**, *189*, 107895. [CrossRef]

30. Jia, Z.; Kong, M.; Yu, B.; Ma, Y.; Pan, J.; Wu, G. Tunable Co/ZnO/C@ MWCNTs Based on Carbon Nanotube-Coated MOF with Excellent Microwave Absorption Properties. *J. Mater. Sci. Technol.* **2022**, *127*, 153–163. [CrossRef]

31. Sun, C.; Jia, Z.; Xu, S.; Hu, D.; Zhang, C.; Wu, G. Synergistic Regulation of Dielectric-Magnetic Dual-Loss and Triple Heterointerface Polarization Via Magnetic MXene for High-Performance Electromagnetic Wave Absorption. *J. Mater. Sci. Technol.* **2022**, *113*, 128–137. [CrossRef]

32. Jang, M.G.; Ryu, S.C.; Juhn, K.J.; Kim, S.K.; Kim, W.N. Effects of Carbon Fiber Modification with Multiwall CNT on The Electrical Conductivity and EMI Shielding Effectiveness of Polycarbonate/Carbon Fiber/CNT Composites. *J. Appl. Polym. Sci.* **2019**, *136*, 47302. [CrossRef]

33. Bai, Y.; Zhou, S.; Lei, X.; Zou, H.; Liang, M. Enhanced Thermal Conductivity of Polycarbonate-Based Composites by Constructing A Dense Filler Packing Structure Consisting of Hybrid Boron Nitride and Flake Graphite. *J. Appl. Polym. Sci.* **2022**, *139*, e52895. [CrossRef]

34. Ren, Y.J.; Zhang, Y.F.; Fang, H.M.; Ding, T.P.; Li, J.L.; Bai, S.L. Simultaneous enhancement on thermal and mechanical properties of polypropylene composites filled with graphite platelets and graphene sheets. *Compos. Part A Appl. Sci. Manuf.* **2018**, *112*, 57–63. [CrossRef]

35. Jin, X.; Wang, J.; Dai, L.; Wang, W.; Wu, H. Largely enhanced thermal conductive, dielectric, mechanical and anti-dripping performance in polycarbonate/boron nitride composites with graphene nanoplatelet and carbon nanotube. *Compos. Sci. Technol.* **2019**, *184*, 107862. [CrossRef]

36. Feng, C.P.; Bai, L.; Bao, R.Y.; Liu, Z.Q.; Yang, M.B.; Chen, J.; Yang, W. Electrically insulating POE/BN elastomeric composites with high through-plane thermal conductivity fabricated by two-roll milling and hot compression. *Adv. Compos. Hybrid Mater.* **2018**, *1*, 160–167. [CrossRef]

37. Zhao, Z.; Shen, S.; Li, Y.; Zhang, X.; Su, J.; Li, H.; Tang, D. Strain-sensing behavior of flexible polypropylene/poly (ethylene-co-octene)/multiwalled carbon nanotube nanocomposites under cyclic tensile deformation. *Polym. Compos.* **2022**, *43*, 7–20. [CrossRef]

38. Zhao, Z.; Zhang, X.; Yang, Q.; Ai, T.; Zhou, S. Crystallization and Microstructure evolution of microinjection molded isotactic polypropylene with the assistance of poly(Ethylene Terephthalate). *Polymers* **2020**, *12*, 219. [CrossRef]

39. Yang, J.; Tang, L.S.; Bao, R.Y.; Bai, L.; Liu, Z.Y.; Xie, B.H.; Yang, M.B.; Yang, W. Hybrid network structure of boron nitride and graphene oxide in shape-stabilized composite phase change materials with enhanced thermal conductivity and light-to-electric energy conversion capability. *Sol. Energy Mater. Sol. Cells* **2018**, *174*, 56–64. [CrossRef]
40. Zhao, X.G.; Tang, Y.L.; Wang, J.; Li, Y.H.; Li, D.K.; Zuo, X.C.; Yang, H.M. Visible Light Locking in Mineral-Based Composite Phase Change Materials Enabling High Photothermal Conversion and Storage. *ACS Appl. Mater. Interfaces* **2023**, *15*, 49132–49145. [CrossRef]
41. Sun, K.Y.; Dong, H.S.; Kou, Y.; Yang, H.N.; Liu, H.Q.; Li, Y.G.; Shi, Q. Flexible graphene aerogel-based phase change film for solar-thermal energy conversion and storage in personal thermal management applications. *Chem. Eng. J.* **2021**, *419*, 129637. [CrossRef]
42. Cai, Y.X.; Zhang, N.; Cao, X.L.; Yuan, Y.P.; Zhang, Z.L.; Yu, N.Y. Ultra-light and flexible graphene aerogel-based form-stable phase change materials for energy conversion and energy storage. *Sol. Energy Mater. Sol. Cells* **2023**, *252*, 112176. [CrossRef]
43. Zheng, J.L.; Deng, Y.; Liu, Y.L.; Wu, F.Z.; Wang, W.H.; Wang, H.; Sun, S.Y.; Lu, J. Paraffin/polyvinyl alcohol/MXene flexible phase change composite films for thermal management applications. *Chem. Eng. J.* **2023**, *453*, 139727. [CrossRef]

Article

Impact Energy Dissipation and Quantitative Models of Injection Molded Short Fiber-Reinforced Thermoplastics

Quan Jiang, Tetsuo Takayama * and Akihiro Nishioka

Graduate School of Organic Materials Science, Yamagata University, 4-3-16 Jonan, Yonezawa 992-8510, Japan; t221291d@st.yamagata-u.ac.jp (Q.J.)
* Correspondence: t-taka@yz.yamagata-u.ac.jp; Tel.: +81-238-26-3085

Abstract: Glass short fiber-reinforced thermoplastics (SGFRTP) are used to reduce carbon dioxide emissions from transportation equipment, especially household vehicles. The mechanical properties required for SGFRTP include flexural strength, impact resistance, etc. In particular, impact resistance is an important indicator of the use of SGFRTP. For this study, a mechanical model was developed to explain the notched impact strength of SGFRTP injection molded products in terms of their interfacial shear strength. The values obtained from the model show good agreement with the experimentally obtained results ($R^2 > 0.95$). Results also suggest that the model applies to different fiber orientation angle and a range of fiber lengths in the molded product that are sufficiently shorter than the critical fiber length.

Keywords: impact strength; injection molding; interfacial shear strength; modeling; short fiber-reinforced thermoplastics

Citation: Jiang, Q.; Takayama, T.; Nishioka, A. Impact Energy Dissipation and Quantitative Models of Injection Molded Short Fiber-Reinforced Thermoplastics. *Polymers* **2023**, *15*, 4297. https://doi.org/10.3390/polym15214297

Academic Editors: Andrew N. Hrymak and Shengtai Zhou

Received: 3 October 2023
Revised: 25 October 2023
Accepted: 30 October 2023
Published: 1 November 2023

1. Introduction

Reducing greenhouse gases such as carbon dioxide is recognized today as a common goal worldwide to help alleviate global warming. Particularly, the reduction in carbon dioxide emissions from transportation equipment such as automobiles and aircraft is required [1].

To reduce the carbon dioxide emissions of transportation equipment, the introduction of non-petroleum-based power systems is being considered in the automotive sector. One example is battery-powered electric vehicles. A battery is heavier than a conventional gasoline engine, thereby increasing the total vehicle weight. An increase in the vehicle weight also engenders increased collision energy. Concern exists that effects on the human body during accidents will be grievous. In the aircraft industry, carbon fiber-reinforced plastic (CFRP) has been applied to hulls to improve fuel efficiency through weight reduction [2]. For automobiles, CFRP has been used on the chassis of high-end cars, achieving a remarkable weight reduction of 100 kg. Suppose CFRP is applied to the chassis of regular passenger cars. In that case, it can be expected to contribute to carbon dioxide reduction significantly. However, this is expected to occur in the distant future because of the high costs of CFRP. To reduce vehicle weight to a practical level, the back door and instrument panel components can be assumed to be replaceable with even 3D-printed fiber-reinforced plastics [3]. This strategy has already been implemented for the back doors of some household automobiles. Currently, short glass fiber-reinforced thermoplastics (SGFRTP) containing more than 40 wt% discontinuous glass fiber (GF) are used to meet the required properties. GF has a higher specific gravity than carbon fiber (CF). Moreover, it requires higher fiber contents because of the discontinuous and random dispersion of the fibers. Therefore, even though the replacement of metallic materials with SGFRTP has been achieved, weight reduction effects have not been as significant as expected. In the case of SGFRTP, the molding is conducted mainly by injection molding, a mass production technique used for thermoplastics. In injection molding, the material is often supplied as pre-made composite pellets. Therefore, the fiber length cannot be greater than the pellet size. Moreover, the shear stress in the resin during the plasticization and inflow processes leads to fiber pressure loss. With injection molding, the fiber length in SGFRTP is invariably shorter than the pellet size. The fiber length in the SGFRTP injection molded products tends to be shorter than the critical fiber length [4]. Furthermore, with injection molding, the fibers flow into the mold cavity in a fountain flow pattern. Therefore, the fiber orientation varies in the thickness dimension.

The mechanical properties of SGFRTP injection molded products vary depending on the thickness and the molded product part. Usually, notched impact tests are used to evaluate the impact resistance of molded products. The impact resistance of SGFRTP injection molded products is also often evaluated according to the notched impact strength. The SGFRTP strength and impact resistance are governed by factors originating from the fiber, the matrix, and the interface between the fiber and the matrix [5]. Interfaces have been modeled in several ways, e.g., a narrow continuum region with graded properties, an infinitely thin surface separated by springs, and cohesive zones with specific traction–separation relations. Table 1 presents the dominant factors affecting the strength and impact resistance of SGFRTP. Among the factors originating from the fiber–matrix interface, interfacial shear strength (IFSS) and interfacial strength (IFS) have attracted attention as factors controlling the mechanical properties of fiber-reinforced plastics.

Table 1. Factors affecting SFRTP strength and impact strength.

SFRTP Structure	Factors
Matrix	Yield stress, Elastic modulus, Poisson's ratio, Density, Linear expansion coefficient, etc.
Fiber	Strength, Elastic modulus, Poisson's ratio, Density, Linear expansion coefficient, etc.
Matrix/Fiber interface	Interfacial shear strength (IFSS), Interfacial strength (IFS), Friction coefficient, Interaction force, Specific surface area, etc.

In fact, pull-out [6], push-out [7], fragmentation [8,9], and pinhole pull-out methods [10] have been proposed to evaluate IFSS. The tensile test [11] has also been proposed to evaluate IFSS. Although these methods are fundamentally effective for capturing the fiber/matrix interface, they cannot be applied to actual molded products because they require the preparation of unique, molded products for evaluation. Given this background, Jiang et al. proposed a method to obtain the IFSS by short beam testing using SGFRTP injection molded products [12]. This method enables the evaluation of IFSS using molded products. The correlation between IFSS and mechanical properties such as strength and toughness can be derived quantitatively. In recent years, it has been reported that there is a positive qualitative correlation between IFSS and impact strength.

T. K. Kallel et al. reported the effects of fiber orientation and interfacial shear strength on the mechanical and structural properties of GF-reinforced polypropylene (PP-g-MAH) and maleic anhydride (MAH) grafted SGFRTP [13]. The obtained results have shown that the interfacial adhesion was substantially improved when the PP-g-MA compatibilizer was added. The enhancement of the mechanical properties such as, impact strength and tensile strength, was affected by the increase in fiber orientation consistency and the enhancement of interfacial shear strength. However, the meso-mechanical model is only discussed for tensile strength.

Lou et al. reported the effects of some parameters, such as fiber length, fraction of GF, and fraction of maleic anhydride grafted PP-g-MAH on the mechanical properties and fracture structure of long fiber-reinforced thermoplastic composites (LFRT) [14]. The obtained results have shown that the mass fraction of GF is the main factor affecting the mechanical properties of the materials. Meanwhile, by studying the fracture mechanism of the 30 wt% GF LFRT, it can be concluded that the GF is subjected to most of the deformation force. When fracture occurs, the fiber is pulled out of the matrix in particular directions or broken. However, due to the inability to quantitatively analyze the failure state of long fibers, there was no discussion about quantitative models.

B.B. Yin et al. reported the effects of some parameters, such as fiber length and interfacial shear strength on the machine learning, and materials informatics approaches for evaluating the interfacial properties of fiber-reinforced composites [15]. The obtained results have shown positive correlations between IFSS and mechanical properties such as strength and toughness. X. Fang et al. reported positive correlations between interfacial shear strength and GF sheet-reinforced thermoplastics [16]. S. Parveen et al. even reported positive correlations between interfacial shear strength and glass fiber-reinforced epoxy composites [17].

From the above existing research, we can conclude that the correlation between IFSS and impact strength has not been clarified. Nevertheless, some correlations between impact properties and related parameters such as fiber length, fiber orientation, interfacial shear strength, and fiber volume fraction have been found. However, a quantitative model expressing impact strength using IFSS and IFS has not been proposed either. Both researchers and product development engineers want to have a simple universal model to predict impact strength. In our previous paper, we considered a quantification model based on pull-out [13]. However, the effects of fiber orientation angle, fiber length, etc., have not yet been discussed in detail. In particular, extreme fiber orientation angles, where all fibers are parallel to the fracture plane, need to be considered.

For this study, the correlation between the notched impact strength and the IFSS calculated using the short beam test was clarified for SGFRTP injection molded products. The model was validated by comparison with experimentally obtained results.

2. Materials and Methods

2.1. Materials

Polypropylene (PP, Novatec MA1B; Japan Polypropylene Corp. Tokyo, Japan, MFR = 21 g/10 min, 230 °C, 2.16 kg, T_m = 160 °C, T_g = 0 °C) and polystyrene (PS, Toyo Styrene GPPS G210C; Toyo-Styrene Co., Ltd. Tokyo, Japan MFR = 10 g/10 min, 230 °C, 2.16 kg, T_g = 105 °C) were used as matrices. Glass fibers (GF, ECS 03 T351; Nippon Electric Glass Co., Ltd. Tokyo, Japan, D = 13 μm, L = 3~5 mm) surface-modified with amino groups were used as fibers. Polypropylene maleic anhydride (MAH-PP, SCONA TSPP10213; BYK Additives & Instruments Co., Ltd., Wesel, Germany) and polystyrene maleic anhydride (MAH-PS, RESISFY R200; Denka Co., Ltd., Tokyo, Japan) were used as additives.

2.2. Sample Preparations

These materials were filled into a twin-screw extruder (IMC0-00; Imoto Machinery Co., Ltd., Kyoto, Japan) and were melt mixed at a 230 °C melting temperature and a 60 rpm screw speed. The configuration of the 15-mm-diameter screw installed in the extruder is portrayed in Figure 1a. The ratio of screw length to screw diameter is 25.

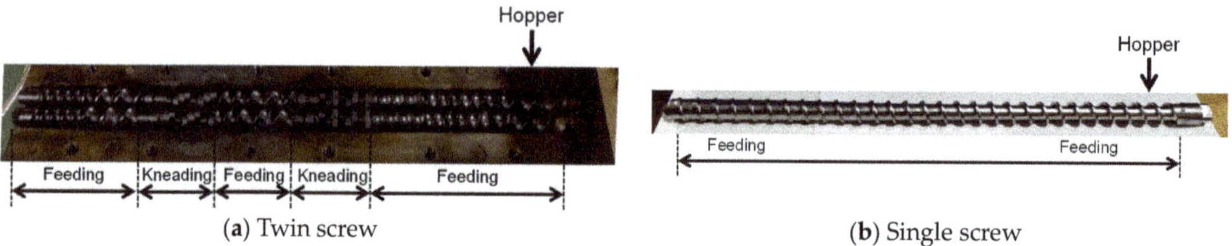

Figure 1. Screw configuration used for the twin and single screw extruder.

To explore the impact of different fiber length distributions, PP, MAH-PP, and GF were filled into a single screw extruder (APEX JAPAN Co., Ltd., Saitama, Japan, AS-1) as portrayed in Figure 1b. All materials were melt mixed at a 200 °C melting temperature and a 20 rpm screw speed. In order to further control the variables, we added the same proportion of materials to the twin-screw extruder, and they were melt mixed at a 200 °C melting temperature and a 60 rpm screw speed. The melt-kneaded strands were pelletized using a pelletizer (cold-cut pelletizer; Toyo Seiki Co., Ltd., Tokyo, Japan) to obtain composite pellets with a 3 mm pellet length. The mixing ratios are shown in Table 2; we fixed the GF content at 30 wt% and MAH-PP and MAH-PS at 1.5 wt%, and used the twin-screw extruder at a 230 °C melting temperature (PP/GF, PP/MAH-PP/GF, PS/GF, PS/MAH-PS/GF). In addition, we fixed the GF content at 10 wt% and MAH-PP at 5 wt% using the single screw and twin-screw extruder at a 200 °C melting temperature (PP/MAH-PP/GF-Single screw (PP/MAH-PP/GF-S) and PS/MAH-PS/GF-Twin-screw (PP/MAH-PP/GF-T)).

Table 2. Mixing ratios for melt-mixing.

Code	Mixing Temp. [°C]	PP [wt%]	MAH-PP [wt%]	PS [wt%]	MAH-PS [wt%]	GF [wt%]
PP/GF		70				30
PP/MAH-PP/GF	230	68.5	1.5			30
PS/GF				70		30
PS/MAH-PS/GF				68.5	1.5	30
PP/MAH-PP/GF-S	200	85	5			10
PP/MAH-PP/GF-T		85	5			10

The obtained composite material pellets were filled into an ultra-compact electric injection molding machine (C, Mobile0813; Shinko Sellbic Co., Ltd., Tokyo, Japan) and were injection molded to obtain beam-shaped molded products. This machine uses a pre-plunger system with a 10-mm-diameter plunger and mold clamping pressure of 29.4 kN. Table 3 shows the injection molding conditions. To clarify the effect of fiber orientation angle on impact performance, we designed two short beam-shaped specimens with different flow conditions, and a molded product thickness of 2 mm (specific size: $50 \times 5 \times 2$ mm). These two types of products are shown in Figure 2. Figure 2a shows that beam specimens were performed in a single flow direction. Figure 2b shows that the beam with weld specimens were performed by adjusting the flow path so that a weld was formed in the center.

Table 3. Injection molding conditions.

Parameter	PP/GF PP/MAH-PP/GF	PS/GF PS/MAH-PS/GF	PP/MAH-PP/GF-S PP/MAH-PP/GF-T	
Specimen shape	Beam	Beam	Beam	Beam with weld
Injection temp. [°C]	230	230	200	
Mold temp. [°C]	50	50	50	
Injection speed [mm/s]	30	20	10	
Holding pressure [MPa]	84	92	84	
Injection time [s]	45	45	45	
Cooling time [s]	15	15	15	
GF content [wt%]	10, 20, 30	10, 20, 30	10	

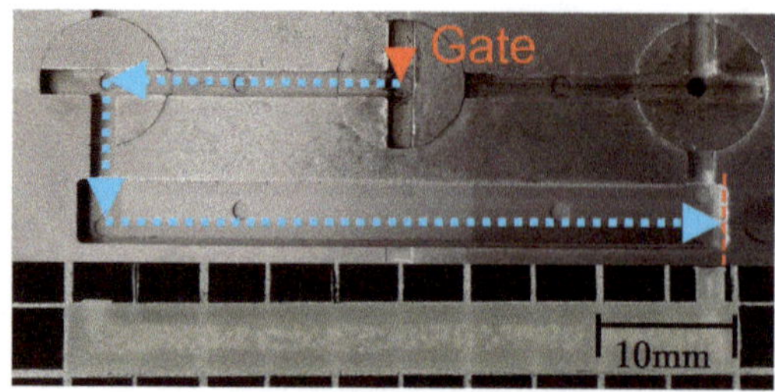

(**a**) Beam specimens

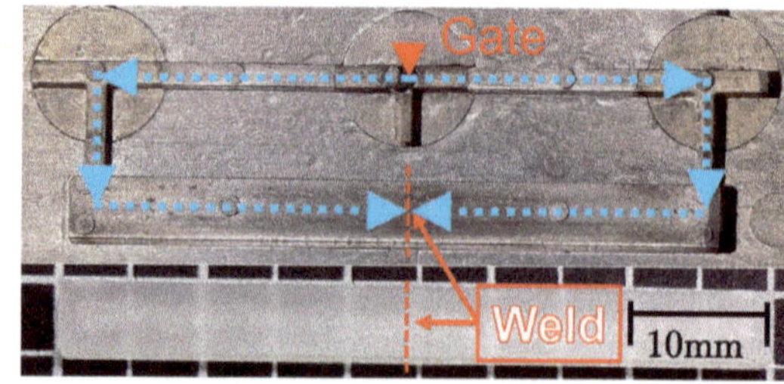

(**b**) Beam with weld specimens

Figure 2. Mold cavity shapes, injection flow path, beam, beam with weld specimen's image. Reprinted with permission from Ref. [18]. 2023. Elsevier.

2.3. Charpy Impact Tests

A notch with a 0.25 mm tip radius was machined in the center of the obtained beam-shaped molded product. The notch depth was approximately 1 mm. The notched beams were subjected to Charpy impact tests using a Charpy impact tester (MYS-Tester Co., Ltd., Osaka, Japan). The loading speed was 2.91 m/s. The spun length was 40 mm. The Charpy impact strength of the specimens was found by calculating the absorbed energy U from the obtained swing angle based on the following Equation (1).

$$a_{iN} = \frac{U}{B(W - a)} \tag{1}$$

Wherein, B stands for the molded product thickness. Also, W denotes the width of the molded product and a represents the notch depth.

2.4. Short Beam Share Testing for Determination of IFSS

The obtained beams were subjected to short beam share testing using a small universal mechanical testing machine (MCT-2150; A&D Co., Ltd., Tokyo, Japan). The loading speed was 10 mm/min and the spun length was 10 mm. The obtained load–deflection curve was differentiated by deflection to obtain the stiffness–deflection curve. Figure 3 presents an example of the stiffness–deflection curve. In the case of SGFRTP injection molded products, the fibers in the molded products are oriented randomly. In other words, most of the fibers are oriented obliquely to the flow direction. However, when a three-point bending load is applied, a bending moment and a shear stress are generated at the loading plane. The shear stresses are conjugate. They reach their maximum at the neutral plane. In the short beam test, high shear stress is generated near the neutral plane by shortening the spun length, which induces slippage at the interface. In the case of obliquely oriented fibers, the interface slippage is regarded as occurring first in either the parallel or perpendicular direction to the loading direction and then in the opposite direction as loading increases. In other words, a discontinuous decrease in stiffness is observed at two points. The point at which the stiffness decrease occurs under small shear stress is designated as Point 1. The point at which the decrease occurs under large shear stress is called Point 2.

In the case of SGFRTP injection molded products, the fibers near the neutral plane are oriented at an angle close to perpendicular to the flow direction. Therefore, the interface slippage is expected to occur first in a direction parallel to the loading direction. In other words, the relation between IFSS and shear stress at Point 1, in this case, can be expressed by Equation (2) below.

$$IFSS = \frac{\tau_1}{\cos \theta_f} = \frac{3P_1}{4BW \cos \theta_f} \tag{2}$$

Wherein, θ_f is the orientation angle of the GF dispersed near the neutral plane. The same relation at Point 2 can be expressed as the following Equation (3).

$$IFSS = \frac{\tau_2}{\sin\theta_f} = \frac{3P_2}{4BW\sin\theta_f} \tag{3}$$

From Equations (2) and (3), θ_f is expressed by Equation (4) below.

$$\theta_f = \tan^{-1}\frac{\tau_2}{\tau_1} = \tan^{-1}\frac{P_2}{P_1} \tag{4}$$

Jiang et al. showed that this angle represents a frequent orientation angle in molded products [12]. The IFSS is obtainable by substituting the θ_f obtained in Equation (4) into Equations (2) or (3). In this study, the *IFSS* was calculated based on the abovementioned method. The correlation between the *IFSS* and the Charpy impact strength was investigated. In this study, all *IFSS* evaluations were conducted using beam specimens without weld.

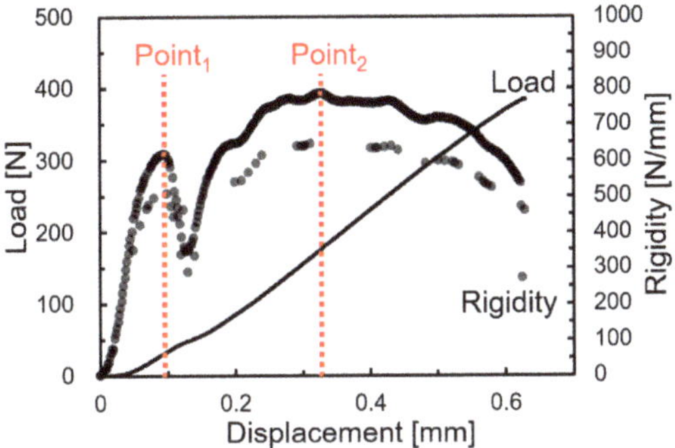

Figure 3. Load–displacement and rigidity–displacement curves were obtained from three-point bending tests using the short beam shear test.

2.5. Tensile Testing for Determination of IFS

The obtained beam with weld specimens was subjected to tensile testing using a small universal testing machine (IMADA Co., Ltd., Toyohashi, Japan, FSA-1KE-1000N-L). The loading speed was 0.5 mm/min; the span distance was 22 mm. The *IFS* of the specimens was found by using the maximum load P_{max} based on the following Equation (5) [11].

$$IFS = \frac{P_{max}}{BW} \tag{5}$$

In this study, the *IFS* was calculated based on the method described above and all *IFS* evaluations were conducted using beam with weld specimens. The correlation between the *IFS* and the Charpy impact strength was also investigated.

2.6. Fiber Length Measurement

To measure the fiber length in the molded product, the area subjected to the Charpy impact test was cut out by machining. The cut-out area was placed in a small electric furnace (FT-HP-100; Full-Tech Co., Ltd., Osaka, Japan) and was fired at 600 °C for 4 h to obtain the residue. Then, the residue was spread out on a glass slide. Details of the residue were photographed using a phase contrast microscope (BA410EPH-1080; Shimadzu Rika Corp., Tokyo, Japan). Then, using image analysis software (WinLOOF ver.7.0; Mitani Corp., Fukui, Japan), the fiber lengths of more than 500 glass fibers in the photographs were measured. The average fiber length and its standard deviation were found.

2.7. Fiber Orientation Measurement

The fiber orientation corresponding to the core layer in the area subjected to the Charpy impact test was photographed using a microfocus X-ray CT system (ScanXmate-D225RSS270; Comscantecno Co., Ltd., Kanagawa, Japan) and a sub-microfocus X-ray CT unit (MARS TOHKEN SOLUTION Co., Ltd., Tokyo, Japan, TUX-3200N). The obtained fiber orientation photographs were used to ascertain the average fiber orientation angle of more than 350 fibers with respect to the flow direction using image analysis software (WinLOOF ver.7.0; Mitani Corp., Fukui, Japan).

2.8. Fracture Surface Observations

The fracture surface obtained after the Charpy impact test was photographed using a phase contrast microscope and a digital microscope (TG200HD2-Me; Shodensha Co., Ltd., Osaka, Japan). The photographic angle was set as 0° for the phase contrast microscope and 45° for the digital microscope, with respect to the side of the specimen. Using image analysis software, the fiber pull-out lengths of more than 300 fibers were measured from the photographs taken using a phase contrast microscope. The average pull-out length and its standard deviation were calculated.

3. Results

3.1. IFSS of Injection Molded SGFRTP

Figure 4a presents the results of using the twin-screw extruder at a melting temperature of 230 °C for IFSS evaluation. The error bars in the figure represent the standard deviation. Results showed a decreasing trend with increasing fiber content. However, the IFSS tended to increase with the addition of maleic anhydride-modified polymer, which is compatible with the matrix. This trend is consistent with the results reported by Kallel et al. for PP/GF-30 wt% [13].

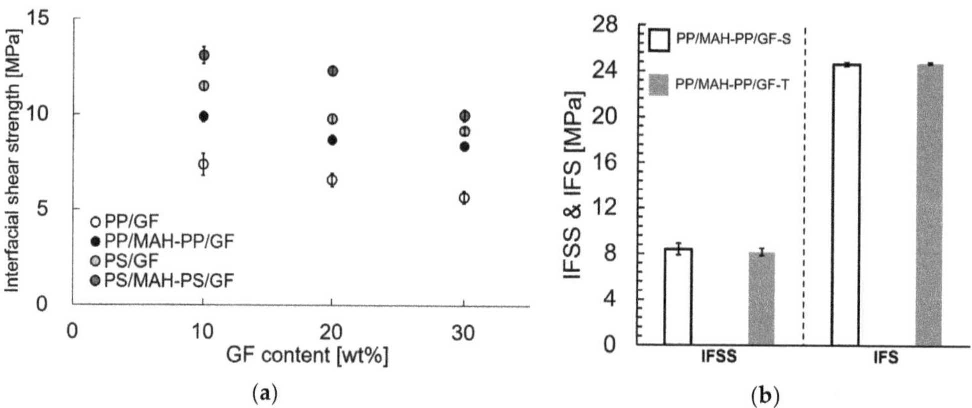

Figure 4. Interfacial shear strengths and interfacial strengths of different melt-mixing condition SGFRTPs. (**a**) Twin-screw extruder with 230 °C melting temperature melt-mixing condition. (**b**) Twin-screw and single-screw extruder with 200 °C melting temperature melt-mixing condition.

Figure 4b presents the results of using the twin-screw and single-screw extruders at a melting temperature of 200 °C for IFSS and IFS evaluation. The error bars in the figure represent the standard deviation. Results showed that the results of different extruders are almost the same.

3.2. Charpy Impact Strength of Injection Molded SGFRTP

Figure 5a portrays the results of using the twin-screw extruder at a melting temperature of 230 °C for evaluation of the Charpy impact strength. The error bars in the figure indicate the standard deviation. The value of a_{iN} tended to increase with the increase in fiber content. Furthermore, the addition of maleic anhydride-modified polymer, which is compatible with the matrix, showed a tendency to increase the value. This trend is consistent with results reported by Kallel et al. for PP/GF-30 wt% [13].

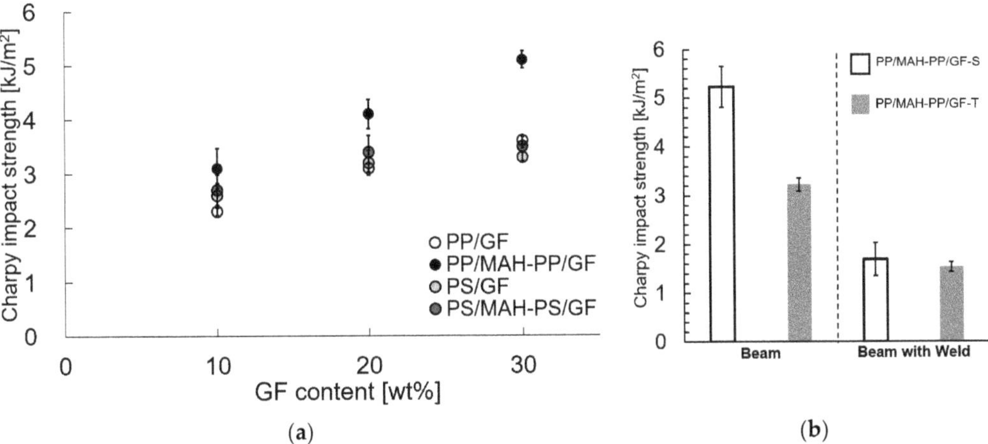

Figure 5. Charpy impact strengths of different melt-mixing condition SGFRTPs. (**a**) Twin-screw extruder with 230 °C melting temperature melt-mixing condition. (**b**) Twin-screw and single-screw extruder with 200 °C melting temperature melt-mixing condition.

Figure 5b portrays results of using the twin-screw and single-screw extruders at a melting temperature of 200 °C for evaluation of the Charpy impact strength. The error bars in the figure indicate the standard deviation. The value of a_{iN} tended to increase obviously when using the single-screw extruder. Furthermore, the beam with weld specimen with the same orientation angle as the impact direction (perpendicular to the flow direction of injection molding) showed almost the same a_{iN} results for different extruders.

4. Discussion

4.1. Quantitative Models of Energy Dissipation at Impact Loading of Beam Specimens' SGFRTP

Figure 6 presents results of phase contrast microscope observations of the notch tip area. As these photographs show, fiber pull-out was observed near the notch tip in all specimens of the compositions examined for this study. As the fiber content increased, the number of pulled-out fibers tended to increase. Figure 7 presents results of digital microscopic observations. These photographs were taken with magnification that accommodates viewing the entire fractured surface. The red dotted wireframe emphasizes the entire fracture surface inside. These photographs indicate that lines are drawn through the entire fractured surface of all the specimens of the compositions examined for this study. These results suggest that the main mechanism of impact energy dissipation in the studied compositions is fiber pull-out. This impact fracture behavior is consistent with the results reported by Quan et al. [18].

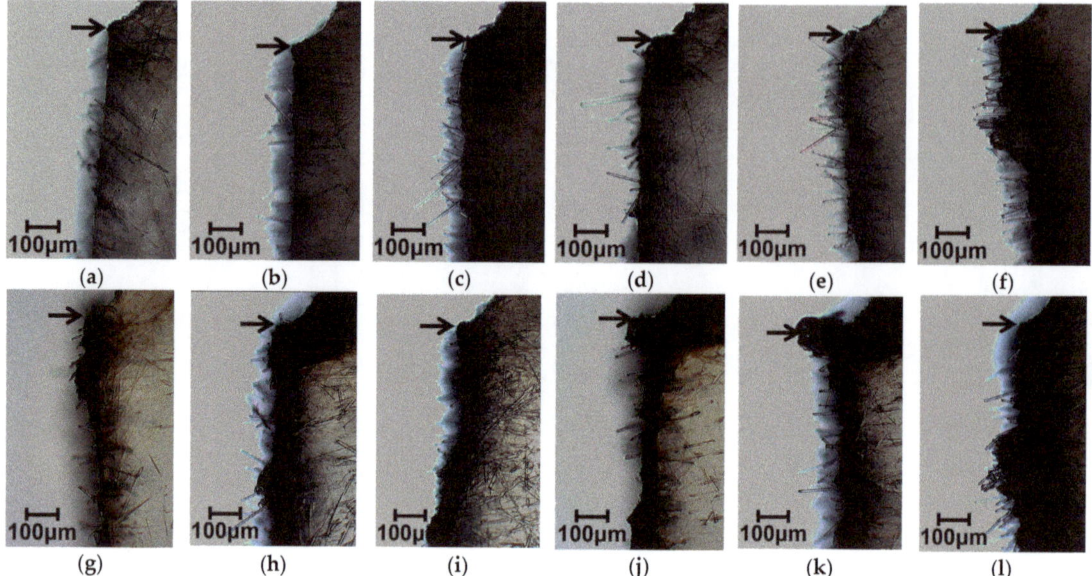

Figure 6. Phase contrast microscope images using the twin-screw extruder at a melting temperature of 230 °C; beam specimens' SGFRTPs observed after Charpy impact tests. Arrows in the figures indicate the notch tip. (**a**) PP/GF-10 wt%. (**b**) PP/GF-20 wt%. (**c**) PP/GF-30 wt%. (**d**) PP/MAH-PP/GF-10 wt%. (**e**) PP/MAH-PP/GF-20 wt%. (**f**) PP/MAH-PP/GF-30 wt%. (**g**) PS/GF-10 wt%. (**h**) PS/F-20 wt%. (**i**) PS/GF-30 wt%. (**j**) PS/MAH-PS/GF-10 wt%. (**k**) PS/MAH-PS/GF-20 wt%. (**l**) PS/MAH-PS/GF-30 wt%.

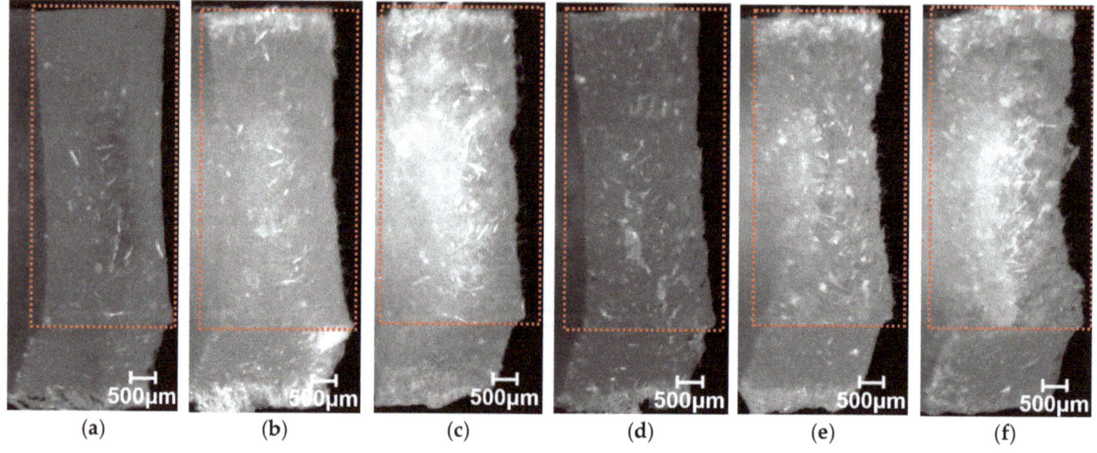

Figure 7. *Cont.*

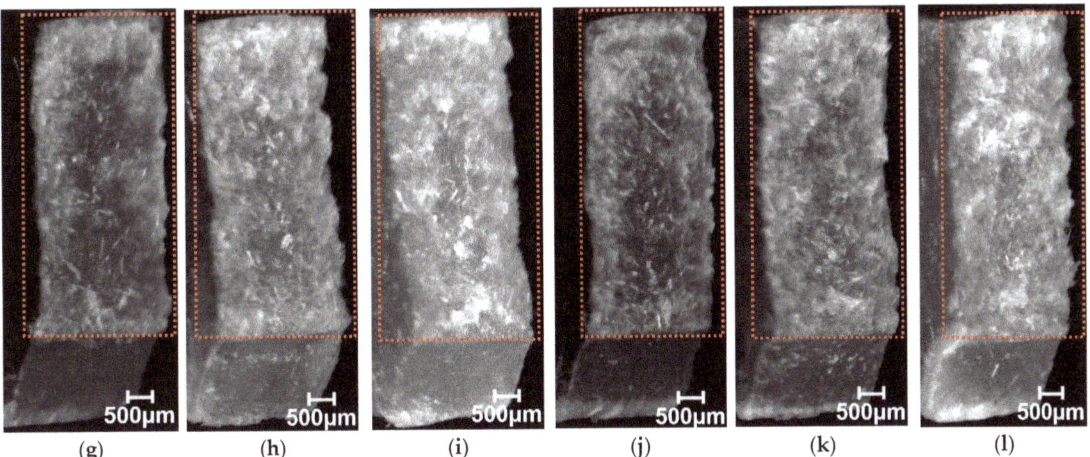

Figure 7. Fractured surface images using the twin-screw extruder at a melting temperature of 230 °C; beam specimens' SGFRTPs observed after Charpy. (**a**) PP/GF-10 wt%. (**b**) PP/GF-20 wt%. (**c**) PP/GF-30 wt%. (**d**) PP/MAH-PP/GF-10 wt%. (**e**) PP/MAH-PP/GF-20 wt%. (**f**) PP/MAH-PP/GF-30 wt%. (**g**) PS/GF-10 wt%. (**h**) PS/GF-20 wt%. (**i**) PS/GF-30 wt%. (**j**) PS/MAH-PS/GF-10 wt%. (**k**) PS/MAH-PS/GF-20 wt%. (**l**) PS/MAH-PS/GF-30 wt%.

Figure 8 shows the average fiber pull-out length, l_p, as measured from the phase contrast microscope observation. The error bars in the figure show the standard deviation. Actually, l_p tended to become shorter with increasing fiber content and interfacial shear strength. The l_p of PS was shorter than that of PP compared to that of matrix.

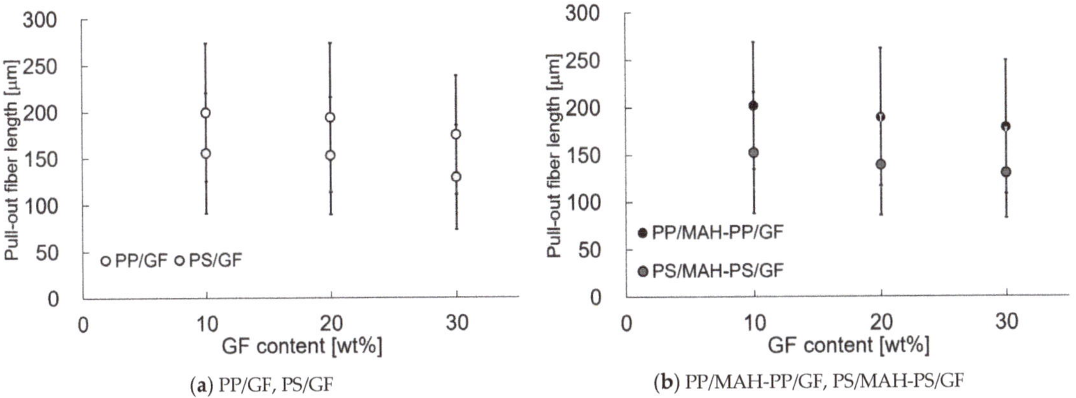

(**a**) PP/GF, PS/GF   (**b**) PP/MAH-PP/GF, PS/MAH-PS/GF

Figure 8. Pull-out fiber lengths of beam specimens' SGFRTPs observed from Charpy impact tests.

Figure 9 shows the fiber content dependence of the average fiber length of the fibers, L_f, present in the molded product. The error bars in the figure show the standard deviation. Results show that L_f tends to become shorter as the fiber content increases. The L_f of PS was shorter than that of PP when compared with that of matrix. At the same time, we should also note that compared to the unadded matrix, the addition of MAH-PP or MAH-PS affected the fiber length and the pull-out fiber's length. With the addition of MAH-PP or MAH-PS, the fiber length was slightly shorter compared to that without addition. Results indicate that an increase in interfacial shear strength will cause damage to the fibers during injection molding. The relation between L_f and l_p in Figures 8 and 9 suggests that some correlation exists between L_f and l_p. Figure 10 presents a relation between L_f and l_p, approximated by the linear function equation shown in the figure. When the dispersed fibers are shorter than the critical fiber length l_c shown in Equation (6) below, fiber pull-out dominates the yield condition of SFRTP.

$$l_c = \frac{\sigma_F d}{2\tau} \tag{6}$$

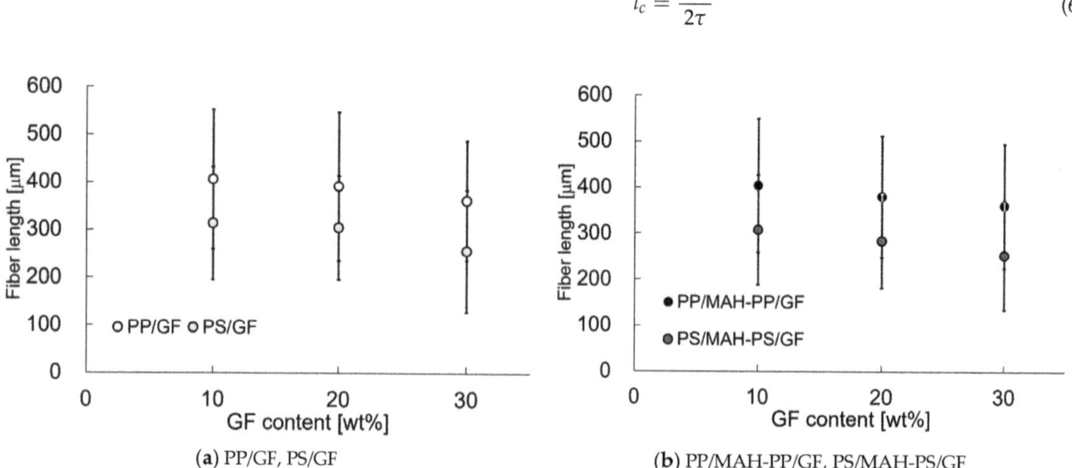

Figure 9. Fiber lengths of beam specimens' injection molded SGFRTPs.

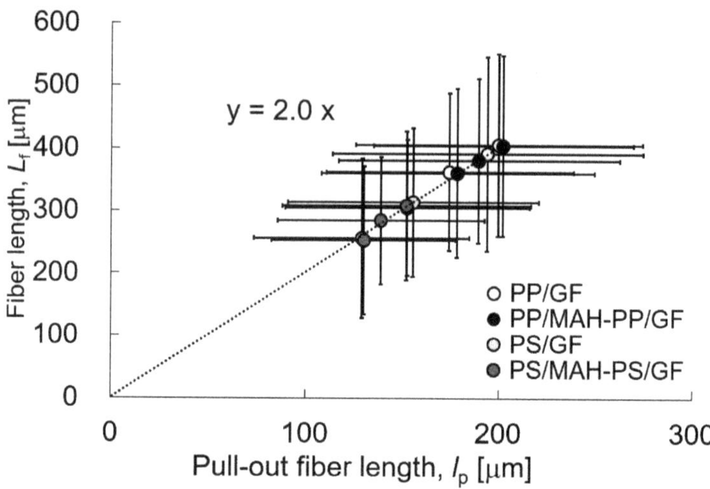

Figure 10. Relations between pull-out fiber length and fiber length of beam specimens' SGFRTPs.

In that equation, d is the fiber diameter and τ denotes the IFSS. The shear stress inside the fiber is maximal at half of the fiber length. In other words, theoretically, a 2:1 relation can be found between L_f and l_p [19]. The relation between L_f and l_p obtained in this study is approximately 2:1, which is equal to the theoretical value. This result also indicates that the addition of MAH-PP or MAH-PS does not affect fiber pull-out. This finding implies that the energy dissipation quantitative models in the Charpy impact test are mostly attributable to fiber pull-out. Figure 11 shows the value of L_f divided by l_c. This value is smaller than 1, which means that the yield condition of SFRTP is dominated by fiber pull-out. All results in the figure are less than 0.5. From these results, one can infer that the energy dissipation quantitative models in the Charpy impact test of the compositions studied in this paper are attributable to fiber pull-out.

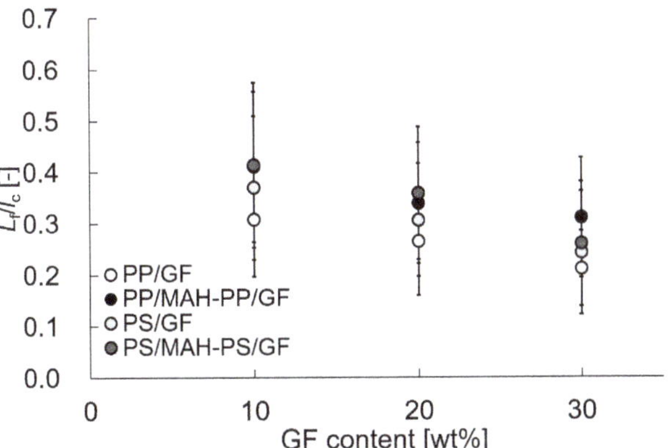

Figure 11. L_f/l_c of beam specimens' SGFRTPs.

4.2. Notched Impact Strength of Fiber Pull-Out

Based on the discussion presented above, we assume for these analyses that most of the energy dissipated in the Charpy impact test comes from fiber pull-out. The amount of energy dissipated by this mechanism is modeled using IFSS. Figure 12 shows the energy dissipation region during the Charpy impact test. The energy dissipation attributable to fiber pull-out occurs only in the region where the fiber is pulled out. Therefore, the energy dissipating region V_P is expressed using Equation (7).

$$V_P - l_p B(W - a) \tag{7}$$

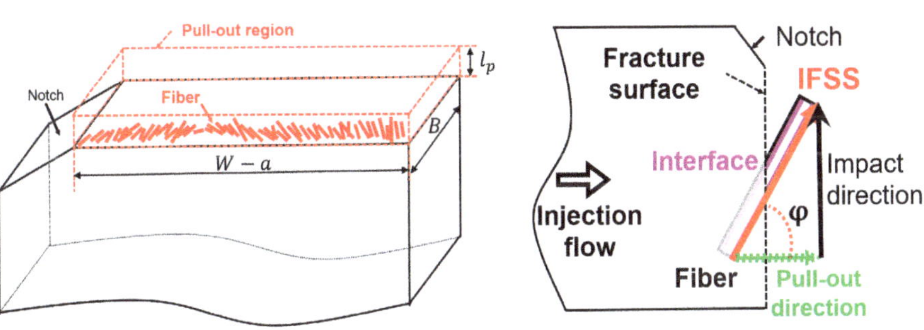

(a) Overviews   **(b)** Enlarged view of fiber pull-out

Figure 12. Pull-out model with an explanation of the notched impact strength. Reprinted with permission from Ref. [18]. 2023. Elsevier.

The energy dissipated by the fiber withdrawal is regarded as coming from the friction at the interface between the fiber and the matrix phase. Therefore, the total amount of energy dissipated U_P can be expressed by Equation (8).

$$U_P = \iota \cos \psi l_p S_f V_P \tag{8}$$

In that equation, φ denotes the orientation angle of the fiber with respect to the loading direction. In the Charpy impact test, the Charpy impact strength is obtained by dividing

U_P by the ligament area. Therefore, the final theoretical value $a_{Model\text{-}P}$ is expressed by the following Equation (9).

$$a_{Model-P} = \frac{U_P}{B(W-a)} = \tau \cos \varphi S_f l_p^2 \tag{9}$$

4.3. Quantitative Models of Energy Dissipation at Impact Loading of Beam with Weld Specimens' SGFRTPs

Figure 13 presents results of phase contrast microscopy observations of the notch tip area. As shown in these photographs, fiber pull-out was observed near the notch tip in all beam specimens. Compared to the beam specimens, no fiber pull-out was observed in the Charpy impact test beam specimens with weld. Figure 14 presents results of digital microscopy observations. These photographs were taken at a magnification that allows the entire fracture surface to be viewed. These photographs show that lines are drawn through the entire fractured surface of beam specimens. Compared to the beam specimens, almost no lines are drawn through the fractured surfaces of beam with weld specimens. These results suggest that interfacial debonding is the main mechanism of impact energy dissipation in the beam with weld specimens.

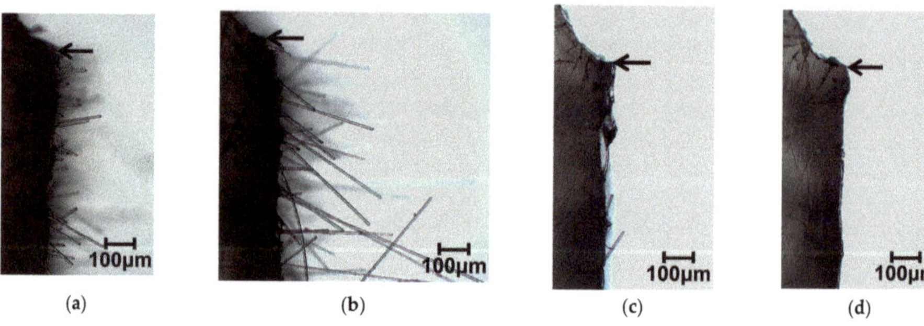

(a) (b) (c) (d)

Figure 13. Phase contrast microscope images of using twin-screw and single-screw extruders at a melting temperature of 200 °C on beam and beam with weld specimens' SGFRTPs after Charpy impact tests. Arrows in the figures indicate the notch tip. (**a**) PP/MAH-PP/GF-T Beam. (**b**) PP/MAH-PP/GF-S Beam. (**c**) PP/MAH-PP/GF-T Beam with Weld. (**d**) PP/MAH-PP/GF-S Beam with Weld.

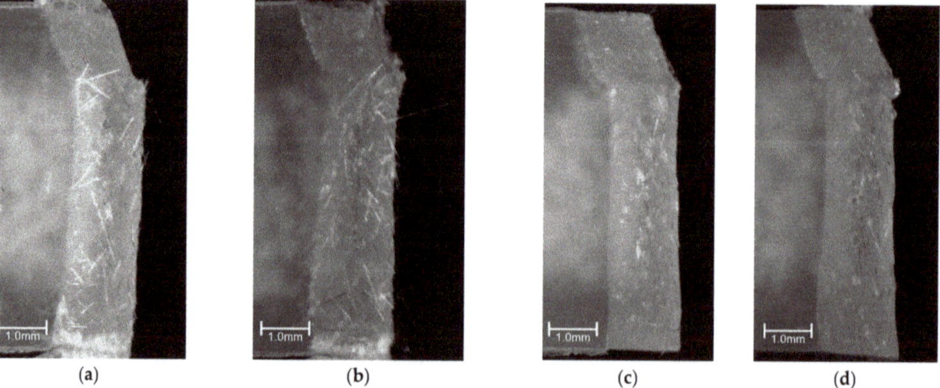

(a) (b) (c) (d)

Figure 14. Fractured surface images using twin-screw and single-screw extruders at a melting temperature of 200 °C on beam and beam with weld specimens' SGFRTPs after Charpy impact tests. (**a**) PP/MAH-PP/GF-T Beam. (**b**) PP/MAH-PP/GF-S Beam. (**c**) PP/MAH-PP/GF-T Beam with Weld. (**d**) PP/MAH-PP/GF-S Beam with Weld.

Based on the discussion presented above, we assume for these analyses that most of the energy dissipated in the Charpy impact test comes from interfacial debonding when the fiber orientation direction is the same as the impact direction. Figure 15 shows the area of energy dissipation during the Charpy impact test. The energy dissipation due to interfacial debonding occurs only in the region of the fiber-to-fiber distance $\langle L_T \rangle$ between the welding point. Therefore, the energy dissipating area V_D is expressed by the following Equation (10).

$$V_D = \langle L_T \rangle B (W - a) \tag{10}$$

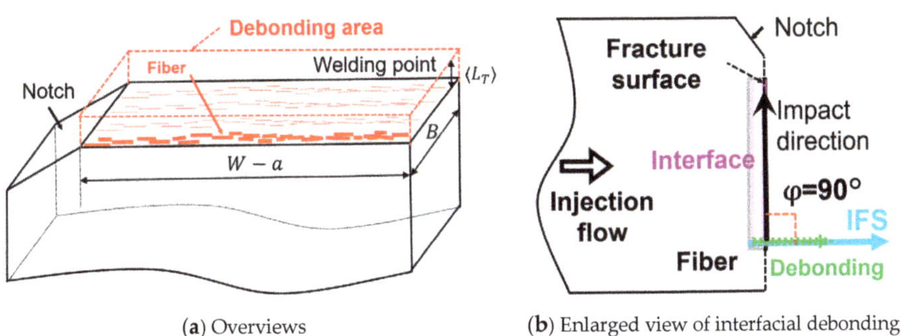

(**a**) Overviews   (**b**) Enlarged view of interfacial debonding

Figure 15. Interfacial debonding model with an explanation of the notched impact strength.

The energy dissipated by the interfacial debonding is considered to come from the IFS between the fiber and the matrix phase. Therefore, the total amount of energy dissipated U_D can be expressed by Equation (11).

$$U_D = IFS \cdot V_D \tag{11}$$

In the Charpy impact test, the Charpy impact strength is obtained by dividing U_D by the ligament area. Therefore, the final theoretical value $a_{Model\text{-}D}$ is expressed by the following Equation (12).

$$a_{Model-D} = \frac{U_D}{B(W - a)} = IFS \cdot \langle L_T \rangle \tag{12}$$

4.4. Validity of the Proposed Theory

To perform calculations using these models, φ and $\langle L_T \rangle$ must be found. Figure 16 presents X-ray CT observations taken from the surface of the molded product to about 1 mm in the thickness direction using a twin-screw extruder with a melting temperature of 230 °C for injection molded beam specimens. Since notches are introduced in the width direction in this study, we extracted more than 350 fibers from the region excluding the notches. We obtained the average orientation angle and its distribution. Figure 17 presents the orientation angle distribution and the average orientation angle. In all compositions, the fibers were mostly oriented perpendicular to the flow direction. As described in this paper, φ was used to obtain $a_{Model\text{-}P}$ using the average orientation angle. Figure 18 presents X-ray CT observations taken from the surface of the molded product to about 1 mm in the thickness direction using a twin-screw and single-screw extruder with a 200 °C melting temperature beam and a beam with weld specimens. We obtained the average orientation angle using the same method as in Section 4.1. We also obtained the average fiber to fiber distance near the welding point region, excluding the notches. As described in this paper, $\langle L_T \rangle$ was used to obtain $a_{Model\text{-}D}$ using the fiber-to-fiber distance.

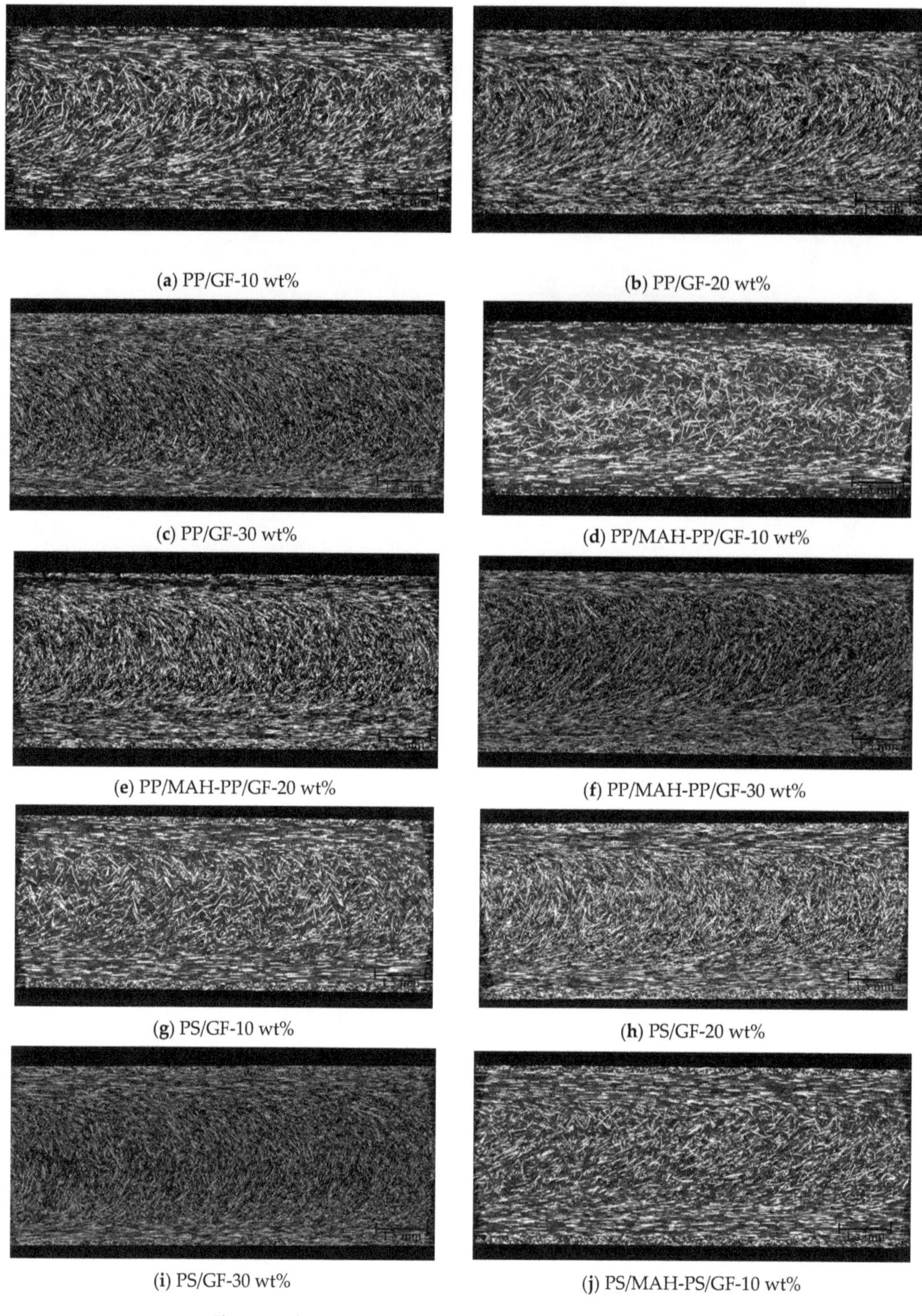

(**a**) PP/GF-10 wt%

(**b**) PP/GF-20 wt%

(**c**) PP/GF-30 wt%

(**d**) PP/MAH-PP/GF-10 wt%

(**e**) PP/MAH-PP/GF-20 wt%

(**f**) PP/MAH-PP/GF-30 wt%

(**g**) PS/GF-10 wt%

(**h**) PS/GF-20 wt%

(**i**) PS/GF-30 wt%

(**j**) PS/MAH-PS/GF-10 wt%

Figure 16. *Cont.*

(**k**) PS/MAH-PS/GF-20 wt%

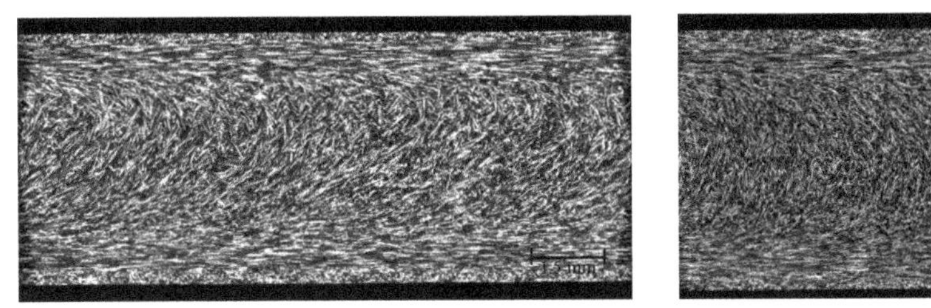

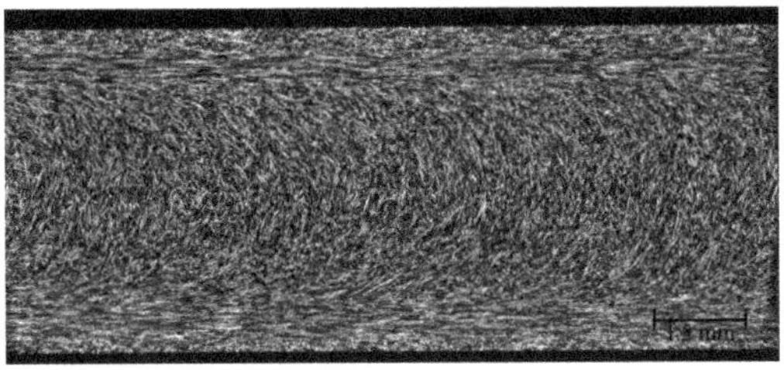

(**l**) PS/MAH-PS/GF-30 wt%

Figure 16. X-ray CT images using the twin-screw extruder at a melting temperature of 230 °C on injection molded beam specimens' SGFRTPs. Reprinted with permission from Ref. [18]. 2023. Elsevier.

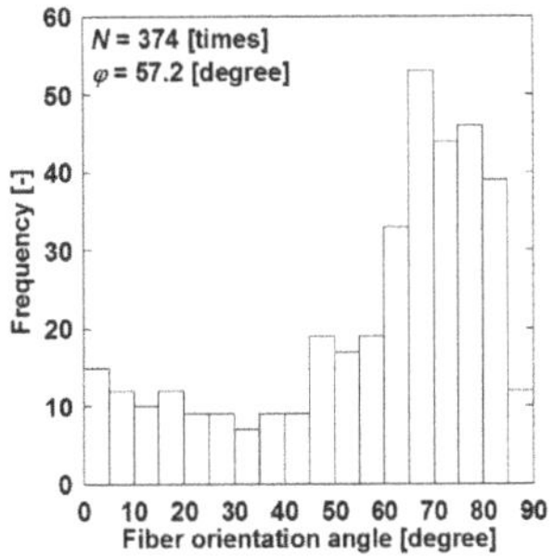

(**a**) PP/GF-10 wt%

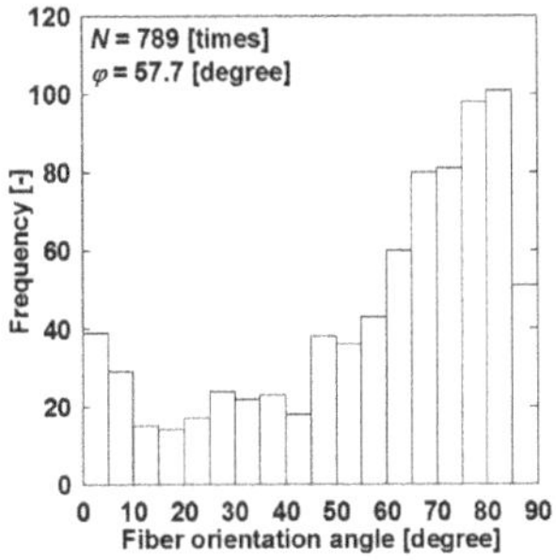

(**b**) PP/GF-20 wt%

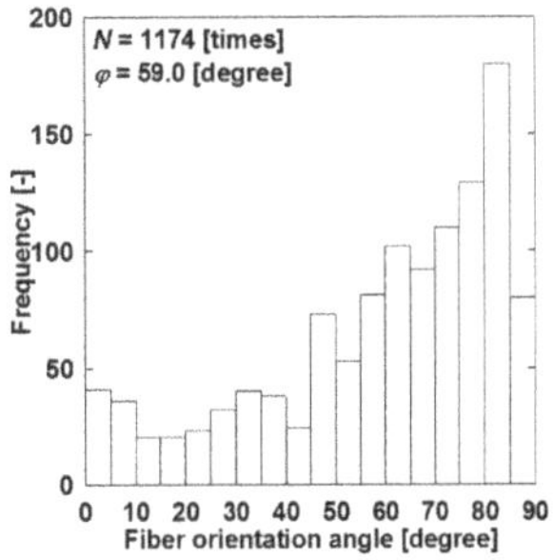

(**c**) PP/GF-30 wt%

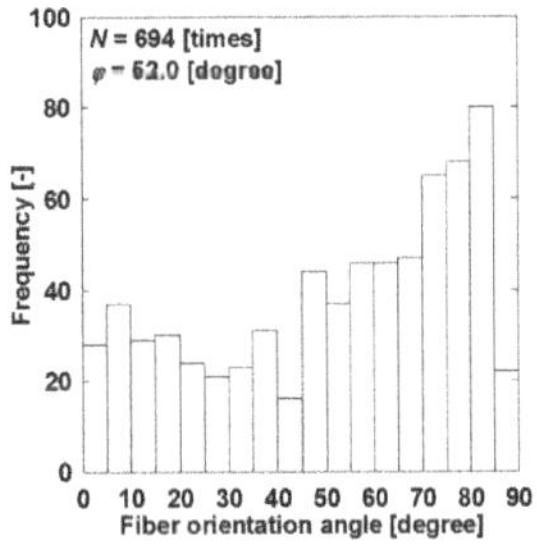

(**d**) PP/MAH-PP/GF-10 wt%

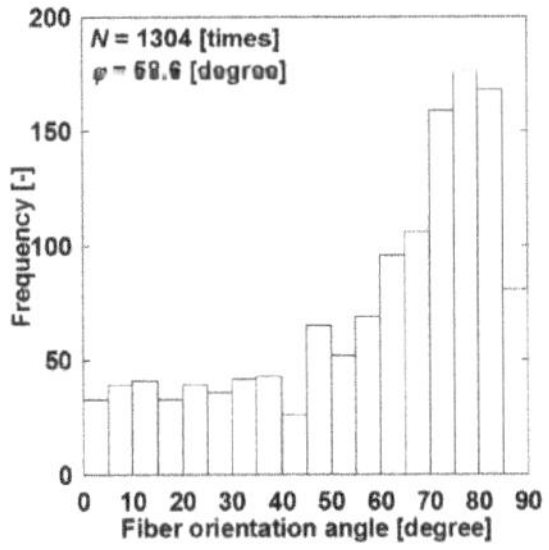

(**e**) PP/MAH-PP/GF-20 wt%

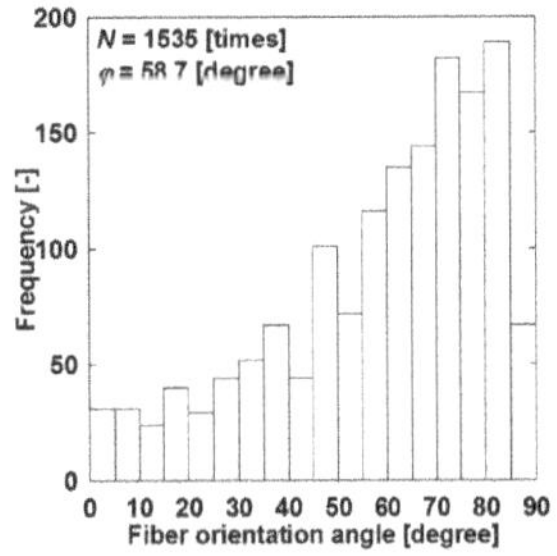

(**f**) PP/MAH-PP/GF-30 wt%

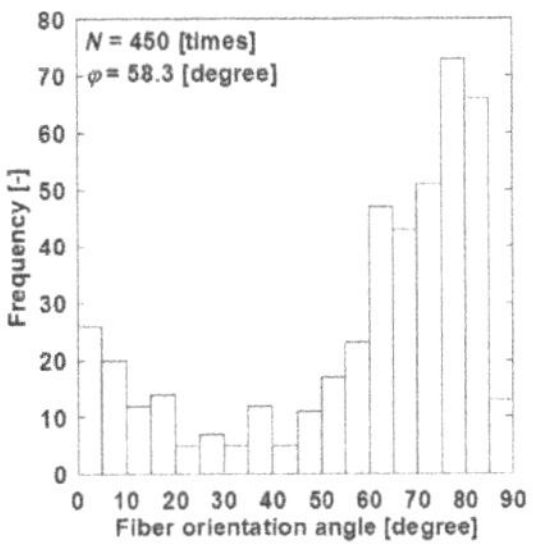

(**g**) PS/GF-10 wt%

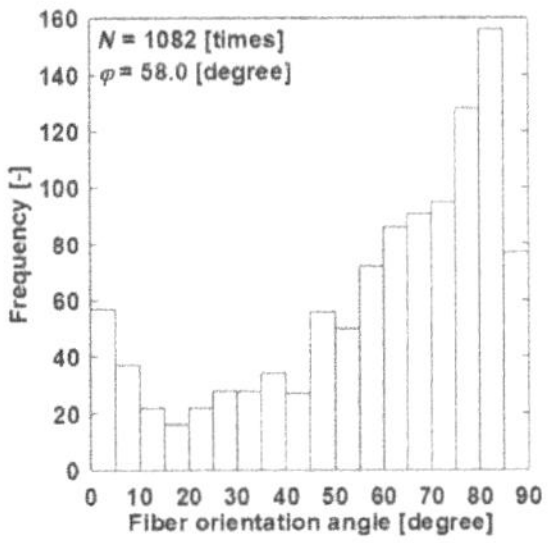

(**h**) PS/GF-20 wt%

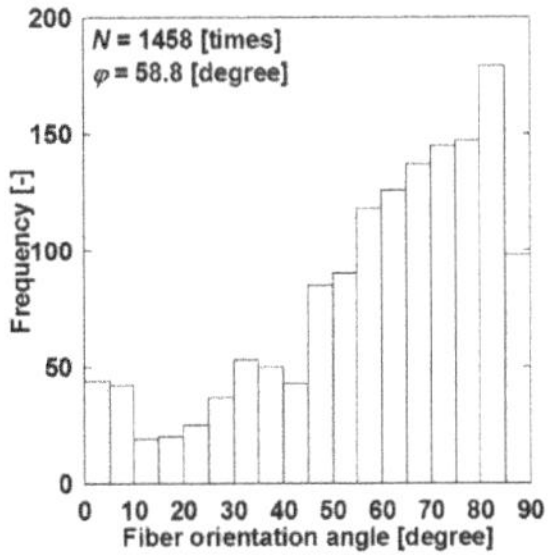

(**i**) PS/GF-30 wt%

Figure 17. *Cont.*

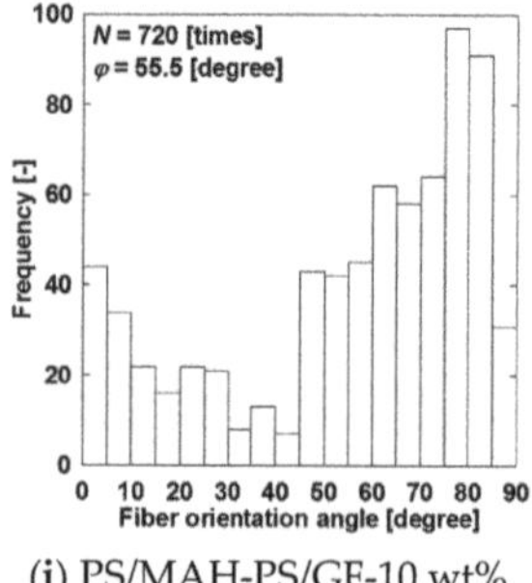
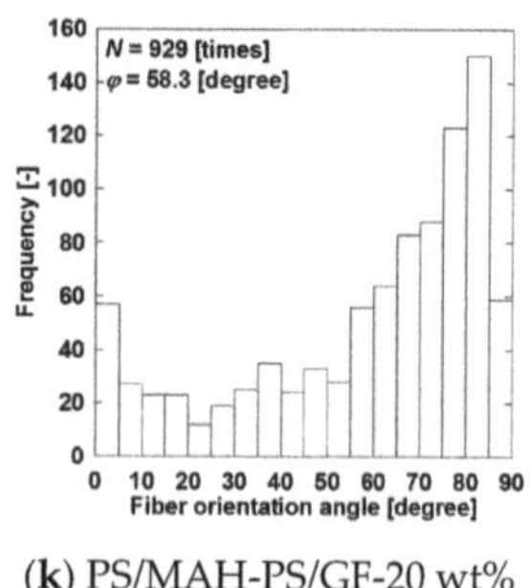
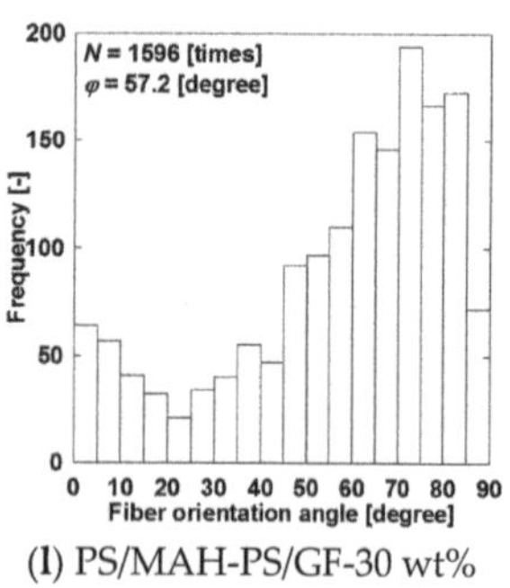

(j) PS/MAH-PS/GF-10 wt% **(k)** PS/MAH-PS/GF-20 wt% **(l)** PS/MAH-PS/GF-30 wt%

Figure 17. Orientation angle distributions and average orientation angles of twin-screw and single-screw extruders at a melting temperature of 200 °C on beam and beam with weld specimens' SGFRTP.

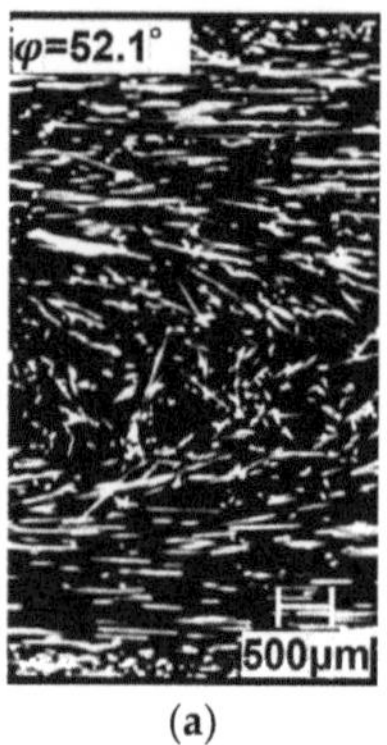
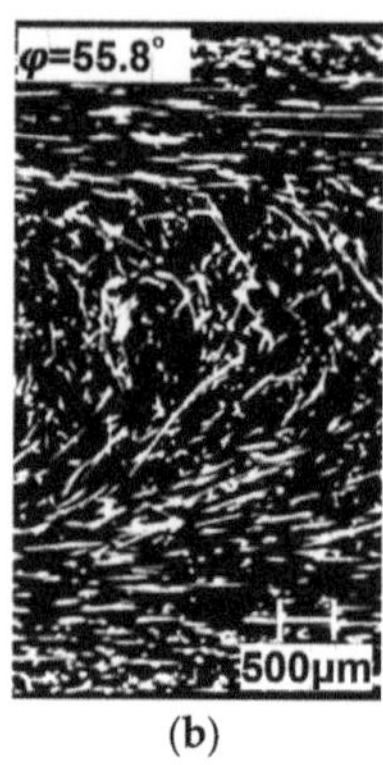
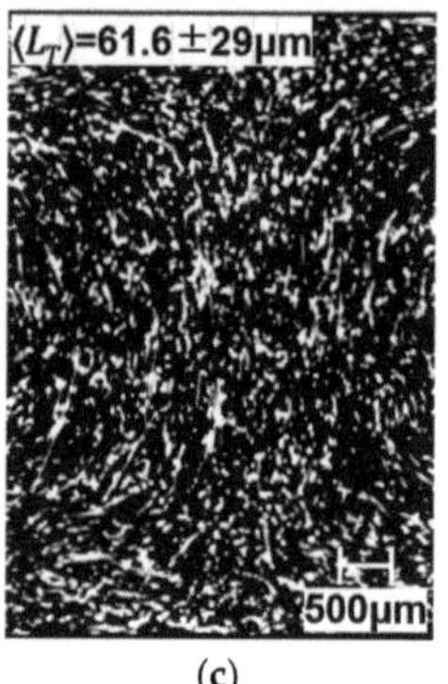
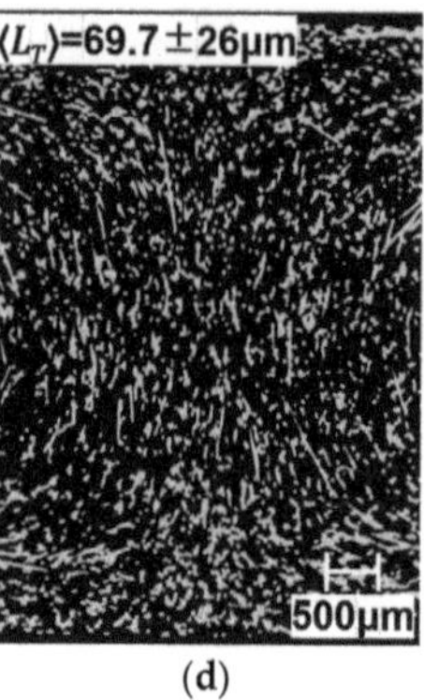

(a) **(b)** **(c)** **(d)**

Figure 18. X-ray CT images, average orientation angles, and average fiber-to-fiber distance using twin-screw and single-screw extruders at a melting temperature of 200 °C on beam and beam with weld specimens' SGFRTPs. (**a**) PP/MAH-PP/GF-T Beam. (**b**) PP/MAH-PP/GF-S Beam. (**c**) PP/MAH-PP/GF-T Beam with Weld. (**d**) PP/MAH-PP/GF-S Beam with Weld.

Figure 19 shows the correlation between a_{iN}, $a_{Model-P}$, and $a_{Model-D}$. The correlation of SGFRTPs shown in this figure is proportional. The coefficient of determination R^2 is close to 0.95, which is a good agreement. Also, the verification results show that the impact strength can be explained by Equation (9) when fiber pull-out is the main source of energy dissipation in quantitative models. In addition, the impact strength can be explained by Equation (12) when interfacial debonding is the main source of energy dissipation in quantitative models.

It is important to note the correlation between the interfacial debonding model and the cohesive force model (CFM) in this study. The cohesive zone approach is used to describe fracture and failure behavior in various material systems [20–22]. In the CFM, a stress-based criterion for debonding and a frictional resistance-based criterion for interfacial sliding were used to capture debonding and sliding [23]. In the Charpy impact test, debonding is postulated to occur under the combined action of normal tensile stress (mode I) at the interface. The physical meaning of $\langle L_T \rangle$ can have the same meaning as crack tip opening displacement (COD). The same physical meaning can also illustrate the reliability of this model.

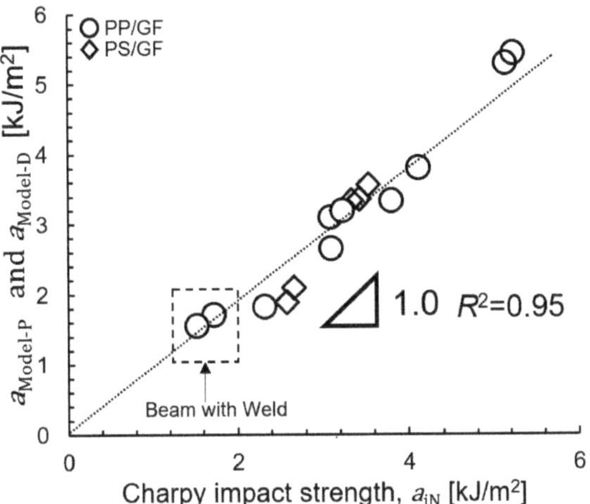

Figure 19. Correlation between a_{iN}, $a_{Model-P}$, and $a_{Model-D}$.

5. Conclusions

This study investigated the correlation between IFSS, IFS, and impact strength of injection molded glass fiber-reinforced thermoplastics made of a polypropylene and polystyrene matrix.

- ► IFSS decreased with the increase in fiber content.
- ► IFSS and IFS showed almost no changes following change in fiber length.
- ► Charpy impact strengths increased with the increase in fiber content or fiber length.
- ► Charpy impact strengths showed almost no changes when the fiber orientation direction was the same as the impact direction.
- ► The relation between L_f and l_p obtained in this study was approximately 2:1, which is equal to the theoretical value. At the same time, both are shorter than the critical fiber length. Usually when the fiber is shorter than the critical fiber length, the fiber break does not happen in SGFRTP.
- ► Fracture surface observation results indicated that fiber pull-out and interfacial debonding were the primary sources of energy dissipation in the quantitative models.

Two theoretical equations with fiber pull-out and interfacial debonding as the main energy dissipation sources in the quantitative models were developed and correlated with the experimental results. Good agreement was obtained ($R^2 > 0.95$). Furthermore, the fiber orientation at the interface will not affect the universality of the model.

This study clarified the quantitative models for interfacial shear strength, interfacial strength, and impact strength of injection molded beam-shaped specimens with different fiber orientation angles. However, this is based on the situation where the dispersed fiber length in injection molded products is less than the critical fiber length. When the fiber length distribution is longer than the critical fiber length in the molded product, especially in the fiber extraction model, the relationship between the fiber extraction length and the critical fiber length needs to be further clarified. Since fibers with weaker interfacial strength or less susceptibility to fracture during molding, such as organic fibers, plant fibers, etc., can be used, the universality of this model needs to be further verified.

Author Contributions: Conceptualization, Q.J. and T.T.; methodology, Q.J. and T.T.; validation, Q.J.; investigation, Q.J.; resources, Q.J.; data curation, Q.J.; writing—original draft preparation, Q.J. and T.T.; writing—review and editing, Q.J. and T.T.; supervision, T.T. and A.N.; project administration, T.T. and A.N. All authors have read and agreed to the published version of the manuscript.

Funding: This research was funded by JST, the establishment of university fellowships towards creating science, technology, and innovation, grant number JPMJFS2104.

Institutional Review Board Statement: Not applicable.

Data Availability Statement: The data presented in this study are available on request from the corresponding author.

Acknowledgments: The X-ray CT images used for this study were taken at the Miyagi Prefectural Industrial Technology Center and Yamagata Research Institute of Technology. This article is a revised and expanded version of a paper entitled [Relationship between interfacial shear strength and impact strength of injection molded short glass fiber-reinforced thermoplastics], which was presented at [Quan Jiang, Tetsuo Takayama, Akihiro Nishioka, Adelaide, Australia, 6–9 Decembe 2022].

Conflicts of Interest: The funders had no role in the design of the study; in the collection, analyses, or interpretation of data; in the writing of the manuscript; or in the decision to publish the results.

References

1. Aminzadegan, S.; Shahriari, M.; Mehranfar, F.; Abramović, B. Factors affecting the emission of pollutants in different types of transportation: A literature review. *Energy Rep.* **2022**, *8*, 2508–2529. [CrossRef]
2. Hegde, S.; Shenoy, B.S.; Chethan, K.N. Review on carbon fiber reinforced polymer (CFRP) and their mechanical performance. *Mater. Today Proc.* **2019**, *19 Pt 2*, 658–662. [CrossRef]
3. Goh, G.D.; Dikshit, V.; Nagalingam, A.P.; Goh, G.L.; Agarwala, S.; Sing, S.L.; Wei, J.; Yeong, W.Y. Characterization of mechanical properties and fracture mode of additively manufactured carbon fiber and glass fiber reinforced thermoplastics. *Mater. Des.* **2018**, *137*, 79–89. [CrossRef]
4. Shimamoto, D.; Tominaga, Y.; Imai, Y.; Hotta, Y. Fiber orientation and flexural properties of short carbon fiber/epoxy composites. *J. Ceram. Soc. Jpn.* **2016**, *124*, 125–128. [CrossRef]
5. Chang, B.; Gu, J.; Long, Z.; Li, Z.; Ruan, S.; Shen, C. Effects of temperature and fiber orientation on the tensile behavior of short carbon fiber reinforced PEEK composites. *Polym. Compos.* **2021**, *42*, 597–607. [CrossRef]
6. Zarges, J.-C.; Kaufhold, C.; Feldmann, M.; Heim, H.-P. Single fiber pull-out test of regenerated cellulose fibers in polypropylene: An energetic evaluation. *Compos. Part A* **2018**, *105*, 19–27. [CrossRef]
7. To, K.; Hiramoto, K.; Matsuda, Y.; Ogawa, J.; Moriwaki, K.; Hamada, H. Study on the Evaluation Methods of the Interfacial Properties of Glass Fiber Reinforced Polypropylene. *Seikeikakou* **2015**, *27*, 434–439. [CrossRef]
8. Mahato, B.; Babarinde, V.O.; Abaimov, S.G.; Lomov, S.V.; Akhatov, I. Interface strength of glass fibers in polypropylene: Dependence on the cooling rate and the degree of crystallinity. *Polym. Compos.* **2020**, *41*, 1310–1322. [CrossRef]
9. Irisawa, T.; Hashimoto, R.; Arai, M.; Tanabe, Y. The Suitability Evaluation of Aromatic Amorphous Thermoplastics as Matrix Resin for CFRTP Having High Thermal Stability. *J. Fiber Sci. Technol.* **2017**, *73*, 61–66. [CrossRef]
10. Kanai, M.; Kageyama, K.; Matsuo, G. Experimental Analysis of Interfacial Debonding on Resin/Fiber by Pinhole-Type Pull-Out Test. *J. Jpn. Soc. Compos. Mater.* **2021**, *47*, 80–87. [CrossRef]
11. Takayama, T. Weld strength of injection molded short fiber reinforced polypropylene. *Mech. Mater.* **2019**, *136*, 103064. [CrossRef]
12. Quan, J.; Takayama, T. Interfacial shear strength evaluation of short fiber reinforced polypropylene by short beam method. *J. Jpn. Soc. Compos. Mater.* **2022**, *48*, 2–9. [CrossRef]
13. Kallel, T.K.; Taktak, R.; Guermazi, N.; Mnif, N. Mechanical and structural properties of glass fiber-reinforced polypropylene (PPGF) composites. *Polym. Compos.* **2018**, *39*, 3497–3508. [CrossRef]
14. Lou, S.; Yin, W.; Ma, X.; Yuan, J.; Wang, Q. Mechanical Properties and Fracture Mechanism of a Glass Fiber Reinforced Polypropylene Composites. *J. Wuhan Univ. Technol.-Mater. Sci. Ed.* **2020**, *35*, 629–634. [CrossRef]
15. Yin, B.B.; Liew, K.M. Machine learning and materials informatics approaches for evaluating the interfacial properties of fiber-reinforced composites. *Compos. Struct.* **2021**, *273*, 114328. [CrossRef]
16. Fang, X.; Shen, C.; Dai, G. The Influence of Compatibilizers on the Structure and Mechanical Properties of Lightweight Reinforced Thermoplastics. *Polym. Compos.* **2016**, *39*, 2212–2223. [CrossRef]
17. Parveen, S.; Pichandi, S.; Goswami, P.; Rana, S. Novel glass fiber reinforced hierarchical composites with improved interfacial, mechanical and dynamic mechanical properties developed using cellulose microcrystals. *Mater. Des.* **2020**, *188*, 108448. [CrossRef]
18. Jiang, Q.; Takayama, T.; Nishioka, A. Relationship between interfacial shear strength and impact strength of injection molded short glass fiber-reinforced thermoplastics. *Procedia Struct. Integr.* **2023**, *45*, 117–124. [CrossRef]
19. Kelly, A.; Tyson, W.R. Tensile properties of fiber-reinforced metals: Copper/tungsten and copper/molybdenum. *J. Mech. Phys. Solids* **1965**, *13*, 329–338, in1–in2, 339–350. [CrossRef]
20. Chandra, N.; Li, H.; Shet, C.; Ghonem, H. Some issues in the application of cohesive zone models for metal–ceramic interfaces. *Int. J. Solids Struct.* **2002**, *39*, 2827–2855. [CrossRef]
21. Li, J.; Zhang, X.B. A criterion study for non-singular stress concentrations in brittle or quasi-brittle materials. *Eng. Fract. Mech.* **2006**, *73*, 505–523. [CrossRef]

22. Huang, Y.; Guo, F. Effect of plastic deformation on the elastic stress field near a crack tip under small-scale yielding conditions: An extended Irwin's model. *Eng. Fract. Mech.* **2021**, *254*, 107888. [CrossRef]
23. Ananth, C.R.; Chandra, N. Evaluation of interfacial shear properties of metal matrix composites from fiber push-out tests. *Mech. Compos. Mater. Struct.* **1995**, *2*, 309–328. [CrossRef]

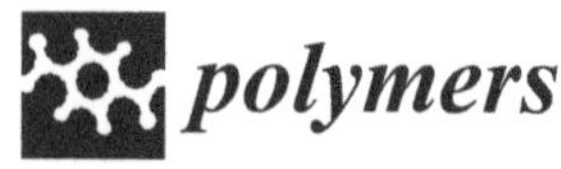

Article

Mechanical Anisotropy of Injection-Molded PP/PS Polymer Blends and Correlation with Morphology

Tetsuo Takayama * and Rin Shibazaki

Graduate School of Organic Materials Science, Yamagata University, Yonezawa 992-8510, Japan
* Correspondence: t-taka@yz.yamagata-u.ac.jp; Tel.: +81-238-26-3085

Abstract: The molecular orientation formed by melt-forming processes depends strongly on the flow direction. Quantifying this anisotropy, which is more pronounced in polymer blends, is important for assessing the mechanical properties of thermoplastic molded products. For injection-molded polymer blends, this study used short-beam shear testing to evaluate the mechanical anisotropy as a stress concentration factor, and clarified the correlation between the evaluation results and the phase structure. Furthermore, because only shear yielding occurs with short-beam shear testing, the yielding conditions related to uniaxial tensile loading were identified by comparing the results with those of three-point bending tests. For continuous-phase PP, the phase structure formed a sea-island structure. The yield condition under uniaxial tensile loading was interface debonding. For continuous-phase PS, the phase structure was dispersed and elongated in the flow direction. The addition of styrene–ethylene–butadiene–styrene (SEBS) altered this structure. The yielding condition under uniaxial tensile loading was shear yielding. The aspect ratio of the dispersed phase was found to correlate with the stress concentration factor. When the PP forming the sea-island structure was of continuous phase, the log-complex law was sufficient to explain the shear yield initiation stress without consideration of the interfacial interaction stress.

Keywords: interfacial interaction force; morphology; short-beam shear tests; yield condition; yield initiation stress

Citation: Takayama, T.; Shibazaki, R. Mechanical Anisotropy of Injection-Molded PP/PS Polymer Blends and Correlation with Morphology. *Polymers* **2023**, *15*, 4167. https://doi.org/10.3390/polym15204167

Academic Editors: Andrew N. Hrymak and Shengtai Zhou

Received: 24 September 2023
Revised: 17 October 2023
Accepted: 19 October 2023
Published: 20 October 2023

1. Introduction

Thermoplastics are used in widely diverse applications, from minor daily necessities to automobiles, because of their lighter weight and superior moldability than those of metals and ceramics [1–3]. Moreover, because thermoplastics have lower melting temperatures than metals or ceramics, they can be melt-molded with the expenditure of low energy costs. Among these molding methods, injection molding is often applied for thermoplastics because it enables near-net-shape molding and because it is excellent for mass production [2,4]. Depending on the required shape of the molded product, extrusion [5] or blow molding [6] may be applied, and depending on the required properties, a polymer blend may be prepared by melt mixing and then melt-molded into a product [7–17]. Examples of polymer blends include polypropylene (PP)/polystyrene (PS) blends [7–9], polyethylene terephthalate (PET)/polyethylene [10], and PET/PP blends [11], high-impact polystyrene (PS-HI) blends [12,13] blended with elastomers to improve the toughness of polystyrene, and polycarbonate (PC)/acrylonitrile-butadiene-styrene (ABS) copolymers, which offer a wide range of controllable processing and mechanical properties [14–17]. Polymer blend is a generic term for materials that are composites of two or more polymers. Polymer blends are prepared to achieve physical properties that cannot be achieved with single polymers. However, it is known that most blends thermodynamically form a phase-separated structure, and this is one of the reasons why it is difficult to obtain the desired physical properties [18]. This is also true for mechanical properties, and, in particular, the mechanical anisotropy of a molded product is strongly dependent on the phase-separated structure that is formed. Therefore, it is very important to quantify the anisotropy when discussing the mechanical properties of molded polymer blends. As for PP/PS, which is the subject of this study, the relationship between its phase structure and mechanical properties has been reported, but descriptions of its mechanical anisotropy could not be found in the authors' investigation [7–9].

Various studies have been conducted to examine the correlation between the phase structure and mechanical anisotropy of polymer blends. For instance, Li et al. performed thin-wall injection molding of a polymer blend of polypropylene (PP) and thermoplastic rubber, and cut specimens from the resulting molded product in the direction parallel (MD) and perpendicular (TD) to the flow direction for tensile testing [19]. They found that elongation at the break of specimens cut in the TD direction was about twice as large as that of specimens cut in the MD direction. The reasons for their finding are discussed in terms of the morphology of the dispersed phase. A correlation between elongation at the break of the polymer blend and the morphology of the dispersed phase has been reported [20]. That study reported that the phase structure of a 20 wt.% blend of PP with PS varies with the melt viscosity of PP and that it can be organized by the viscosity ratio of PS to PP.

Such reports discuss the mechanical properties and the anisotropy of the morphology of the dispersed phase which is formed via injection molding. Nevertheless, quantitative analysis of morphology effects of the dispersed phase on the mechanical properties of polymer blends is difficult, as is clarification of the factors contributing to these effects. For some areas, conventional evaluation methods are inadequate for analyzing the mechanical anisotropy which occurs in polymer blend molded products.

The evaluation of mechanical anisotropy of polymer molded products is mainly performed via the cutout method [19]. The cutout method evaluates mechanical anisotropy by performing mechanical tests on specimens cut in a specific direction and comparing the results between the cutout directions. This method is effective for evaluating the mechanical anisotropy of two-dimensional flat surfaces such as films, but is not suitable for evaluating the mechanical anisotropy of three-dimensional objects such as injection-molded products. Residual stresses in the molded product may relax when the injection-molded product is machined, leaving the possibility that the cut-out method may not correctly evaluate the mechanical anisotropy of the molded product [21]. Another method for evaluating mechanical anisotropy is indentation hardness testing [22], with results based on the ratio of indentation lengths obtained from testing. This simple method requires no preparation of specimens, as is necessary for the cut-out method.

Although the reports explained above present methods for quantitatively evaluating mechanical anisotropy, they cannot be characterized as methods for quantitatively analyzing the correlation between a polymer blend's phase structure and its mechanical properties. Moreover, they do not clarify the factors causing such anisotropy. As demonstrated by the discussion presented above, no study has provided evidence clarifying the mechanisms of mechanical anisotropy in polymer blend injection-molded products.

Against this background, the authors have proposed a method for evaluating the mechanical anisotropy of polymer injection-molded parts in three dimensions by applying the method for evaluating the interfacial shear strength via a short-beam shear test proposed by Quan et al. [23,24]. Using this method, the stress concentration factor derived from the phase structure of the molded product can be quantified in the triaxial direction. In addition, since this method generates only shear stress, the yield condition can be limited to shear yield. When mechanical evaluation is performed via the tensile testing of polymer blend molded products, the two yield conditions are interface delamination and shear yield, and it is difficult to identify the factor of stress at yield initiation, but when combined with the results of short-beam shear testing, the yield condition due to tensile loading of the polymer blend can be identified. Based on the discussion presented above, the primary objective of this study of polymer blend injection-molded parts is using short-beam shear testing for the evaluation of mechanical anisotropy as a stress concentration factor, and clarifying the correlation between the obtained values and the phase structure. Furthermore, because only shear yielding occurs in short-beam shear testing, the yielding conditions attributable to uniaxial tensile loading were identified by comparing the results with those obtained from three-point bending testing. Because multiple phase structures were obtained in this study, the stress at shear yield initiation obtained for each phase structure was also modeled to clarify the correlation between the phase structure and the stress at yield initiation.

2. Materials and Methods

2.1. Materials

Table 1 presents the materials used for this study. The table also shows the melt flow rate (MFR), an index of melt viscosity. Two types of polypropylene were used: homo-type polypropylene (H-PP) and block-type polypropylene (B-PP). The block-type PP used in this paper is a propylene-ethylene block copolymer containing 16 wt.% ethylene-propylene rubber with an average molecular weight of 450,000 g/mol [25]. Styrene–ethylene–butadiene–styrene (SEBS) copolymers of two types were used to control the phase structure: SEBS with a low styrene ratio was designated as L-SEBS, and SEBS with a high styrene ratio was designated as H-SEBS.

Table 1. Material information.

Material	Code	Manufacturer	Name	MFR (g/10 min)
PP	H-PP	Japan Polypropylene Corp., Tokyo, Japan	Novatec-PP MA1B	21@230 °C, 2.160 kgf
	B-PP	Japan Polypropylene Corp., Tokyo, Japan	Novatec-PP BC03B	30@230 °C, 2.160 kgf
PS	PS	Toyo Styrene Co., Ltd., Tokyo, Japan	Toyo styrene G210C	10@230 °C, 2.160 kgf
SEBS	L-SEBS	Asahi Kasei Corp., Tokyo, Japan	Tuftec H1052	13@230 °C, 2.160 kgf
	H-SEBS	Asahi Kasei Corp., Tokyo, Japan	Tuftec H1043	2@230 °C, 2.160 kgf

2.2. Sample Preparation

After these materials were poured into a twin-screw extruder (IMC-00 type, L/D = 25; Imoto Machinery Co., Ltd., Kyoto, Japan) with a screw diameter of 15 mm, they were melt-kneaded. Table 2 presents the studied compositions. SEBS was added at the ratio shown in that table to promote a fine dispersion of the dispersed phase. For this study, the melt-kneading temperature was set as 230 °C. The screw speed was fixed as 60 rpm. The resulting strands were cut into granular pieces using a pelletizer; they were used as pellets. The resulting pellets were filled into an ultra-compact electric injection molding machine (C. Mobile 0813; Shinko Sellbic Co., Ltd., Tokyo, Japan) and were injection molded to obtain strip-shaped molded products with the dimensions presented in Figure 1. Table 3 shows the injection molding conditions. The molding adopted the same geometry as in the literature [24]. All temperatures and times for injection molding were fixed. The holding pressure was varied to provide a good molded product.

Table 2. Material compositions.

H-PP (vol.%)	B-PP (vol.%)	PS (vol.%)	L-SEBS (vol.%)	H-SEBS (vol.%)
100.0				
77.0		23.0		
69.2		25.2	4.0	1.6
33.3		66.7		
31.4		63.4	1.7	3.5
		100.0		
	100.0			
	77.0	23.0		
	69.2	25.2	4.0	1.6
	33.3	66.7		
	31.4	63.4	1.7	3.5
		100.0		

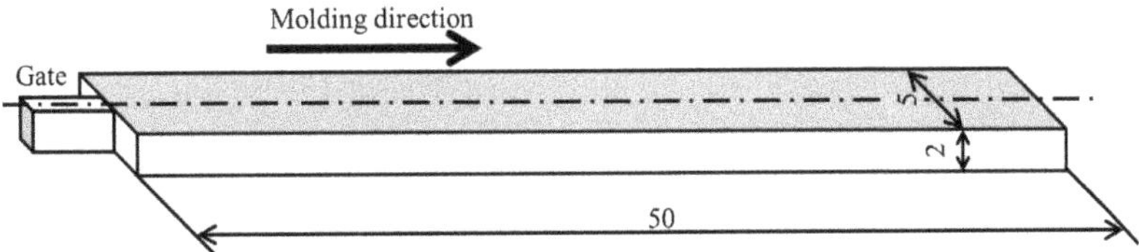

Figure 1. Beam specimen geometry (unit: mm) [24].

Table 3. Injection molding conditions.

H-PP (vol.%)	B-PP (vol.%)	PS (vol.%)	L-SEBS (vol.%)	H-SEBS (vol.%)	T_{inj} (°C)	T_{mold} (°C)	V_{inj} (mm/s)	P_{hold} (MPa)	T_{inj} (s)	T_{cool} (s)
100					230	50	30	46	10	15
77.0		23.0			230	50	30	46	10	15
69.2		25.2	4.0	1.6	230	50	30	49	10	15
33.3		66.7			230	50	30	56	10	15
31.4		63.4	1.7	3.5	230	50	30	56	10	15
		100.0			230	50	30	46	10	15
	100.0				225	50	30	46	10	15
	77.0	23.0			230	50	30	46	10	15
	69.2	25.2	4.0	1.6	230	50	30	49	10	15
	33.3	66.7			230	50	30	56	10	15
	31.4	63.4	1.7	3.5	230	50	30	56	10	15
		100.0			230	50	30	46	10	15

2.3. Mechanical Anisotropy Determination via Short-Beam Shear Testing [24]

Short-beam shear tests were conducted on a compact universal mechanical testing machine (MCT-2150; A&D Co., Ltd., Tokyo, Japan) using strips of molded products obtained via injection molding. The distance between spans was 10 mm. The loading speed was 10 mm/min. The obtained load–deflection curve is differentiated by the deflection to obtain the stiffness. The average shear stress τ was obtained from the load at each time point based on Equation (1).

$$\tau = \frac{3P}{4A} \tag{1}$$

where P represents the load; and A stands for the cross sectional area of the specimen. The stiffness-averaged shear stress curve was finally obtained using the calculations shown above. An example of a stiffness-averaged shear stress curve obtained from this test is presented in Figure 2. All of the materials examined for this study exhibited curves similar to this example. They show a rapid increase in stiffness at the beginning of loading followed by a discontinuous decrease. Subsequently, the stiffness becomes stable, but upon closer inspection, several points are apparent at which the stiffness decreases. As described in this paper, shear yielding is regarded as initiated at these points. Because shear stress is conjugate, when only shear stress occurs, this stress acts equally in the triaxial direction. Therefore, there would be, at most, three points of stiffness reduction. These points are τ_s, τ_m, and τ_l, starting from the lowest value, where τ_s signifies the shear stress acting in the specimen thickness direction, τ_m in the specimen width direction, and τ_l in the flow direction. Using these values, mechanical anisotropy was determined as the minimum mechanical anisotropy A_s shown in Equation (2), the intermediate mechanical anisotropy A_m shown in Equation (3), and the maximum mechanical anisotropy A_l shown in Equation (4).

$$A_s = 1 - \frac{\tau_m}{\tau_l} \tag{2}$$

$$A_m = 1 - \frac{\tau_s}{\tau_m} \tag{3}$$

$$A_l = 1 - \frac{\tau_s}{\tau_l} \tag{4}$$

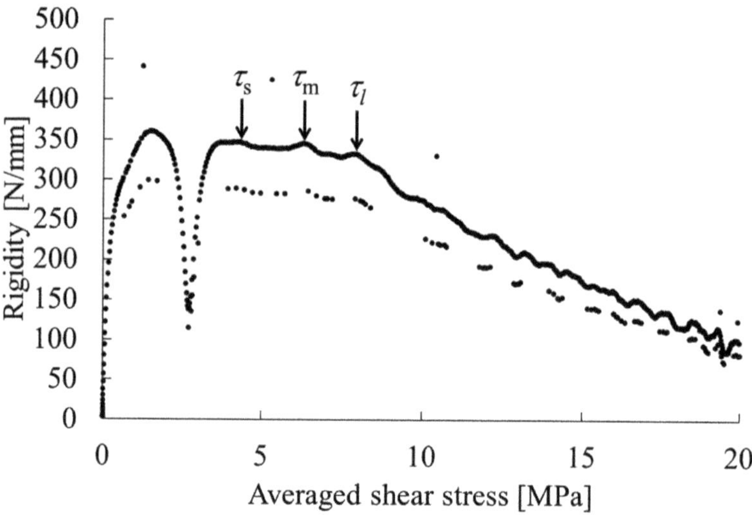

Figure 2. Rigidity—averaged shear stress curve obtained from short-beam shear tests [24].

These variables take values from 0 to 1, with a maximum of 1. In addition, from the definition, $A_l \geq A_m$.

Using these values, the stress concentration factors were determined via Equations (5)–(7) as γ_s, γ_m, and γ_l, from largest to smallest.

$$\gamma_s = \frac{\tau_y}{\sqrt{3}\tau_s} = \frac{1}{\sqrt{3}}\sqrt{1 + \frac{1}{(1 - A_m)^2} + \frac{1}{(1 - A_l)^2}} \tag{5}$$

$$\gamma_m = \frac{1}{\sqrt{3}}\sqrt{1 + \frac{1}{(1 - A_s)^2} + (1 - A_m)^2} \tag{6}$$

$$\gamma_l = \frac{1}{\sqrt{3}}\sqrt{1 + (1 - A_s)^2 + (1 - A_l)^2} \tag{7}$$

Their minimum value is 1 in the isotropic case. Their maximum value is theoretically infinite. Furthermore, using the stress concentration factor obtained from Equation (5) and τ_s, the stress at yield initiation in the MD direction was obtained using Equation (8).

$$\sigma_{y,MD} = 3\gamma_s\tau_s \tag{8}$$

The stress at yield initiation obtained here is the value obtained at shear yielding. As described in this paper, the correlation between these stress concentration factors, the stress at yield initiation, and the phase structure is investigated.

2.4. Morphology Observation

Sections were prepared from the injection-molded specimen using a microtome (RX-860; Yamato Kohki Industrial Co., Ltd., Asaka, Japan). The phase structure was observed using a phase contrast microscope (BA410EPH-1080; Shimadzu Rika Corp., Tokyo, Japan). Compositions containing a substantial amount of PS were examined using scanning electron microscopy (SEM, Technex Co., Ltd., Tokyo, Japan, Tiny-SEM 510) on the cut surfaces created during section preparation to confirm the phase structure due to the difficulty in confirming it via phase contrast observation. Figure 3 shows the locations used for observation of the phase structure. We attempted to observe the phase structure three-dimensionally by observing the area corresponding to the core layer of the molded product from two angles: the MD-TD plane and the TD-ND plane. From the MD-TD plane image, 100 dispersed phases were extracted from the phase difference image. The aspect ratio was calculated as the ratio of lengths in the MD and TD directions. We attempted to achieve the purposes of this study by clarifying the degrees of correlation between the obtained phase structures and aspect ratios and the stress concentration factors in each direction.

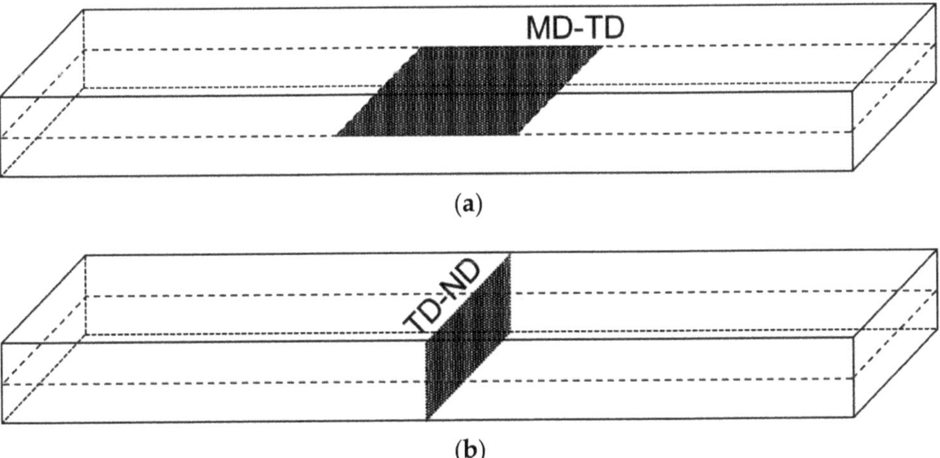

Figure 3. Morphology observation area. (**a**) MD-TD plane. (**b**) TD-ND plane.

3. Results and Discussions

3.1. Relation of Morphology and Mechanical Anisotropy of Injection-Molded PP-Rich Polymer Blends

Figure 4 shows the phase contrast microscopy results for the PP-rich compositions; both MD-TD and TD-ND cross sections of the PP-rich compositions show a sea-island structure [7–9], leading to the conclusion that the phase structure in the PP-rich compositions is a sea-island structure. The addition of SEBS to this composition also resulted in a sea-island structure; the size of the dispersed phase appeared to be smaller and the particle size distribution narrower than that of the product without SEBS. Figure 5 portrays examples of stiffness-averaged shear stress curves for compositions with high PP content. The addition of SEBS tended to shift the shear stress at the onset of the yield in each direction toward the low stress side. Table 4 shows the shear stress at yield, mechanical illegality and stress concentration factors in each direction obtained via short-beam shear tests for the PP-rich compositions. This indicates that the morphology of the dispersed phase of the PP-rich composition is almost unchanged depending on the type of PP and the presence or absence of SEBS.

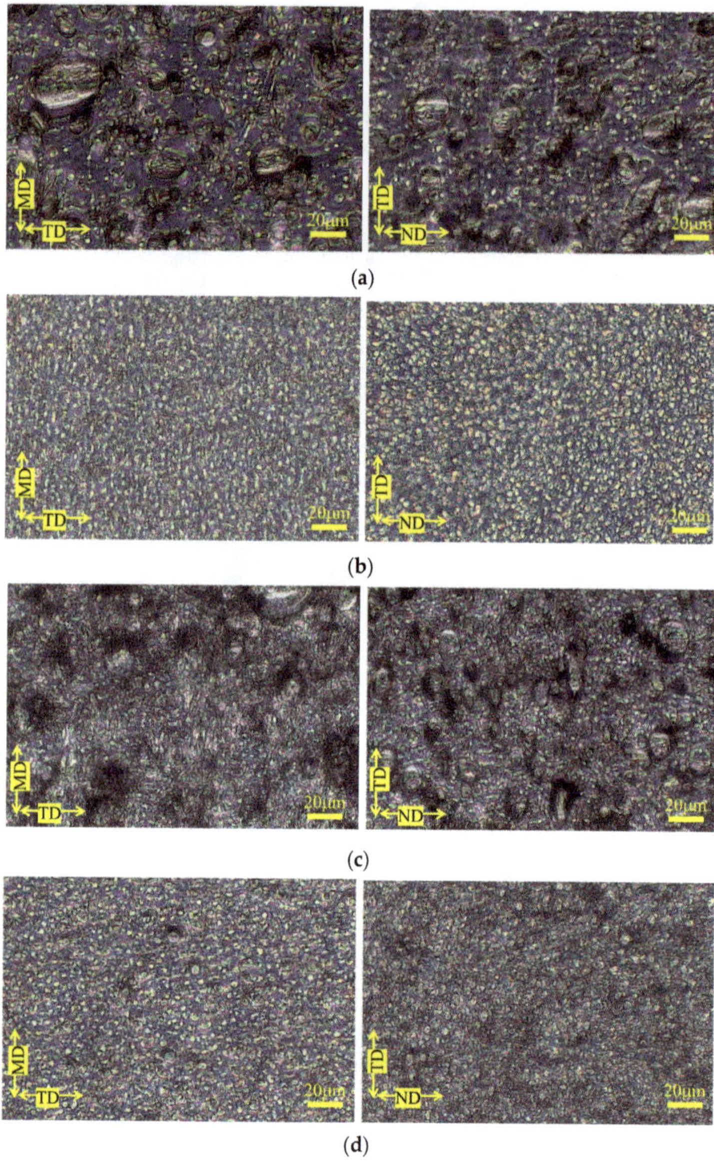

Figure 4. Phase contrast microscopic observations of the composition with high PP content. (**a**) H-PP/PS = 77.0/23.0 vol.%. (**b**) H-PP/PS/SEBS = 69.2/25.2/5.6 vol.%. (**c**) B-PP/PS = 77.0/23.0 vol.%. (**d**) B-PP/PS/SEBS = 69.2/25.2/5.6 vol.%.

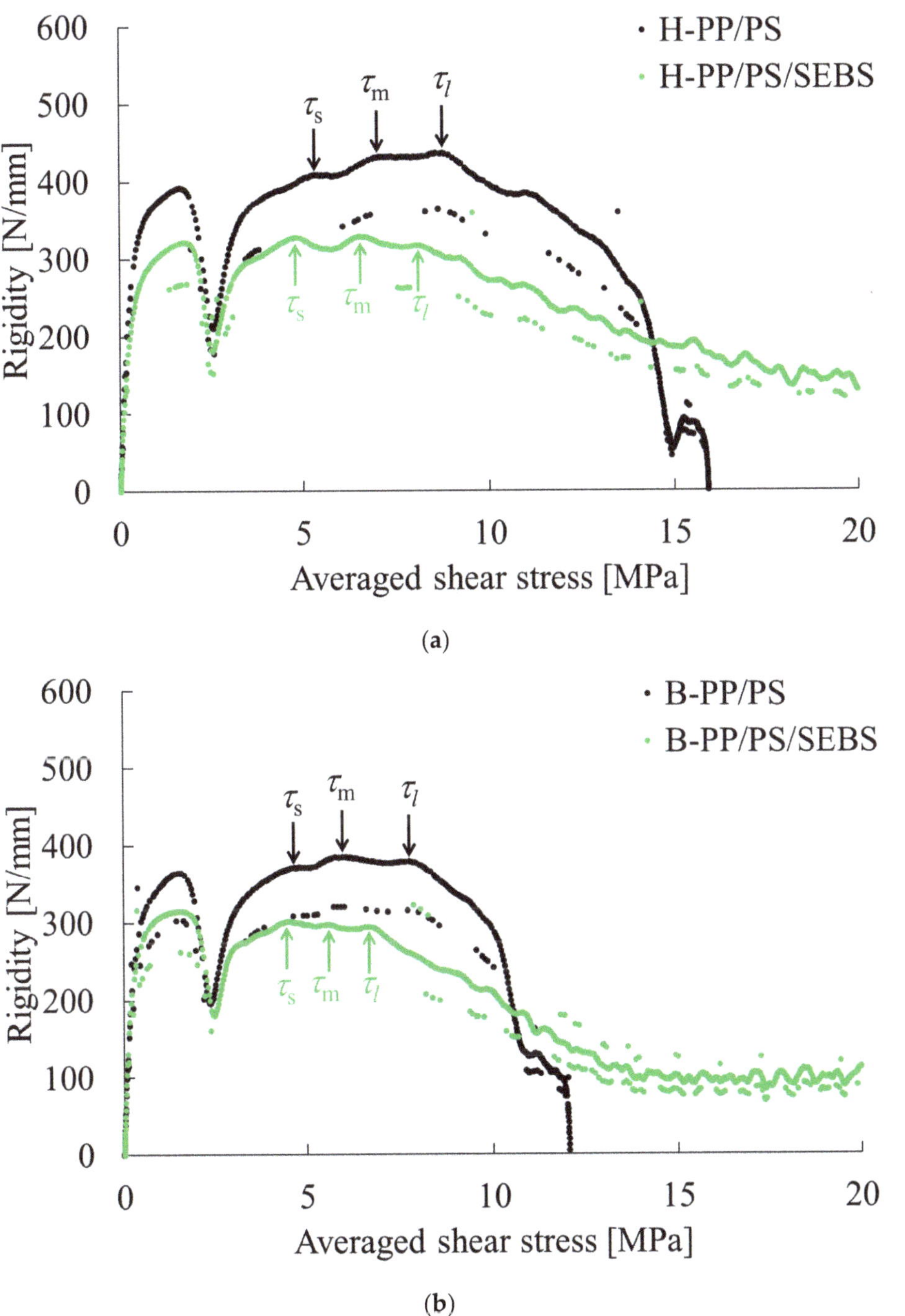

Figure 5. Rigidity—averaged shear stress curves of PP-rich polymer blends obtained via short-beam shear testing. (**a**) H-PP/PS = 77.0/23.0 vol.% and H-PP/PS/SEBS = 69.2/25.2/5.6 vol.%. (**b**) B-PP/PS = 77.0/23.0 vol.% and B-PP/PS/SEBS = 69.2/25.2/5.6 vol.%.

Table 4. Mechanical anisotropy of PP/PS polymer blends with high PP content.

(a) Mechanical Anisotropy Factor										
H-PP (vol.%)	B-PP (vol.%)	PS (vol.%)	L-SEBS (vol.%)	H-SEBS (vol.%)	τ_s (MPa)	τ_m (MPa)	τ_l (MPa)	A_s (-)	A_m (-)	A_l (-)
100					4.2	6.1	8.1	0.25	0.31	0.48
77.0		23.0			5.5	7.0	8.3	0.16	0.21	0.34
69.2		25.2	4.0	1.6	5.2	6.7	7.7	0.13	0.22	0.32
	100				4.5	5.6	6.6	0.15	0.20	0.32
	77.0	23.0			5.2	6.6	7.9	0.16	0.21	0.34
	69.2	25.2	1.7	3.5	4.4	5.4	6.5	0.17	0.19	0.32
		100.0			13.1	16.2	18.8	0.14	0.19	0.30

(b) Stress concentration factor and aspect ratio								
H-PP (vol.%)	B-PP (vol.%)	PS (vol.%)	L-SEBS (vol.%)	H-SEBS (vol.%)	γ_s (-)	γ_m (-)	γ_l (-)	A.R. (-)
100					1.51	1.04	0.78	-
77.0		23.0			1.28	1.00	0.85	1.2
69.2		25.2	4.0	1.6	1.27	0.99	0.86	1.3
	100				1.25	1.01	0.85	-
	77.0	23.0			1.28	1.01	0.84	1.3
	69.2	25.2	4.0	1.6	1.25	1.02	0.85	1.2
		100.0			1.24	1.00	0.86	-

3.2. Relation of Morphology and Mechanical Anisotropy of Injection-Molded PS-Rich Polymer Blends

The phase contrast microscopy and scanning electron microscopy results of the PS-rich composition are shown in Figures 6 and 7, respectively. For the SEBS-free variant, in the MD-TD cross section, both phase contrast and SEM images displayed an elliptical dispersed phase elongated towards the MD direction. The SEM image of TD-ND cross section also presented an elliptical dispersed phase elongated towards the TD direction. From these results, it was inferred that the dispersed phase in the PS-rich composition without SEBS is dispersed in an elongated disk shape. Upon the addition of SEBS to PS/H-PP, the phase contrast image exhibited a dispersed phase elongated in the MD direction, with a rod-like structure, while the TD-ND cross section showed a sea-island pattern. Furthermore, via the SEM image of the TD-ND cross section, it was revealed that the dispersed phase was finely dispersed. Thus, we can conclude that the addition of SEBS to PS/H-PP resulted in a phase morphology with a cylindrical dispersed phase with a relatively small diameter. Adding SEBS to PS/B-PP resulted in the dispersed phase arranging into a network, as observed in the phase contrast image of the MD-TD cross section. The TD-ND cross section's phase contrast image and SEM image further confirmed a sea-island structure. These findings suggest that incorporating SEBS into PS/B-PP results in the dispersed phase forming a two-dimensional network structure. Figure 8 presents an example of stiffness-averaged shear stress curves for a composition with a high PS content. The addition of SEBS to PS/H-PP tended to shift the shear stress at the onset of yield in each direction toward the high stress side, whereas the addition of SEBS to PS/B-PP tended to shift the shear stress at the onset of yield in each direction toward the low stress side.

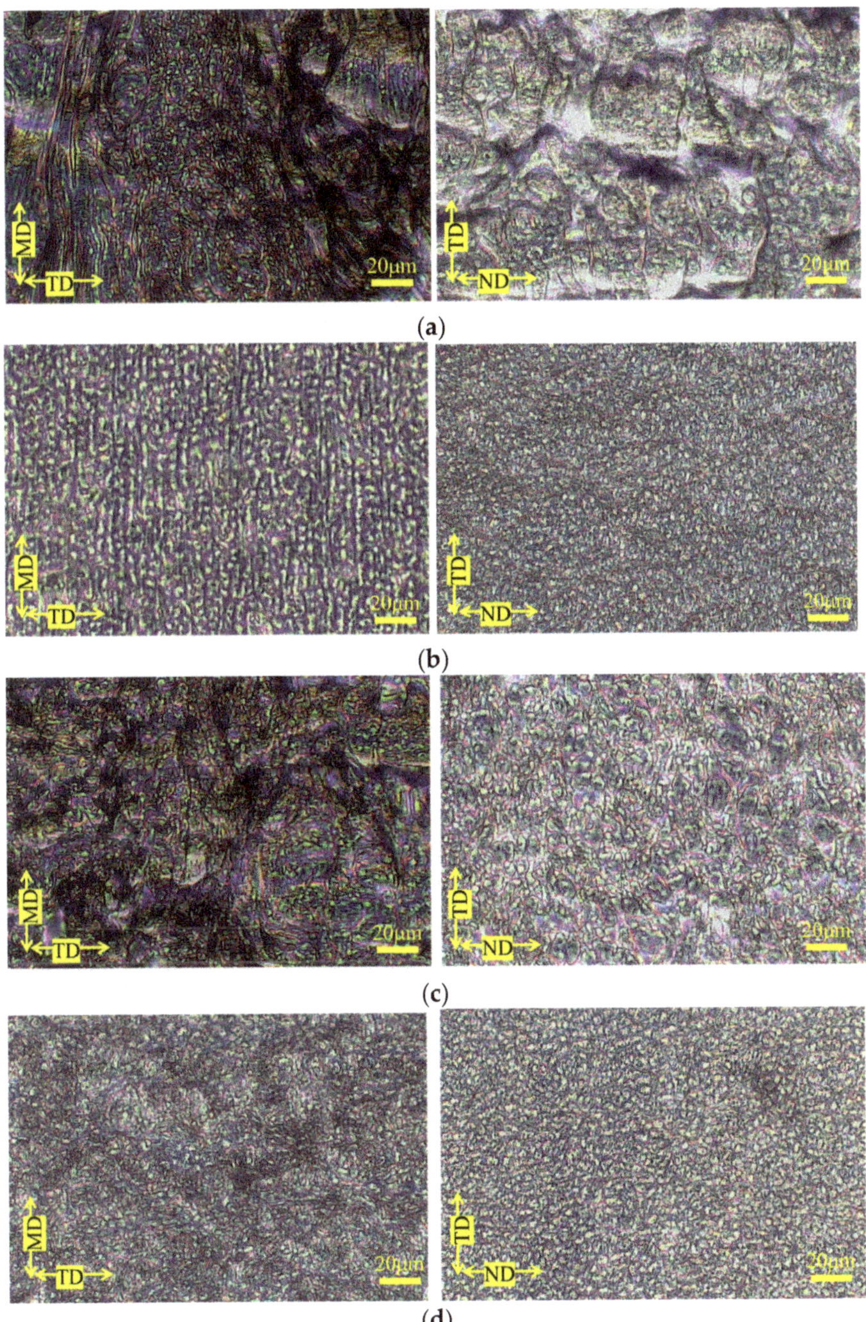

Figure 6. Phase contrast microscopic observations of the composition with high PS content. (a) H-PP/PS = 33.3/66.7 vol.%. (b) H-PP/PS/SEBS = 31.4/63.4/5.2 vol.%. (c) B-PP/PS = 33.3/66.7 vol.%. (d) B-PP/PS/SEBS = 31.4/63.4/5.2 vol.%.

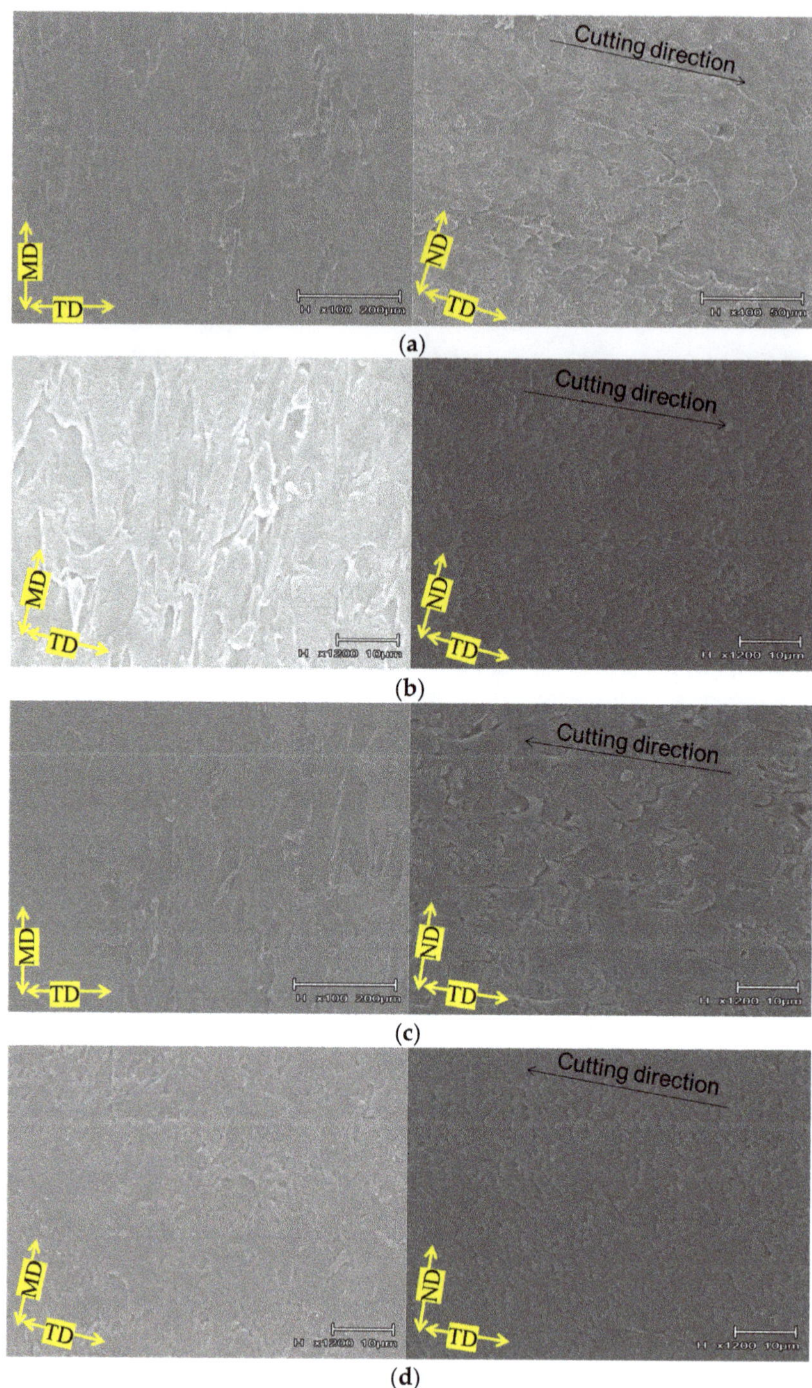

Figure 7. Scanning electron microscope observations of the composition with high PS content. (**a**) H-PP/PS = 33.3/66.7 vol.%. (**b**) H-PP/PS/SEBS = 31.4/63.4/5.2 vol.%. (**c**) B-PP/PS = 33.3/66.7 vol.%. (**d**) B-PP/PS/SEBS = 31.4/63.4/5.2 vol.%.

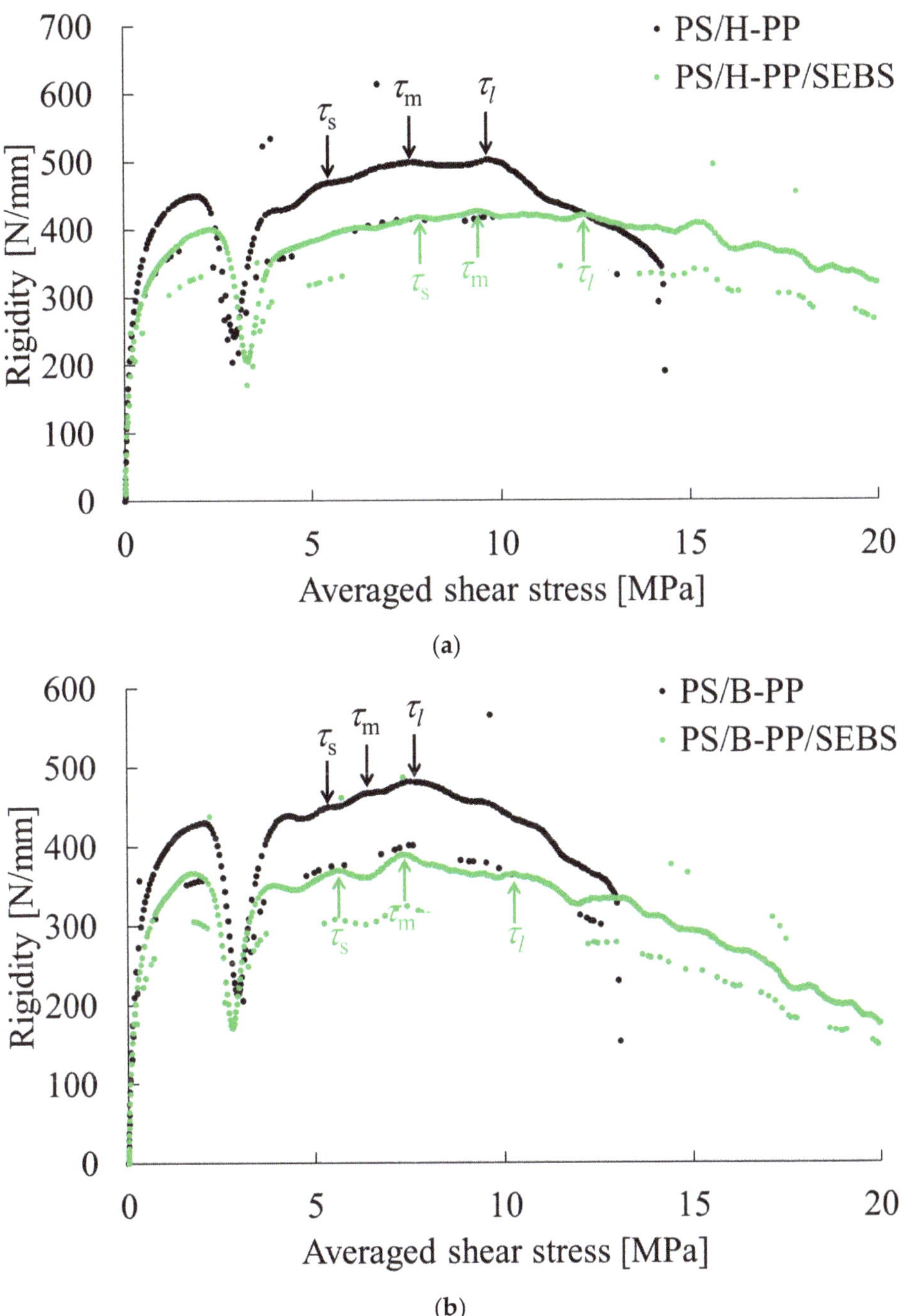

(a)

(b)

Figure 8. Rigidity—averaged shear stress curves of PS-rich polymer blends obtained via short-beam shear testing. (a) H-PP/PS = 33.3/66.7 vol.% and H-PP/PS/SEBS = 31.4/63.4/5.2 vol.%. (b) B-PP/PS = 33.3/66.7 vol.% and B-PP/PS/SEBS = 31.4/63.4/5.2 vol.%.

Table 5 presents the shear stress at yield initiation, mechanical anisotropy factors, and stress concentration factors in each direction obtained from short-beam shear tests for the PS-rich compositions. The stress concentration factors of the PS-rich compositions did not change with the type of PP when SEBS was not added; they were almost constant. When SEBS was added to these compositions, γ_s and γ_m tended to increase, although γ_l tended to decrease. This result suggests that the morphology of the dispersed phase is almost identical in the compositions with high PS content without SEBS, but that the morphology differs when SEBS is added.

Table 5. Mechanical anisotropy of PP/PS polymer blends with high PS content.

(a) Mechanical Anisotropy Factor										
H-PP (vol.%)	B-PP (vol.%)	PS (vol.%)	L-SEBS (vol.%)	H-SEBS (vol.%)	τ_s (MPa)	τ_m (MPa)	τ_l (MPa)	A_s (-)	A_m (-)	A_l (-)
100.0					4.5	5.6	6.6	0.15	0.20	0.32
33.3		66.7			5.2	7.6	8.9	0.15	0.32	0.42
31.4		63.4	1.7	3.5	6.4	9.3	11.9	0.22	0.31	0.46
	100.0				4.5	5.6	6.6	0.15	0.20	0.32
	33.3	66.7			4.8	7.0	8.0	0.13	0.31	0.40
	31.4	63.4	1.7	3.5	5.6	8.2	10.5	0.22	0.32	0.47
		100.0			13.1	16.2	18.8	0.14	0.19	0.30

(b) Stress concentration factor and aspect ratio								
H-PP (vol.%)	B-PP (vol.%)	PS (vol.%)	L-SEBS (vol.%)	H-SEBS (vol.%)	γ_s (-)	γ_m (-)	γ_l (-)	A.R. (-)
100.0					1.25	1.01	0.85	-
33.3		66.7			1.42	0.97	0.83	2.5
31.4		63.4	1.7	3.5	1.48	1.02	0.80	6.7
	100.0				1.25	1.01	0.85	-
	33.3	66.7			1.40	0.96	0.84	2.3
	31.4	63.4	1.7	3.5	1.49	1.02	0.79	∞
		100.0			1.24	1.00	0.86	-

These results indicate that the mechanical anisotropy of PP/PS polymer blends depends on the morphology of the dispersed phase and indicate that it is only slightly affected by its size or its distribution. Based on these results, we investigated the degree of correlation between the aspect ratio, which is related to the morphology of the dispersed phase, and each stress concentration coefficient. Figure 9 shows the relation between the aspect ratio and each stress concentration coefficient obtained from phase contrast microscopic observations. Tables 4b and 5b show the average values of the aspect ratios. The compositions which form a two-dimensional network are assumed to have continuity in the dispersed phase. The aspect ratio is assumed to be infinite, as represented by the dashed line in that figure. The γ_s and γ_m values tend to increase, although γ_l tends to decrease, as the aspect ratio increases. The γ_s and γ_l values change asymptotically toward the value of the aspect ratio at infinity. These results indicate that the mechanical anisotropy of PP/PS polymer blends is correlated with the aspect ratio of the dispersed phase. When comparing particle-dispersed and fiber-dispersed composites, it has been found that the latter display greater mechanical anisotropy. As the aspect ratio increases, these composites develop mechanical anisotropy comparable to that of continuous fibers [26–28]. This correlation is believed to be applicable to blends of PP and PS polymers.

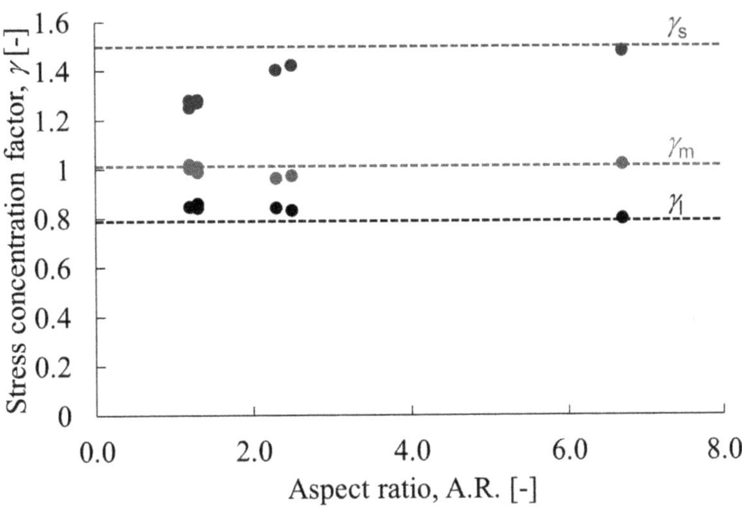

Figure 9. Relations between the aspect ratio and stress concentration factors. The figure displays the value when A.R. is infinite with a dashed line.

3.3. Yield Conditions for PP/PS Polymer Blends

From the obtained values of mechanical anisotropy, the yield initiation stress in the MD direction was obtained using Equation (8). Table 6 presents the results obtained from comparing the yield initiation stress in the MD direction with the yield initiation stress $\sigma_{y,exp}$ obtained from a three-point bending test under the same conditions as those reported in the literature [19]. $\sigma_{y,exp}$ is expressed as in Equation (9).

$$\sigma_{y,exp} = \frac{2\sigma_f}{3(1+v)} \tag{9}$$

Table 6. Results of comparing the yield initiation stress in the MD direction with the yield initiation stress $\sigma_{y,exp}$ obtained from a three-point bending test.

H-PP (vol.%)	B-PP (vol.%)	PS (vol.%)	L-SEBS (vol.%)	H-SEBS (vol.%)	τ_s (MPa)	γ_s (-)	$\sigma_{y,MD}$ (MPa)	$\sigma_{fy,exp}$ (MPa)	v (-)	$\sigma_{y,exp}$ (MPa)
100					4.2	1.51	19.0	28.0	0.407	19.9
77.0		23.0			5.5	1.28	21.1	25.5	0.392	18.3
69.2		25.2	4.0	1.6	5.2	1.27	19.8	23.6	0.393	16.9
33.3		66.7			5.2	1.42	22.2	31.5	0.363	23.1
31.4		63.4	1.7	3.5	6.4	1.48	28.4	38.0	0.366	27.8
	100.0				4.5	1.25	16.9	24.1	0.413	17.1
	77.0	23.0			5.2	1.28	20.0	22.2	0.398	15.9
	69.2	25.2	4.0	1.6	4.4	1.25	16.5	19.9	0.398	14.3
	33.3	66.7			4.8	1.40	20.2	28.0	0.366	20.5
	31.4	63.4	1.7	3.5	5.6	1.49	25.0	32.2	0.368	23.5
		100.0			13.1	1.24	48.6	65.5	0.342	48.8

In that equation, σ_f represents the flexural strength: v is Poisson's ratio. The value v of the polymer blend is obtained using Equation (10).

$$v = \sum_{i=1}^{n} v_i V_i \tag{10}$$

In that equation, V stands for the volume content; and n denotes the number of compositions. The fact that $\sigma_{y,MD}$ is greater than $\sigma_{y,exp}$ suggests that the yielding conditions generated by the short-beam shear test differ from those generated by the three-point bending test. Because the short-beam shear test imparts only shear stress, the only applicable yield condition is shear yielding. By contrast, the three-point bending test imparts mainly vertical stress, which results in expansion stress. Under these stress conditions, yielding might occur because of debonding at the interface in addition to shear yielding. Table 7 presents the yield conditions for the compositions examined for this study. To confirm this point, a model was constructed for this study to ascertain the yielding initiation stress when yielding occurs under the conditions of interface debonding.

Table 7. Yield conditions for the compositions.

H-PP (vol.%)	B-PP (vol.%)	PS (vol.%)	L-SEBS (vol.%)	H-SEBS (vol.%)	$\sigma_{y,MD}$ (MPa)	$\sigma_{y,exp}$ (MPa)	YC_{MD}
77.0		23.0			21.1	18.3	Debonding
69.2		25.2	4.0	1.6	19.8	16.9	Debonding
33.3		66.7			22.2	23.1	Shear yield
31.4		63.4	1.7	3.5	28.4	27.8	Shear yield
	77.0	23.0			20.0	15.9	Debonding
	69.2	25.2	4.0	1.6	16.5	14.3	Debonding
	33.3	66.7			20.2	20.5	Shear yield
	31.4	63.4	1.7	3.5	25.0	23.5	Shear yield

When obtaining a molded product through the melt forming process, the product always undergoes a heating and cooling process. Therefore, thermal strain is expected to remain inside the molded product. For polymer blends, multiple phases with different coefficients of thermal expansion are mixed together. Therefore, thermal strain corresponding to the difference is assumed to occur at the interface. This strain is expressed as ε_v in Equation (11).

$$\varepsilon_v = 3\left(\alpha_m V_m - \sum_{j=1}^{n-1} \alpha_j V_j\right)\Delta T \tag{11}$$

where α is the coefficient of linear expansion. Also, ΔT expresses the difference between the molding process temperature and the test temperature. This strain is positive in the direction of contraction. When the expansion strain is generated by an external force and the strain reaches ε_v, the interface will delaminate. In the case of a linear elastic body, if the necessary expansion stress at that time is σ_v, it can be expressed as in Equation (12).

$$\sigma_v = \varepsilon_v K \tag{12}$$

In that equation, K denotes the bulk modulus. Equation (13) is used to ascertain the bulk modulus K_p of the polymer blend.

$$K_p = \sum_{i=1}^{n} K_i V_i \tag{13}$$

When the strain is small, the expansion stress produced by uniaxial tensile loading is $1/3$ of the applied vertical stress. Therefore, the vertical stress σ_d required to produce an expansion strain of ε_v can be expressed as shown in Equation (14).

$$\sigma_d = \frac{\sigma_v}{3} = \left(\alpha_m V_m - \sum_{j=1}^{n-1} \alpha_j V_j\right)\Delta T K \tag{14}$$

For this study, σ_d was obtained using the theory above. One-third of the difference from $\sigma_{y,exp}$ was obtained as σ_i. The values of α, v, and K of each polymer were obtained using the method described in the literature [29]. Table 8 presents those results and density ρ. The volume fraction of each material was obtained using the ρ and weight ratio of each material. Table 9 presents the obtained volume fractions and $\sigma_{y,exp}$, σ_v from Equation (12), and σ_d from Equation (14). In other words, if yielding occurs because of interface debonding, $\sigma_{y,exp}$ is obtainable via Equation (15).

$$\sigma_{y,exp} = \sigma_d + 3\sigma_i \tag{15}$$

Table 8. Physical properties of polymers.

	FS (MPa)	FM (MPa)	α (10^{-5}/K)	v (-)	E (MPa)	K (MPa)	ρ (g/cm^3)
H-PP	42.1	1728	9	0.407	767	1374	0.9
B-PP	36.3	1469	10	0.412	623	1181	0.9
PS	98.2	3237	7	0.342	2100	2215	1.04
L-SEBS	1.6	40	61	0.481	4	38	0.89
H-SEBS	46	1200	11	0.367	693	868	0.97

Table 9. Interfacial interaction stress for compositions yielding caused by interfacial debonding.

H-PP (vol.%)	B-PP (vol.%)	PS (vol.%)	L-SEBS (vol.%)	H-SEBS (vol.%)	YC_{MD}	$\sigma_{y,exp}$ (MPa)	K (MPa)	σ_d (MPa)	σ_i (MPa)
77.0		23.0			Debonding	18.3	1567	163.8	−48.5
69.2		25.2	4.0	1.6	Debonding	16.9	1524	198.9	−60.6
	77.0	23.0			Debonding	15.9	1419	164.3	−49.5
	69.2	25.2	4.0	1.6	Debonding	14.3	1375	193.4	−59.7

σ_i denoting a simple blend of PP and PS takes a negative value. Therefore, an interaction force is generated in the direction of interface separation. The addition of SEBS causes an increase in σ_i, which means that the addition of SEBS generates a stronger force in the direction of interface debonding. This increase in force is thought to have promoted the micro-dispersion of the phase structure.

However, when comparing $\sigma_{y,exp}$ and $\sigma_{y,MD}$ for compositions with more PS, both values were almost identical, which indicates that the yield condition is shear yield. Although the yield initiation stress increased with the addition of SEBS, no change occurred in the yield condition. The reason for this lack of change is discussed by modeling the yield initiation stress initiated by shear yielding and by comparing it with the experimentally obtained results.

When the distributed phases are loaded in the direction of elongation, strain occurs equally in each phase. When the yield initiation strain of either phase is reached, the yield initiation of the entire system is regarded as occurring. In this case, the shear yield initiation stress of the polymer blend σ_{yp} is expressed as in Equation (16).

$$\sigma_{yp} = 3\sqrt{3}\left(\sum_{i=1}^{n} \alpha_i V_i\right)\Delta T K_p(1-2v)\cos\theta + \sqrt{3}\sigma_i\cos\theta = \sigma_{y,s} + \sqrt{3}\sigma_i\cos\theta \tag{16}$$

where θ represents the shear angle. The shear yield initiation stress obtained by excluding the σ_i component in Equation (16) above is presented in Table 10 as $\sigma_{y,s}$. The calculated value of $\sigma_{y,s}$ is greater than $\sigma_{y,MD}$. This difference can be attributed to σ_i. In this case, σ_i was obtained by transforming Equation (16) into Equation (17).

$$\sigma_i = \frac{\sigma_{yp} - \sigma_{y,s}}{\sqrt{3}\cos\theta} \tag{17}$$

Table 10. Experimentally obtained and calculated shear yield initiation stress.

H-PP (vol.%)	B-PP (vol.%)	PS (vol.%)	L-SEBS (vol.%)	H-SEBS (vol.%)	$\sigma_{y,MD}$ (MPa)	$\sigma_{y,p}$ or $\sigma_{y,s}$ (MPa)	σ_i (MPa)	Structure
77.0		23.0			21.1	23.6	−48.5	Sea-island
69.2		25.2	4.0	1.6	19.8	21.0	−60.6	Sea-island
33.3		66.7			22.2	36.4	−9.5	Elongated Disc
31.4		63.4	1.7	3.5	28.4	38.0	−7.3	Cylinder
	77.0	23.0			20.0	21.4	−49.5	Sea-island
	69.2	25.2	4.0	1.6	16.5	19.2	−59.7	Sea-island
	33.3	66.7			20.2	34.5	−10.0	Elongated Disc
	31.4	63.4	1.7	3.5	25.0	36.1	−9.0	2D network

Table 10 also shows the value of σ_i obtained via Equation (17). The value of σ_i of the phase structure elongated in the loading direction is smaller than that found for the sea-island structure. The change caused by the addition of SEBS is slight. The reason for this slight change might be that the phase structure changes with the addition of SEBS into PS-rich compositions. Also, σ_i can be expressed as in Equation (18) using the interfacial interaction force and the specific surface area at the interface.

$$\sigma_i = F_i S_i \tag{18}$$

In that equation, F_i stands for the interfacial interaction force; and S_i denotes the specific surface area of the interface. S_i depends on the volume content of the dispersed phase and its morphology. For example, for the same volume fraction and the same diameter, a 1.5:1 relation exists between the sizes of the specific surface area when dispersed in a spherical form and when dispersed in a rod form. In other words, even though the addition of SEBS acted in the direction of increasing F_i, the change in σ_i was regarded as minute and as a result of the decrease in the specific surface area at the interface because of the change in phase structure. These results indicate that the shear yield initiation stress of a polymer blend with a dispersed phase elongated in the loading direction is obtainable from the shear yield initiation stress considering the interfacial interaction force.

The shear yield initiation stress of the PP-rich compositions is also discussed. Figures 6 and 7 show that in these compositions, a sea-island structure is formed with and without SEBS. In this structure, the interfacial interaction forces are isotropic and might cancel each other out. Furthermore, because the dispersed phase is spherical, we inferred that the logarithmic complex law shown in Equation (19) is valid [30].

$$\sigma_{y,p} = e^{\{\sum_{i=1}^{n} (V_i \ln \sigma_{yi})\}} \tag{19}$$

The interfacial interaction forces are not considered in σ_{yi} here. The shear yield initiation stress obtained using Equation (19) above is shown in Table 9 as $\sigma_{y,p}$. The calculated $\sigma_{y,p}$ shows good agreement with $\sigma_{y,MD}$. These results indicate that the shear yield initiation stress of a polymer blend with a spherical dispersed phase is obtainable by averaging the shear yield initiation stress without considering the interfacial interaction force using the logarithmic compound law.

As described in this paper, short-beam shear tests were performed on PP/PS polymer blends to evaluate their mechanical anisotropy with respect to yield initiation stress. The results showed a correlation between the phase structure and the evaluated mechanical anisotropy, suggesting that there is anisotropy in fracture toughness and other mechanical properties related to yield initiation stress. To assess fracture toughness, an anisotropic evaluation of notched impact strength has been performed on 3D molded parts [31]. Additional studies will also be conducted for injection-molded products. Anisotropy associated with the fracture toughness of polymer blends is more pronounced than yield initiation stress.

For industrial considerations, elucidating this mechanism is particularly important. The mechanisms of anisotropy affecting fracture toughness will be further clarified in future studies based on the findings presented here.

4. Conclusions

For this study, short-beam shear tests were performed on injection-molded PP/PS polymer blends to evaluate their anisotropy with respect to the shear yield initiation stress. The correlation between the obtained anisotropy and the phase structure inside the injection-molded products was investigated. The correlation between the phase structure and the shear yield initiation stress was modeled together with the yield conditions identified via a comparison with the three-point bending test results. The relevant findings are presented below.

- When PP is a continuous phase, the phase structure forms a sea-island structure. The yield condition under uniaxial tensile loading was interface debonding.
- When PS is a continuous phase, the phase structure has a dispersed phase that is elongated in the flow direction. This structure was changed by the addition of SEBS. The yielding condition under uniaxial tensile loading was shear yielding.
- The aspect ratio of the dispersed phase was correlated with the stress concentration factor.
- When the PP forming the sea-island structure is a continuous phase, the shear yield initiation stress is explainable by the log-complex law without considering the interfacial interaction stress.
- When the PS forming the structure with stretched dispersed phase is a continuous phase, the shear yield initiation stress is explainable by the shear yield initiation stress considering the interfacial interaction force.

Author Contributions: Conceptualization, T.T. and R.S.; methodology, T.T.; validation, T.T. and R.S.; formal analysis, R.S.; investigation, T.T.; resources, R.S.; data curation, T.T.; writing—original draft preparation, T.T.; writing—review and editing, T.T.; visualization, R.S.; supervision, T.T.; project administration, T.T. All authors have read and agreed to the published version of the manuscript.

Funding: This research received no external funding.

Institutional Review Board Statement: Not applicable.

Data Availability Statement: The data presented in this study are available on request from the corresponding author.

Conflicts of Interest: The authors declare no conflict of interest.

References

1. McCrum, N.G.; Buckley, C.P.; Bucknall, C.B. *Principles of Polymer Engineering*; Oxford University Press: New York, NY, USA, 1988.
2. Juster, H.; Aar, B.; Brouwer, H. A Review on Microfabrication of Thermoplastic Polymer-Based Microneedle Arrays. *Polym. Eng. Sci.* **2019**, *59*, 877–890. [CrossRef]
3. Chaudhary, B.; Li, H.; Matos, H. Long-term mechanical performance of 3D printed thermoplastics in seawater environments. *Results Mater.* **2023**, *17*, 100381. [CrossRef]
4. Ge, J.; Luo, M.; Catalanotti, G.; Falzon, B.G.; Higgins, C.; McClory, C.; Thiebot, J.-A.; Zhang, L.; He, M.; Zhang, D.; et al. Process characteristics, damage mechanisms and challenges in machining of fibre reinforced thermoplastic polymer (FRTP) composites: A review. *China Plast.* **2023**, *36*, 140–149.
5. Thakkar, R.; Thakkar, R.; Pillai, A.; Ashour, E.A.; Repka, M.A. Systematic screening of pharmaceutical polymers for hot melt extrusion processing: A comprehensive review. *Int. J. Pharm.* **2020**, *576*, 118989. [CrossRef] [PubMed]
6. Zhou, Y.; Yu, F.; Deng, H.; Huang, Y.; Li, G.; Fu, Q. Morphology evolution of polymer blends under intense shear during high speed thin-wall injection molding. *J. Phys. Chem. B* **2017**, *8*, 6257–6270. [CrossRef] [PubMed]
7. Wang, Y.; Xiao, Y.; Zhang, Q.; Gao, X.-L.; Fu, Q. The morphology and mechanical properties of dynamic packing injection molded PP/PS blends. *Polymer* **2003**, *44*, 1469–1480. [CrossRef]
8. Omonov, T.S.; Harrats, C.; Groeninckx, G.; Moldenaers, P. Anisotropy and instability of the co-continuous phase morphology in uncompatibilized and reactively compatibilized polypropylene/polystyrene blends. *Polymer* **2007**, *48*, 5289–5302. [CrossRef]

9. Mao, Z.; Zhang, X.; Jiang, G.; Zhang, J. Fabricating sea-island structure and co-continuous structure in PMMA/ASA and PMMA/CPE blends: Correlation between impact property and phase morphology. *Polym. Test.* **2019**, *73*, 21–30. [CrossRef]

10. Li, Z.-M.; Yang, W.; Yang, S.; Huang, R.; Yang, M.-B. Morphology-tensile behavior relationship in injection molded poly(ethylene terephthalate)/polyethylene and polycarbonate/polyethylene blends (I). *J. Mater. Sci.* **2004**, *39*, 413–431. [CrossRef]

11. Friedrich, K.; Evstatiev, M.; Fakirov, S.; Evstatiev, O.; Ishii, M.; Harrass, M. Microfibrillar reinforced composites from PET/PP blends: Processing, morphology and mechanical properties. *Compos. Sci. Technol.* **2005**, *65*, 107–116. [CrossRef]

12. Sajjadi, S.A.; Ghasemi, F.A.; Rajaee, P.; Fasihi, M. Evaluation of fracture properties of 3D printed high impact polystyrene according to essential work of fracture: Effect of raster angle. *Addit. Manuf.* **2022**, *59*, 103191. [CrossRef]

13. Haile, L. Study of Recycling of Waste High Impact Polystyrene (PS-HI) and Polystyrene with Flame Retardant Additives (PS-FR) Co-Polymers. *J. Text. Sci. Fash. Technol.* **2021**, *7*, 1–11. [CrossRef]

14. Nishino, K.; Shindo, Y.; Takayama, T.; Ito, H. Improvement of impact strength and hydrolytic stability of PC/ABS blend using reactive polymer. *J. Appl. Polym. Sci.* **2017**, *134*, 44550. [CrossRef]

15. Kannan, S.; Ramamoorthy, M. Mechanical characterization and experimental modal analysis of 3D Printed ABS, PC and PC-ABS materials. *Mater. Res. Express* **2020**, *7*, 015341. [CrossRef]

16. Krache, R.; Debbah, I. Some Mechanical and Thermal Properties of PC/ABS Blends. *Mater. Sci. Appl.* **2011**, *2*, 404–410. [CrossRef]

17. Tan, Z.Y.; Xu, X.F.; Sun, S.L.; Zhou, C.; Ao, Y.H.; Zhang, H.X.; Han, Y. Influence of rubber content in ABS in wide range on the mechanical properties and morphology of PC/ABS blends with different composition. *Polym. Eng. Sci.* **2006**, *46*, 1476–1484. [CrossRef]

18. Macosko, C.W. Morphology development and control in immiscible polymer blends. *Macromol. Symp.* **2000**, *149*, 171–184. [CrossRef]

19. Li, S.; Tian, H.; Hu, G.-H.; Ning, N.; Tian, M.; Zhang, L. Effects of shear during injection molding on the anisotropic microstructure and properties of EPDM/PP TPV containing rubber nanoparticle agglomerates. *Polymer* **2021**, *229*, 124008. [CrossRef]

20. Hammani, S.; Moulai-Mostefa, N.; Samyn, P.; Bechelany, M.; Dufresne, A.; Barhoum, A. Morphology, Rheology and Crystallization in Relation to the Viscosity Ratio of Polystyrene/Polypropylene Polymer Blends. *Materials* **2020**, *13*, 926. [CrossRef]

21. Guevara-Morales, A.; Figueroa-López, U. Residual stresses in injection molded products. *J. Mater. Sci.* **2014**, *49*, 4399–4415. [CrossRef]

22. Sargent, P.M.; Page, T.F. Factors affecting the measurement of hardness and hardness anisotropy. *J. Mater. Sci.* **1985**, *20*, 2388–2398. [CrossRef]

23. Jiang, Q.; Takayama, T. Relationship between interfacial shear strength and impact strength of injection molded short glass fiber-reinforced thermoplastics. *Procedia Struct. Integr.* **2023**, *45*, 117–124. [CrossRef]

24. Takayama, T.; Shibazaki, R. Evaluating Mechanical Anisotropy of Injection Molded Polymer Products using Short Beam Shear Testing. *Results Mater.* **2023**, *19*, 100434. [CrossRef]

25. Fukunaga, S. Propylene-Based Resin Composition, Propylene-Based Resin Mixture and Molded Body. JP2015193695A, 5 November 2015.

26. Kim, Y.; Park, O.O. Effect of Fiber Length on Mechanical Properties of Injection Molded Long-Fiber-Reinforced Thermoplastics. *Macromol. Res.* **2020**, *28*, 433–444. [CrossRef]

27. Zhang, Y.; Wu, M.; Zheng, S.; Xie, Y.; Jin, Z. Effect of process parameters on the mechanical properties of fiber-reinforced thermoplastics fabricated by direct fiber feeding injection molding. *Polym. Eng. Sci.* **2023**, *63*, 2457–2467. [CrossRef]

28. Cathelin, J. Through-Process Modelling for Accurate Prediction of Long Term Anisotropic Mechanics in Fibre Reinforced Thermoplastics. *MATEC Web Conf.* **2018**, *165*, 17002. [CrossRef]

29. Takayama, T.; Motoyama, Y. Injection molding temperature dependence of elastic coefficients obtained using three-point bending tests to ascertain thermoplastic polymer coefficients. *Mech. Eng. J.* **2021**, *8*, 20–00414. [CrossRef]

30. Gray, R.W.; McCrum, N.G. Origin of the γ relaxations in polyethylene and polytetrafluoroethylene. *J. Polym. Sci. Part A-2 Polym. Phys.* **1969**, *7*, 1329–1355. [CrossRef]

31. Spreeman, M.E.; Stretz, H.A.; Dadmun, M.D. Role of compatibilizer in 3D printing of polymer blends. *Addit. Manuf.* **2019**, *27*, 267–277. [CrossRef]

Article

Experimental Development of an Injection Molding Process Window

Mason Myers [1], Rachmat Mulyana [1], Jose M. Castro [1,*] and Ben Hoffman [2]

[1] Industrial and Systems Engineering, College of Engineering, Columbus Campus, The Ohio State University, Columbus, OH 43210, USA; myers.1828@osu.edu (M.M.); mulyana.1@osu.edu (R.M.)
[2] Honda of America Mfg Inc., Raymond, OH 43067, USA; ben_hoffman@na.honda.com
* Correspondence: castro.38@osu.edu; Tel.: +1-(624)-688-8233

Abstract: Injection molding is one of the most common and effective manufacturing processes used to produce plastic products and impacts industries around the world. However, injection molding is a complex process that requires careful consideration of several key control variables. These variables and how they are utilized greatly affect the resulting polymer parts of any molding operation. The bounds of the acceptable values of each Control Process Variable (CPV) must be analyzed and delimited to ensure manufacturing success and produce injected molded parts efficiently and effectively. One such method by which the key CPVs of an injection molding operation can be delimited is through the development of a process window. Once developed, operating CPVs at values inside the boundaries of the window or region will allow for the consistent production of parts that comply with the desired Performance Measures (PM), promoting a stable manufacturing process. This work proposes a novel approach to experimentally developing process windows and illustrates the methodology with a specific molding operation. A semicrystalline material was selected as it is more sensitive to process conditions than amorphous materials.

Keywords: injection molding; process windows; simulation; controllable process variables; performance measures

Citation: Myers, M.; Mulyana, R.; Castro, J.M.; Hoffman, B. Experimental Development of an Injection Molding Process Window. *Polymers* **2023**, *15*, 3207. https://doi.org/10.3390/polym15153207

Academic Editors: Luigi Sorrentino, Andrew N. Hrymak and Shengtai Zhou

Received: 1 June 2023
Revised: 5 July 2023
Accepted: 25 July 2023
Published: 28 July 2023

1. Introduction

Injection molding is a manufacturing process for plastic parts, that can be found in almost any industry, whether it be electronics, healthcare, consumer goods, or automotive. The global plastic injection molding market is expected to reach over 266 billion dollars by 2030 [1]. The goal of Injection Molding, like any other manufacturing process, is to produce parts or products efficiently and effectively. Several factors involved in this process can significantly affect the resulting products. One such way to increase the performance of the injection molding process is to construct an operating envelope in which key controllable factors, or variables, have been delimited. With the overall goal of producing acceptable values of the relevant performance measures of the specific injection molding process [2–6]. This operating envelope is commonly called a process window. This paper focuses on discussing a novel approach based on experiments through which an injection molding process window can be constructed.

The following steps were followed to develop the process window:

1. Identify and establish key Controllable Process Variables (CPV) for the process, including material attributes, machine settings, and mold/part conditions;
2. Identify the relevant performance measures for the specific process, such as part quality indicators and mechanical property values;
3. Develop a process window to delimit the controllable process variables such as packing time, injection speed, mold temperature, melt temperature, and packing pressure;

4. Supplement the experimental results via simulation to illustrate conditions for which the use of a less desirable location in the process window has to be selected. The software utilized in this study was Moldex3D (2021R2OR 64-bit).

These steps are the basis of the methodology developed, and each of them is discussed in detail in the following sections.

Each stage of the injection molding process critically influences the performance measures of the final product. It was desired to identify a region where the relevant controllable process variables within these stages have a range or envelope of operation in which acceptable parts are produced. This is the so-called process window. This research focused on developing a novel, industrially relevant, organized approach to define this acceptable range or process window. This window's exact shape and size will depend not only on the polymer part to be molded but also on the machine and the quality of the mold used. The specific values of the process window are also material dependent; however, the approach presented here towards developing the process window should be applicable to other materials. We selected semicrystalline materials as they are more sensitive to process conditions than amorphous materials.

2. Materials and Methods

The machine used throughout this research was a Sumitomo 180-ton Injection molding Machine (SG180M-HP, Tokyo, Japan). The cavity of the mold used in this study produced the parts utilized for measuring mechanical properties as indicated by the American Society for Testing Materials (ASTM). The mold material was tool steel, with several cooling lines running throughout both halves of the mold. When ejected from the mold, a sprue and cold runner system connect the 4 test samples utilized for mechanical testing, as shown in Figure 1.

Figure 1. Picture of a molded ASTM polymer part.

The 4 test samples produced include a tensile bar, two flexural bars with differing thicknesses of 3 mm and 6 mm, and a disk with a diameter of 50 mm. These samples can be used to conduct a variety of tests, including tensile testing, 3-point bending, and impact testing such as the Izod or Dupont tests. The material used in this work was polypropylene, produced by Advanced Composites. The application of this material in industry was for automotive body panels.

The mechanical properties were measured with an INSTRON 5569 Dual Column Table Top Load Frame. The INSTRON has a maximum speed of 500 mm/min and the minimum speed that can be used is 0.005 mm/min. The load capacity is 50 kN [7]. A speed of 50 mm/min was used, which is larger than the one recommended by the ASTM but is the speed used by automotive manufacturers. A TA Q20 Differential Scanning Calorimeter (DSC) was used to measure the crystallinity of the modified polypropylene samples [8–10].

When operating an injection molding machine, several settings must be selected to produce parts successfully. The definition of a "successful" part depends on the desired performance measures for the specific molding operation and will be outlined later, before the development of the final process windows. The settings that will be discussed here are those that are most important to the overall process and are used as the controllable variables of the process window to be developed. These key variables are melt temperature, mold temperature, injection screw speed, packing pressure, packing/cooling time, clamping force, and shot size. These variables are important in any injection molding operation. This research found it convenient and beneficial to separate these variables into tiers to help streamline and focus the development of the process window. Table 1 presents this concept.

Table 1. Control Variable Tiers.

Primary Control Variable	Mold Temperature (T_w) Melt Temperature (T_m) Packing Pressure (P_{pack})
Secondary Control Variable	Injection Screw Speed Packing Time (t_{pack}) Cooling Time (t_{cool})
Tertiary Control Variable	Shot Size Clamping Force

The Primary Control Variables of mold temperature, melt temperature, and packing pressure were deemed most important for this molding operation as they were the variables used to construct the boundaries of the process window.

The control variables of melt and mold temperature and their effects on the resulting part and operation closely interact with each other. The melt temperature must be high enough to allow the material to fill the mold completely. On the other hand, the mold temperature can be adjusted to benefit cycle time. Most likely, both variables affect the properties of the final part. Simply put, a balance must be achieved to produce parts effectively and efficiently. The last primary variable is packing pressure. This pressure ensures the mold cavity is fully filled and reduces thermal shrinkage as the material cools in the mold. It is important to note that the packing pressure values reported in this work are not the pressure on the polymer inside the mold cavity but the pressure setting entered into the machine, which corresponds to the hydraulic pressure. We measured the corresponding pressure inside the cavity to relate the "real values" to the machine settings. For example, a machine packing pressure of 2.07 MPa corresponds to a cavity pressure of about 41.37 MPa [11]. The approximate multiplier to calculate part pressure from machine pressure settings is twenty [11].

Due to the effects of these 3 primary variables on the overall injection molding process and part quality, understanding their relationship is critical to delimiting a process window to produce acceptable, defect-free parts. For the secondary control variables, that is, injection screw speed and packing/cooling time, the values chosen can have serious implications on the resulting polymer part and were analyzed before the primary control variables were studied. However, after these variables were delimited, specific settings or values were selected and kept constant throughout the process window's development. Lastly, the tertiary control variables are those that can affect a particular molding operation but are easier to establish. The clamping force must be large enough to hold the two mold halves together but not too high to damage them. The shot size was determined by the volume of the molded part. These tertiary control variables required little analysis and were held at the same settings for the entirety of the process window's development.

As the focus of this research was to develop a method to establish a process window to produce thermoplastic injection molded parts, it is necessary to understand the concept of such a window. A process window is a region that delimits the controllable process variables so that inside the boundaries of this region or window-like shape, acceptable parts, that is, parts that comply with the desired performance measures, can be molded. Operating outside of the process window will produce molded parts that are unacceptable due to defects such as flash, sink marks, and short shots. Examples of such defects are shown in Figure 2 [7].

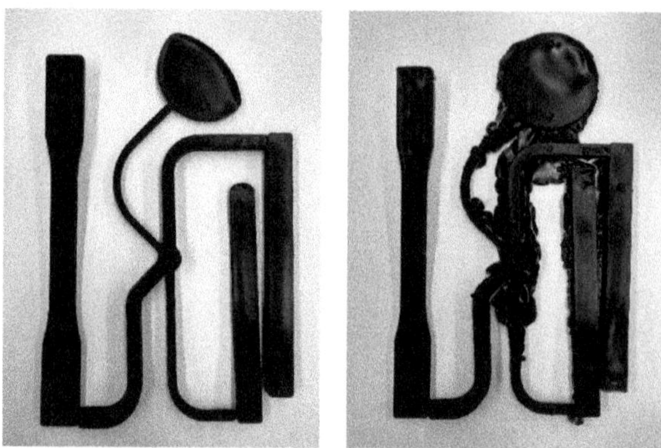

Figure 2. Examples of defect types: short shot, flash, and sink marks.

Figure 3 visually illustrates a process window that considers the controllable process variables of melt temperature and packing pressure. If the melt temperature is too low, the plastic will solidify before the mold is filled, producing a short shot. On the other hand, if the temperature is too high, plastic degradation may occur. If the packing pressure is too low, a short shot could occur, and if the packing pressure is too high, leakage (flash) will occur.

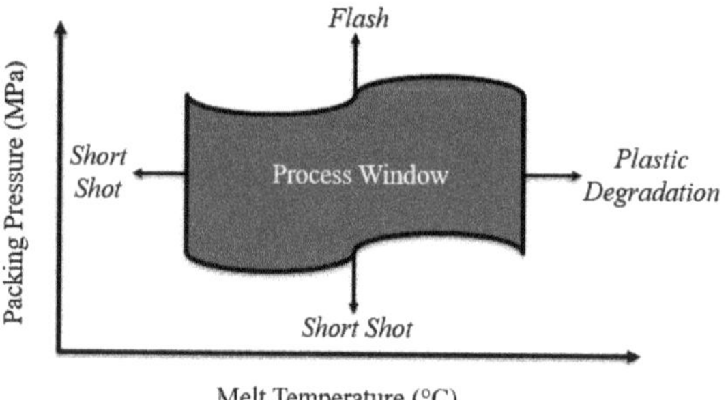

Figure 3. Example of a process window considering melt temperature and packing pressure.

When the selected values of the controllable process variables are near the center of the defined window, the part quality will be less influenced by undesired variations in the process variables. Variations due to unforeseen conditions likely caused by the molding environment. Therefore, the use of a process window in injection molding promotes a stable and more predictable manufacturing process [12]. Although a particular process window will vary from one production setting to another, the method by which these windows can be found has the same steps and considerations. A method that can be used for constructing such a window will be presented throughout.

3. Experimental

The first step in developing an injection molding process window was to identify, define, and determine the base values of the controllable process variables. The manufacturing of polymeric parts via injection molding is a complex problem. Each Controllable Process Variable (CPV) can have an impact on the resulting polymer part. Because of this, it is best to simplify and determine which variables will be changed and which will remain constant based on the goals of the process window being developed. This is the reason we divided them into primary, secondary, and tertiary tiers. It is best to only consider settings critical to the specific molding operation, as too many variables will greatly complicate the development method. We will discuss what CPVs were identified, how their values were found, and if they were kept as constants or varied as part of the process window.

Several values or ranges of the CPV were defined based on known information about the specific molding material, mold, and injection molding machine, as these factors will not change throughout the development process. Table 2 provides the reference for each of these variables and lists their selected ranges or values.

Table 2. Known Variables and Their Origin.

Variable	Origin	Value/Setting
Mold Temperature (T_w)	Material Supplier	80–120 °F (26.7–48.9 °C)
Melt Temperature (T_m)	Material Supplier/Molder Experience	355–410 °F (179.4–210 °C)
Shot Size	Part Volume	2.10 in (53.34 mm)
Clamping Force	Molder Experience	120 Ton (1067.6 kN)

Other variables can also be delimited, but it took more analysis to properly determine their values with respect to the specific operation. These variables are listed in Table 3 and will be defined and discussed in more detail.

Table 3. Variables Needing Further Analysis.

Variable
Mold Closed Time
Injection Screw Speed
Preliminary Packing Pressure (P_{pack})
Packing (t_{pack}) and Cooling (t_{cool}) Time

The CPVs of mold closed time, injection screw speed, preliminary packing pressure, and packing/cooling time required further analysis to determine their respective values. These variables will likely have the most variation from one injection molding operation to another. Therefore, to determine the best settings for these variables, a specific analysis was conducted to isolate the influence of each variable and determine its value regarding the unique molding operation. The values found by this analysis will be different from one operation to another; however, the methods by which they were found will be similar. Although most of these values were later treated as constants in the development of the process window, to find their best values, they were each treated as independent variables during their specific analysis step. A discussion on how each of these variables was defined and the method used to determine their values is presented.

Table 4 presents the various settings that each of the CPVs were set to during this analysis stage [7].

Table 4. CPV Settings for Injection Screw Speed Trials.

CPV	Value
Mold Temperature (T_w)	80 °F (26.7 °C)
Melt Temperature (T_m)	380 °F (193.3 °C)
Packing Pressure (P_{pack})	10–750 psi (0.07–5.2 MPa)
Injection Screw Speed	0.2–10 in/s (5.1–254 mm/s)
Packing Time (t_{pack})	0–52 s
Cooling Time (t_{cool})	0–52 s
Shot Size	2.10 in (53.34 mm)
Clamping Force	120 Ton (1067.6 kN)

The first analysis conducted was to delimit the value needed for mold closure time, that is the time needed for the part to become solid enough so that it can be demolded without any blemishes. The approach we recommend and that is used in our group [2,7] is to start by calculating the conduction time (t_{CT}), that is:

$$t_{CT} = \frac{h^2}{\alpha} \tag{1}$$

where: $h = \frac{partthickness}{2}$, $\alpha = ThermalDiffusity$.

Part thickness is the largest thickness of the polymer part to be molded. This time is a naturally occurring time when making the heat conduction equation dimensionless and represents the time needed for the temperature in the center of the part to be such that its value minus the mold temperature becomes 10% of the maximum temperature difference (melt temperature minus mold temperature) if one assumes one-directional heat transfer and constant mold wall temperature. We suggest and have used it both in our lab and in our interactions with industry as an initial value, in general, a conservative number that can be decreased using experiments. For this case, we kept the calculated value as the mold closed time. This value will later be subdivided between packing time and cooling time. Once the packing pressure does not affect the part quality, the part is kept in the mold without packing pressure. That way, the screw can start retracting for the next cycle, and the mold stresses are minimized.

The injection screw speed of an injection molding machine is the speed at which the screw or plunger moves during the injection stage. Correctly determining the value of this setting is important, as injecting too fast will cause flash, but injecting too slowly will cause the melt front to solidify before the part is completely filled, also known as a short shot. The goal of this analysis was to find a maximum injection speed before flashing occurs to reduce the total cycle time but also produce an acceptable part. This concept is represented and discussed below (Figure 4).

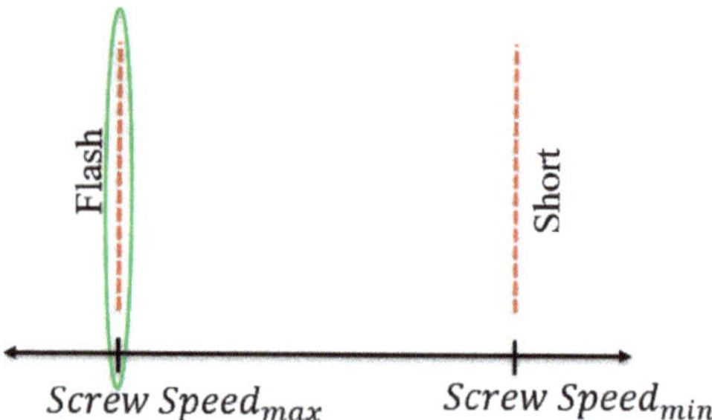

Figure 4. Injection screw speed spectrum with defect boundaries.

Again, the goal of these experiments was to find the injection screw speed just on the boundary before flashing occurs. With regards to this research, it was unnecessary to explore the right-hand side of the spectrum, as injecting the material slowly into the mold has no real benefit and would slow cycle times. However, slower injection times may, in some cases, produce better results, such as surface finish, and therefore should be explored if applicable [13].

Several runs were conducted to establish the proper setting for injection screw speed. To focus on the injection of the material and the screw speed, values associated with packing were minimized by selecting a very low packing pressure of 10 psi (0.07 MPa) and allocating zero seconds of the calculated conduction time to packing (the entire 52 s was set to cooling). Our group has found that using a cushion position that is ten percent of the total shot size (0.210 inches (53.34 mm)) is good practice. Using the settings above, molded parts were produced and visually analyzed. Screw speeds ranging from 0.2 in/s to 10 in/s (5.1–254 mm/s) were tested as shown in Table 4.

After the completion of these trials, a visual inspection [7] of the parts was conducted to determine the best value for this setting. It was found that at a speed less than or equal to 2.0 in/s (50.8 mm/s), the mold would be filled without flash. The setting of injection screw speed was thus established at this constant value during the process window development trials and considered a secondary control variable.

Although packing pressure will be used as a primary CPV during the development of the process window itself, it was necessary to find a "preliminary" setting for this CPV to select values for the other settings. The procedure to delimit this initial packing pressure was to run injection molding trials and vary the packing pressure. By doing so, a preliminary process window related to packing pressure was found in which acceptable maximum (P_{max}) and minimum (P_{min}) packing pressures were established. Pressure outside this window would either cause the polymer part to flash or produce a short shot [7]. From this process window, an average of P_{min} and P_{max} was calculated. This P_{avg} was the value used as the preliminary packing pressure in the remaining analysis trials. The experiments and analysis conducted to preliminarily delimit the packing pressure are as follows:

To delimit and isolate the packing pressure, several of the other molding variables were set to specific values. For example, the melt temperature was set to the middle value of the range provided by the material supplier. With regard to time, the overall mold closing time was kept equal to the calculated conduction time of 52 s. At this stage, this entire length of time was allocated to packing time (t_{pack}), and the cooling time (t_{cool}) was set to zero, again shown in Table 4. During this analysis, samples were produced with packing pressures ranging from 100 psi to 750 psi (0.69–5.17 MPa). Each polymer part produced was then analyzed to find defects and determine the P_{max} and P_{min} of the preliminary packing pressure window. To find this range, a combination of visual inspection and measurements was utilized.

To determine the minimum boundary, both visual inspection for defects and the measurement of the surfacing profile of the molded parts using a profilometer were utilized. The results of these measurements are shown in Figure 5.

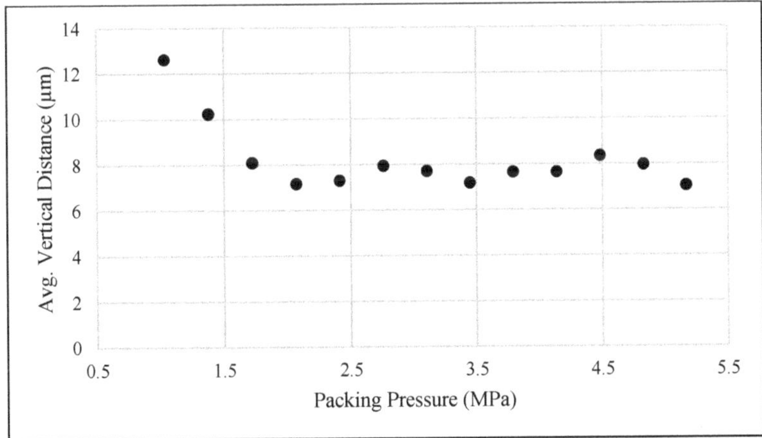

Figure 5. Average vertical distance versus packing pressure measured by a profilometer.

As can be seen from Figure 5, as packing pressure increased, the vertical distance from the highest outside edge of the measured piece to the lowest middle point became stable. From these measurements, it can be seen that 1.7 MPa is the initial packing pressure value where the surface profile began to level out, indicating the part was completely filled. Additionally, visually inspecting the molded parts found that all corners and end pieces were completely filled, and no sink marks were seen, beginning at this value. Therefore, a minimum packing pressure of 1.7 MPa was determined.

To determine the maximum limit for the preliminary packing pressure, several other measurements were utilized. One such measurement was the total part weight. These results are shown in Figure 6.

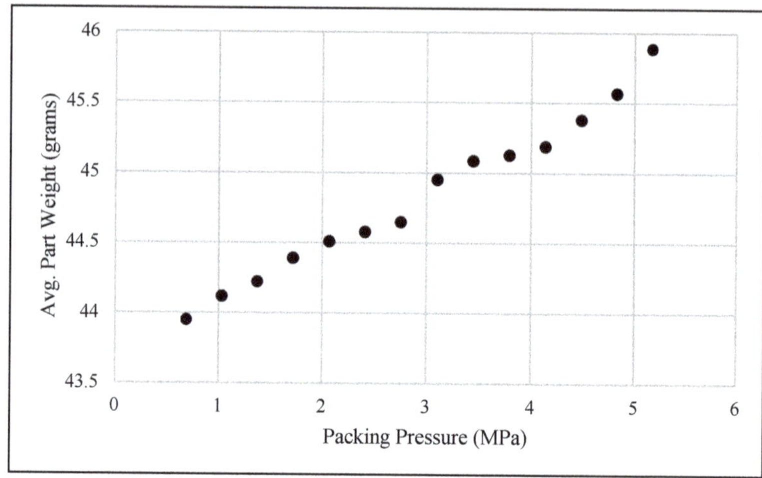

Figure 6. Average molded part weight versus packing pressure.

Looking at Figure 6, which displays the total weights of the mold parts in grams, it can be seen that from about 2.76 to 3.79 MPa, there was an increasing trend in the weights of the molded parts. This was taken as an indication that the flash being seen was becoming more significant and therefore unacceptable, as indicated in Figure 2. It was also observed that as the packing pressure increased, the final cushion position of the injection screw significantly changed. At around 2.76 MPa, a large drop-off from the machine setting of 0.210 inches (5.33 mm) was seen. This indicated that the screw was going well past its desired final position and pushing extra material into the mold, causing significant flash when higher packing pressures were introduced. Combining these results with visual inspection, a maximum value of 3.1 MPa was determined.

From the above analysis, the resulting process window shown in Figure 7 was developed.

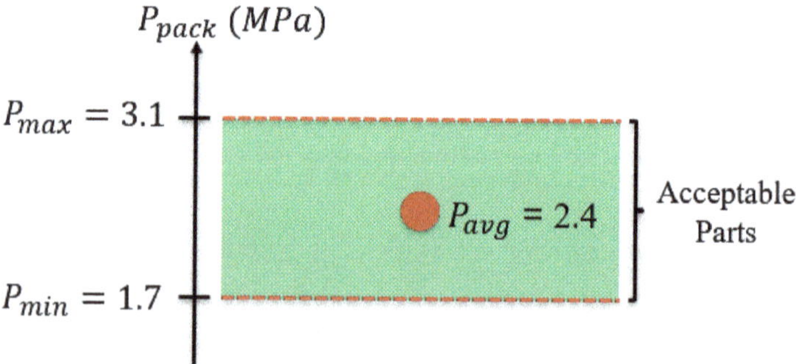

Figure 7. Determined preliminary packing pressure process window.

From these findings, an average packing pressure of 2.4 MPa was calculated. This P_{avg} was used in the additional trials to determine the values of the remaining non-primary controllable variables. Once these values were established, packing pressure was treated as a primary controllable process variable during the development of the final process window.

The calculated conduction time of 52 s is subdivided into packing time and cooling time. Determining the ratio between these two periods was important because, after the point when the packing pressure has no effect, it is desirable to switch from packing to just cooling so the screw can start retracting, as well as to avoid unnecessary stresses on the mold. The process undertaken to delimit this ratio, its justification, and the resulting times are discussed below.

First, a range of ratios between t_{pack} and t_{cool} was selected with respect to the total conduction time. At each of these ratios and the settings shown in Table 4, injection molded samples were collected, inspected, and measured.

Utilizing both the measurement of part weight and surface profile via profilometer [7], it was found that the packing-to-cooling time ratio must be greater than 20 to 80, or 10.4 s and 41.6 s, respectively. To support the results from part weight and shrinkage measurements [7], tensile testing of the parts was conducted. For each of the six-time ratios, tensile testing was completed. These results are presented in Table 5.

Table 5. Tensile Test Mechanical Property Results (Average of 5 Samples Each).

t_{pack}/t_{cool} (%)	Tensile Stress at Maximum Load [MPa]	Tensile Strain (Displacement) at Yield (Zero Slope) [%]	Tensile Stress at Yield (Offset 0.2%) [MPa]	Tensile Stress at Break (Standard) [MPa]	Tensile Strain (Displacement) at Break (Standard) [%]	Modulus Young [MPa]
100/0	19.279	12.122	8.843	14.271	145.491	909.661
80/20	19.353	11.950	8.510	14.864	254.9	992.231
60/40	19.367	11.447	8.443	13.596	118.440	1013.861
40/60	19.311	12.040	8.891	15.230	299.206	914.458
20/80	19.515	10.776	8.397	13.621	128.562	1019.808
0/100	18.169	7.246	8.831	15.54	38.87	941.908

With a combination of part weights, surface profiles, and tensile tests, the ratios were delimited. As stated, the results of the weights and profilometer indicated that the use of ratios less than 20:80 was not desirable [7], and this is supported by the tensile test results. The ratio of one hundred percent packing time and zero percent cooling is also not acceptable, as it did not allow the screw to start retracting early enough to prepare material for the next cycle, thus slowing production. Additionally, by completing a conduction time analysis on the part's cylindrical sprue, it was found that packing for more than approximately 45 s would not be beneficial as a large percent of the sprue cross-section, or entry into the mold, will have already solidified.

Combining all the experimental evidence, it was determined that the best packing time to cooling time ratio was 40 to 60 percent, respectively, as shown in Table 6. This conclusion was reached based on several factors. Firstly, the samples produced under this ratio visually appeared to be the best. The weight and profilometer measurements also supported this ratio, as the samples were shown to be completely filled. Additionally, this ratio produced the best mechanical properties, with the highest tensile stress at yield and break, along with increased ductility (tensile strain at break).

Table 6. Established Packing and Cooling Time Ratio.

t_{pack} (%)	t_{cool} (%)	t_{pack} (%)	t_{cool} (%)
40	60	20.8	31.2

These secondary settings would remain constant throughout the development of the process window discussed in the remaining sections.

4. Results and Discussion

With the secondary and tertiary process variables established, we proceeded to develop the process window. As previously stated, mold temperature, melt temperature, and packing pressure were the primary CPVs that defined the boundaries of the process window. Table 7 summarizes all the controllable variables and their values as found previously.

Table 7. Key Molding Variables and Determined Values.

Control Variable Tiers	Variable	Setting
Primary Control Variable	Mold Temperature Melt Temperature Packing Pressure	Controllable Process Variable (CPV) CPV CPV
Secondary Control Variable	Injection Screw Speed Conduction Time Packing Time Cooling Time	2.0 in/s (50.8 mm/s) 52 s 20.8 s 31.2 s
Tertiary Control Variable	Shot Size Clamp Force	2.10 in (53.34 mm) 120 Ton (1067.6 kN)

As the primary CPV defined the boundaries of the process window, their effect on the performance measures was evaluated next. A full factorial for both temperatures at three levels was conducted. Ten samples at each setting were collected. The values of the melt temperature used are 179.4, 193.3, and 210 °C. The values for the mold temperatures are 26.7, 37.8, and 48.9 °C. Packing pressure was varied at each temperature combination until a minimum and maximum value were found for each. Nine unique temperature combinations were tested, all with various levels of packing pressure. To thoroughly construct the process window, a total of more than 750 parts were modeled and analyzed. The first window that was developed was based on visual inspection.

Quality standards [7] were established in order to define what constitutes an acceptable part. These standards will likely vary on a case-by-case basis, but they must be defined to keep the process repeatable and measurable. In the case of this research, visual quality standards were developed to focus on key areas or zones of the part being produced. Figure 8 highlights these areas.

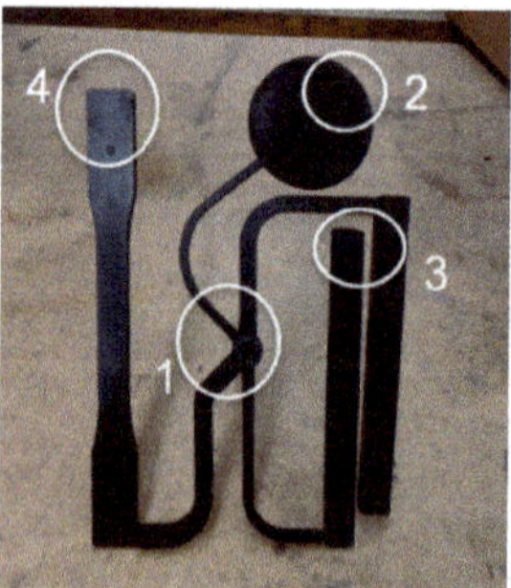

Figure 8. Visual quality standard zones (labeled 1–4) on molded parts.

Within each of the respective locations of the polymer part, a visual inspection for defects was conducted, looking for such things as flash and sink marks. As depicted in Figure 8, zones 2, 3, and 4 were inspected for complete fill and a lack of shrinkage in corners and edges. Additionally, zones 1 and 3 were inspected for significant flashing. With these quality standards defined, they were used to analyze parts produced at various packing pressures and temperatures. If the quality standards defined were not met by particular molded samples, then the associated molding condition was deemed unacceptable. From this analysis, the boundaries, or minimum and maximum acceptable packing pressure, for each of the nine temperature combinations were found, therefore producing a visual process window.

The resulting visual process window was developed by utilizing the defined quality standards and inspecting the molded parts produced [7]. Based on the standards, a minimum and maximum packing pressure value for each temperature combination—mold temperature and melt temperature—were found. The visual process window was produced by overlaying these ranges onto a single plot. Figure 9 represents the visual process window of this particular molding process.

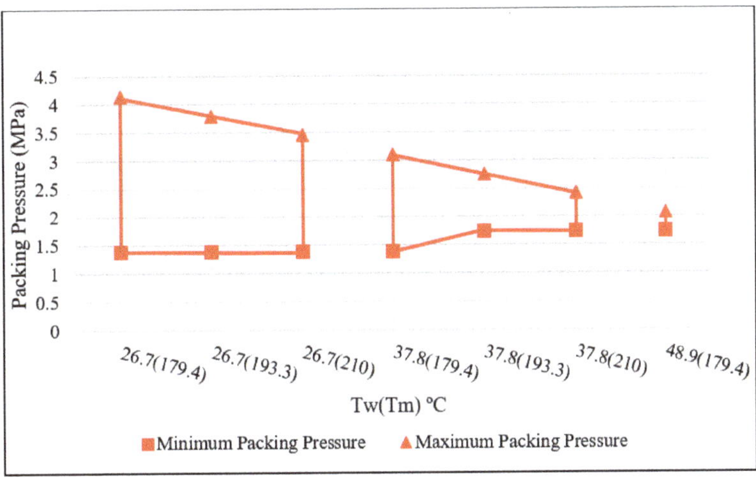

Figure 9. Visual process window.

Several features of this resulting process window must be discussed, along with a justification of the resulting window based on fundamental principles. Firstly, it may be noted that only seven temperature combinations were included within the developed window when nine such combinations were molded. This occurred because no parts produced above the mold and melt temperature settings of 48.9 °F and 179.4 °C were found to be acceptable based on the quality standards. Because these molded parts were found to be visually defective, they were also filtered out or eliminated from the later mechanical property analysis.

In terms of fundamentals, the resulting visual processing window was justified. The steady decrease in acceptable maximum packing pressure (the upper bounds of the window) was supported by the fact that as the temperature increased, the thermoplastic material became less viscous and therefore increased potential leakage, causing the defect of flashing. This was observed on molded parts above the maximum boundary. The acceptable minimum packing pressure was affected by the increasing temperature as well. The increased temperature also caused increased shrinkage as the part cooled in the mold. This explains the increase in the minimum packing pressure values once mold temperatures were at or above 37.8 °C. Parts below the designated minimum were observed to have sink marks or were short in filling.

From the trend seen in the process window, it can be noted that mold temperature appears to have a greater effect on the developed window and part acceptability. The packing pressure range for each temperature combination was greatly reduced as mold temperatures increased. This relationship is most likely material and part dependent.

The left side, the side with a lower mold temperature, is more robust; that is, it would allow for more uncontrollable variations in your molding environment and still enable the production of acceptable parts. Additionally, because the mold temperature is low, the molded part cools faster and can be ejected sooner, promoting a shorter cycle time. For this particular molding operation, the right side of the visual process window is less desirable and may only be used as a boundary of limitation. However, depending on the complexity of the part, the right side or higher mold and melt temperatures may be necessary, as illustrated with an example later.

The process window, based on visual inspection, is a great tool to delimit the relevant CPV, and select the most robust region of the CPV domain. However, it was believed that the visual inspection used to construct the injection molding process window was only part of the whole development process, and more analysis was needed to promote and produce a more robust solution. This is particularly true for semicrystalline materials and less important for amorphous materials [8].

Since our material is semicrystalline, to develop a more robust process window, analysis of the mechanical properties of the parts produced within the visual process window was necessary. Although the visual process window developed indicates to a molder how to operate their machine in order to produce visually acceptable parts, it does not include the effect on the mechanical properties of the polymer part. The approach in industry is to define acceptable parts based on visual inspection; mechanical properties are rarely considered for specific parts after they are molded [14]. This could lead to unacceptable parts, in particular for semicrystalline materials [8]. This research aimed to further the analysis beyond just the visual process window and develop a more refined process window that took mechanical properties into consideration. Mechanical properties, in particular ductility, may be affected by process conditions for semicrystalline materials [8].

The desired mechanical properties of a particular injection molded part will vary from one product to another. In the case of this analysis, experiments were conducted to determine the tensile properties of the samples produced. With the visual process window already defined, the molding conditions and the number of samples for the tensile test were decreased, as polymer parts deemed visually unacceptable were not tested mechanically. The two properties that will be discussed to generate a more refined process window are tensile strain at yield and tensile strain at break, or ductility. The results of these two properties are presented and discussed below (Tables 8 and 9).

Table 8. Tensile Strain (Displacement) at Yield (%).

$T_w(^\circ C)$	26.7			37.8			48.9		
T_m (°C)	179.4	193.3	210	179.4	193.3	210	179.4	193.3	210
1.38	12.691	11.897	11.897	12.000	11.420	10.722	12.125	11.679	
1.72	13.323	11.917	11.917	12.444	11.934	11.234	12.680	11.726	
2.07	13.365	12.183	12.183	13.012	12.719	11.578	13.122		
2.41	14.201	12.425	12.425	13.838	12.971	11.716			
2.76	14.944	12.848	12.848	14.223	12.948	12.225			
3.10	15.466	13.234	13.234	14.758	13.388				
3.45	15.018	13.324	13.324	15.005					
3.79	15.849	13.005	13.005						
4.14	16.617	13.284							

(Row label column: Packing Pressure (MPa))

Looking at the tensile strain at yield results, the effects the CPV has on this property become clear, as higher values indicate that the parts deformed to a greater degree before yielding or the start of plastic/permanent deformation. In terms of packing pressure, the samples produced with increasing pressure result in higher tensile strain in all combinations of mold and melt temperatures. Additionally, with regards to temperature, lower melt temperatures produce samples with increased tensile strain. In all three cases of different mold temperatures, the lower range of the melt temperature (179.4 °C) produces the highest strain values. Lastly, it was observed that overall, the mold temperature of 26.7 °C produced higher tensile strain results. In conclusion, these results indicated that the mechanical property of tensile strain at yield benefited from higher packing pressure and lower mold and melt temperatures, as it takes longer or more deformation to cause parts molded at these conditions to yield.

Table 9 presents the results of the tensile strain at break or the ductility of the various samples that were tested mechanically. These results provide similar conclusions to those found for the tensile strain at yield. Higher ductility is achieved by samples that were both molded at higher packing pressures and at lower mold and melt temperature combinations. With high ductility, parts molded under these conditions will deform but not break or fail easily.

Table 9. Tensile strain (Displacement) at Break (%).

	T_w (°C)	26.7			37.8			48.9		
	T_m (°C)	179.4	193.3	210	179.4	193.3	210	179.4	193.3	210
Packing Pressure (MPa)	1.38	118.557	180.703	165.851	150.313	148.742	188.623	146.582	175.032	
	1.72	134.699	182.579	141.534	177.650	155.772	151.928	160.629	158.783	
	2.07	136.120	181.355	151.462	140.261	155.257	149.869	135.287		
	2.41	125.384	247.614	262.226	145.024	187.168	145.354			
	2.76	299.574	371.187	271.889	158.577	234.021	210.268			
	3.10	320.251	340.745	370.961	198.609	316.248				
	3.45	344.143	373.189	371.243	237.441					
	3.79	371.349	336.061	400.026						
	4.14	325.522	377.526							

The analysis of tensile strain at yield and at break indicates that the percent crystallinity of the samples at lower temperatures is most likely lower, and thus they are more ductile. This has been corroborated by Differential Scanning Calorimetry (DSC) [8–10]. These results justify the conclusion that further analysis of the mechanical proprieties benefits the creation of a more refined injection molding process window. The findings discussed indicated that improvements could be made to the visual process window originally found to produce a window that promotes both visual and mechanical success. This may not be the case for amorphous materials [8,9].

Recall that it was found that using lower mold and melt temperatures produced better tensile strain properties. Combining both the findings of the visual inspection process window and the tensile strain testing (mechanical properties), a final refined process window was produced.

This process window shown in Figure 10 suggests operating at a mold temperature of 26.7 °C and a melt temperature of 179.4 °C to 210 °C. Additionally, packing pressure should be set between 2.41 and 4.14 MPa, depending on the temperature. This process window will promote a more stable and predictable injection molding operation by reducing the impact of undesirable variation in the molding environment and also improving the resulting parts' visual and mechanical properties.

For complicated parts, in particular parts with thin sections, it may be necessary to mold at higher temperatures in order to completely fill the part or to use the least desirable region of the process window. We illustrated this with a simple example, where we analyzed the filling of a flat plate with increasing length using a fan gate, as shown schematically in Figure 11. The Computer Aided Engineering (CAE) software used in this case is Moldex3D.

The specifics of this example were to construct a flat plate of a certain length. Then we ran a filling stage analysis using Moldex3D to determine if the temperature allowed the flat plate to successfully fill. If the plaque was successfully filled, a new plaque of longer length was then analyzed. If the temperatures used did not allow the mold to fill completely, then the temperature values were increased. This process was repeated for a variety of flat plate lengths and used the mold and melt temperature ranges recommended by the material supplier and used in the original visual processing window.

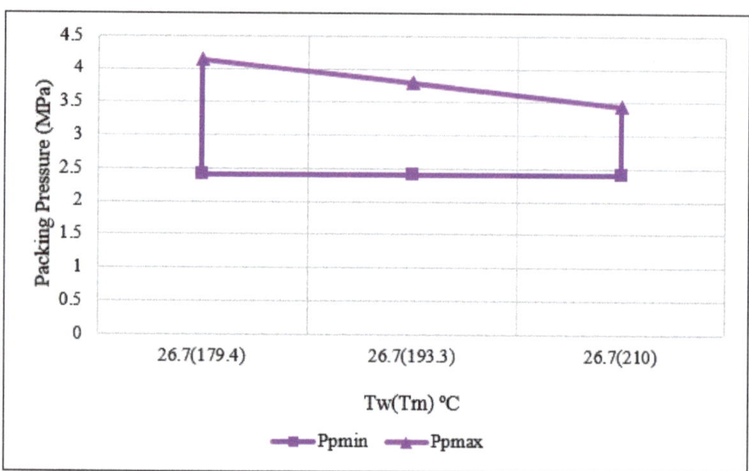

Figure 10. Final refined process window.

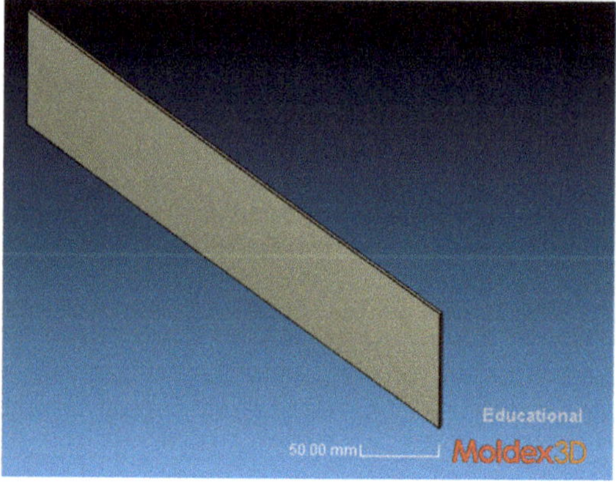

Figure 11. A thin plate model (0.375 m × 0.1 m × 0.002 m) visualized with software Moldex3D.

With the case study defined, several simulations were completed. The results of these runs are shown in Table 10.

Table 10. Simulated Variables and Run Results.

Plate Length (m)	T_w (°C)	T_m (°C)	T_f (s) Calculated	T_f (s) Simulated	Filled?
0.2	26.7	179.4	1.4	1.422	Y
0.3	26.7	179.4	2.1	2.148	Y
0.35	26.7	179.4	2.45	3.205	N
	26.7	193.3	2.45	2.606	Y
0.375	26.7	193.3	2.625	3.479	N
	26.7	210	2.625	2.764	Y
	37.8	179.4	2.625	3.462	N
	37.8	193.3	2.625	2.963	Y
0.4	37.8	193.3	2.8	3.881	N
	37.8	210	2.8	3.132	Y
	48.9	179.4	2.8	3.735	N

As can be seen from the table, plates of shorter lengths filled successfully at lower mold and melt temperatures. However, as the plate length increased, higher temperatures, specifically melt temperatures, were required to fill the mold completely.

The results indicated that the temperature ranges found to be the best for promoting visual and mechanical success during the experimental development of the process windows did not allow for complete filling in certain cases of the thin plate molds. Although it was found earlier that visual inspection and mechanical properties benefited from lower melt and mold temperatures, the simulations show that such temperatures may need to be increased in order to fill the mold when considering, large thin-walled parts. The right side, or the side representing higher temperatures of the visual process window shown in Figure 12, was better justified by this case study. We selected a flat plate for simplicity and easier discussion; however, we could have also performed the same analysis with the ASTM mold. Whereas before the right side of the process window was only used as a boundary of limitation, the simulations conducted indicate that although this area of the process window is not as robust, it may be necessary in order to produce fully filled parts. Again, it must be stated that the shape and size of a particular process window and the values of the key CPV will vary from one operation to another.

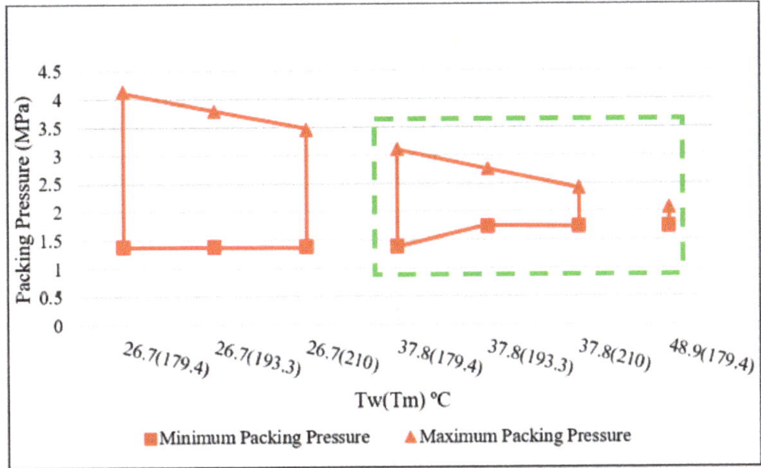

Figure 12. Visual process window is highlighted with simulation results.

Molding thermoplastic parts that are both visually appealing and mechanically sound is desirable. However, the specific material and part dimensions of a particular operation may limit these quality factors. The simulations conducted as a part of this case study indicated that in order to successfully fill a thin plate, areas of the developed process window with higher mold and melt temperatures may have to be utilized, as indicated in Figure 12.

5. Conclusions

This research presented a method for the experimental development of an injection molding process window. By following each stage of the injection molding process, key controllable process variables were able to be isolated and analyzed. Using this method and defined quality standards, a visual process window was first found. This window focused on obtaining parts with acceptable appearance. In addition, this work proposed that this visual process window was only part of the solution, for semicrystalline materials in particular, and that there was a need to include mechanical testing as part of developing the process window. Through tensile testing, a more refined process window was developed.

This refined window allowed for parts with both an acceptable appearance and adequate mechanical properties to be produced [8].

A special case study utilizing simulation was also presented, which helped to justify the use of specific regions of the experimental process windows, even if they were less robust from the visual and mechanical properties point of view.

The proposed approach can be summarized as follows. First, identify and establish key controllable process variables, including material attributes, machine settings, and mold/part conditions. Once this initial information is collected, determine the key performance measures and develop quality/measurement standards for the particular operation. Next, using part and material characteristics, calculate your conduction time (t_{CT}). Then, by following the order of each of the stages in the injection molding process, isolate and analyze each CPV to determine its range or value. Vary relevant controllable process variables such as mold temperature, melt temperature, and packing pressure to produce a large sample of molded parts. Using the produced molded parts and the defined performance measures, develop a visual process window. Lastly, refine the process via mechanical property analysis.

The seven steps above represent the general method used to experimentally construct the process windows presented and discussed in this work. It is suggested that the use of these steps will allow for the development of process windows in other injection molding operations. Again, it is important to note that the specific values presented during this work correspond to the particular molding process investigated (mold/part, machine, and material). Such values will likely differ from one molding process to another. This work is intended to present and provide a more standardized and thorough procedure for experimentally developing injection molding process windows. The steps highlighted above summarize the detailed research completed, and the results presented throughout showcase the success of the method developed.

6. Future Work

Firstly, it is suggested that the experimental method for developing injection molding process windows be tested further. Research using other materials or molding operations will help test the robustness of the method presented. This research would also help improve process window development strategies.

Due to the time commitment of conducting experimental research, it is also recommended that further analysis of simulations be pursued. In the case of this research, simulation is used as a supplementary tool to help better understand the experimental results. However, simulation software has promising potential to help develop injection molding process windows in a more direct way. In the future, with the help of the experimental results presented, strides in conducting more detailed simulation analysis to develop injection molding process windows will be possible. Utilizing the steps developed to construct an injection molding process window in this work as a guide, simulation can become the main tool used in order to avoid the limitations of physical trials. Process window development via simulation will allow us to test real parts where it is not possible to do experimental trials, such as an automotive production line. Work is currently being conducted in this area.

Author Contributions: Conceptualization, M.M., R.M., J.M.C. and B.H.; Methodology, M.M.; Formal analysis, M.M.; Investigation, M.M.; Resources, R.M., J.M.C. and B.H.; Data curation, M.M.; Writing—original draft, M.M.; Writing—review & editing, M.M., R.M. and J.M.C.; Visualization, M.M.; Supervision, R.M. and J.M.C.; Project administration, R.M. and J.M.C.; Funding acquisition, J.M.C. and B.H. All authors have read and agreed to the published version of the manuscript.

Funding: Funding was provided by Honda Manufacturing grant number GR129167. Injection molding process window development through simulation APC was funded by Release Time Account of corresponding author, The Ohio State University.

Institutional Review Board Statement: Not applicable.

Data Availability Statement: Data available from the authors.

Conflicts of Interest: The authors declare no conflict of interest.

References

1. Plastic Injection Molding Market (by Raw Material Type: Polypropylene, Acrylonitrile Butadiene Styrene, Polystyrene, High Density Polyethylene, and Others; and by Application: Automobile, Consumer Goods, Electronics, packaging, Building & Construction, and Healthcare)—Global Industry Analysis, Size, Share, Growth, Trends, Regional Outlook, and Forecast 2021–2030. Precedence Research. Available online: https://www.precedenceresearch.com/plastic-injection-molding-market (accessed on 17 January 2023).
2. Villarreal-Marroquín, M.G.; Chen, P.-H.; Mulyana, R.; Santner, T.J.; Dean, A.M.; Castro, J.M. Multiobjective optimization of injection molding using a calibrated predictor based on physical and simulated data. *Polym. Eng. Sci.* **2016**, *57*, 248–257. [CrossRef]
3. Marroquín, M.G.; Castro, J.M.; Mondragón, Ó.L.; Ríos, M.C. Optimisation via simulation: A metamodelling-based method and a case study. *Eur. J. Ind. Eng.* **2013**, *7*, 275–294. [CrossRef]
4. Rodríguez, A.B.; Niño, E.; Castro, J.M.; Suarez, M.; Cabrera, M. Injection molding process windows considering two conflicting criteria: Simulation results. In Proceedings of the ASME International Mechanical Engineering Congress and Exposition, Houston, TX, USA, 9–15 November 2012; Volume 3, pp. 1447–1453. [CrossRef]
5. Villarreal-Marroquin, M.G.; Cabrera-Rios, M.; Castro, J.M. A multicriteria simulation optimization method for injection molding. In Proceedings of the 2011 Winter Simulation Conference (WSC), Phoenix, AZ, USA, 11–14 December 2011; pp. 2390–2402. [CrossRef]
6. Villarreal-Marroquín, M.G.; Mulyana, R.; Castro, J.M.; Cabrera-Ríos, M. Selecting process parameter in injection molding via simulation optimization. *J. Polym. Eng.* **2011**, *31*, 387–395. [CrossRef]
7. Myers, M.; Castro, J.M.; Mulyana, R.; Hoffman, B. Experimental Development of an Injection Molding Process Window. *Preprints* **2023**, 2023060641. [CrossRef]
8. Cai, K. Effect of Processing Parameters on Mechanical Properties During Injection Molding. Ph.D. Thesis, The Ohio State University, Columbus, OH, USA, 2018.
9. Bartlett, L.P. A Preliminary Study of Using Plastic Molds in Injection Molding. Ph.D. Thesis, The Ohio State University, Columbus, OH, USA, 2017.
10. Bartlett, L.; Grunden, E.; Mulyana, R.; Castro, J.M. A preliminary study on the performance of additive manufacturing tooling for injection molding. SPE ANTEC.:100-104. Available online: https://www.injectionmoldingdivision.org/2019/07/16/a-preliminary-study-on-the-performance-of-additive-manufacturing-tooling-for-injection-molding/ (accessed on 17 January 2023).
11. Mulyana, R. Design of Experiments and Data Envelopment Analysis Based Optimization in Injection Molding. Ph.D. Thesis, The Ohio State University, Columbus, OH, USA, 2006.
12. Yu, S.; Zhang, Y.; Yang, D.; Zhou, H.; Li, J. Offline Prediction of Process Windows for Robust Injection Molding. *J. Appl. Polym. Sci.* **2014**, *131*, 40804. [CrossRef]
13. Ebnesajjad, S. 10—Injection molding. In *Fluoroplastics*; William Andrew: Norwich, NY, USA, 2015; pp. 236–281. [CrossRef]
14. Zhang, D.; Bartlett, L.; Cai, K.; Hoffman, B.; Mulyana, R.; Herderick, E.; Castro, J.M. Difference in Mechanical Properties Between Plastic and Metal Molds for Injection Molding a Talc Reinforced Rubber Modified Polypropylene. *Polym. Eng. Sci.* **2022**, *62*, 3470–3475. [CrossRef]

 *polymers*

Article

High-Temperature Response Polylactic Acid Composites by Tuning Double-Percolated Structures

Haiwei Yao [1], Rong Xue [2], Chouxuan Wang [2], Chengzhi Chen [2], Xin Xie [2], Pengfei Zhang [3], Zhongguo Zhao [2,*] and Yapeng Li [2]

[1] Textile and Clothing, College of Chemical Engineering, Shaanxi Polytechnic Institute, Xianyang 712000, China
[2] National and Local Engineering Laboratory for Slag Comprehensive Utilization and Environment Technology, School of Materials Science and Engineering, Shaanxi University of Technology, Hanzhong 723000, China
[3] School of Textile Science and Engineering, Xi'an Polytechnic University, Xi'an 710048, China
* Correspondence: zhaozhonggo@snut.edu.cn; Tel.: +86-916-2641711

Abstract: Due to the properties of a positive temperature coefficient (PTC) effect and a negative temperature coefficient (NTC) effect, electrically conductive polymer composites (CPCs) have been widely used in polymer thermistors. A dual percolated conductive microstructure was prepared by introducing the polybutylene adipate terephthalate phase (PBAT) into graphene nanoplatelets (GNPs)-filled polylactic acid (PLA) composites, intending to develop a favorable and stable PTC material. To achieve this strategy, GNPs were selectively distributed in the PBAT phase by injection molding. In this study, we investigated the crystallization behavior, electrical conductivity, and temperature response of GNP-filled PLA/PBAT composites. The introduction of GNPs into PLA significantly increased PLA crystallization capacity, where the crystallization onset temperature (T_o) is raised from 116.7 °C to 134.7 °C, and the crystallization half-time ($t_{1/2}$) decreases from 35.8 min to 27.3 min. The addition of 5 wt% PBAT increases the electrical conductivity of PLA/PBAT/GNPs composites by almost two orders of magnitude when compared to PLA/GNPs counterparts. The temperature of the heat treatment is also found to play a role in affecting the electrical conductivity of PLA-based composites. Increasing crystallinity is favorable for increasing electrical conductivity. PLA/PBAT/GNPs composites also show a significant positive temperature coefficient, which is reflected in the temperature–electrical resistance cycling tests.

Keywords: crystallization; electrical conductivity; temperature response behavior; positive temperature coefficient

Citation: Yao, H.; Xue, R.; Wang, C.; Chen, C.; Xie, X.; Zhang, P.; Zhao, Z.; Li, Y. High-Temperature Response Polylactic Acid Composites by Tuning Double-Percolated Structures. *Polymers* **2023**, *15*, 138. https://doi.org/10.3390/polym15010138

Academic Editors: Andrew N. Hrymak and Shengtai Zhou

Received: 2 December 2022
Revised: 21 December 2022
Accepted: 24 December 2022
Published: 28 December 2022

1. Introduction

In recent years, due to the high conductivity, lightweight properties, corrosion resistance, and good processability of electrically conductive polymer composites (CPCs) [1–3], they have been widely used in electromagnetic shielding materials (EMI) [4–6], sensors [7,8], capacitors [9,10], and other fields [11–13]. Due to the non-degradable properties of petroleum-based polymers, bio-based polymers have been extensively studied, and CPCs prepared with degradable polymers as a matrix have become a research hotspot [14–17]. CPCs have a positive temperature coefficient (PTC) of CPCs, that is, the resistance of the composite increases with temperature, and the negative temperature coefficient (NTC) of CPCs, the resistance of the composite, decreases with an increase in temperature. These two important temperature-related characteristics have great theoretical significance for application in self-protection fuses and temperature-sensitive sensors [18–22]. However, CPCs are usually accompanied by two effects of PTC and NTC in a certain temperature range, which limits their application. Therefore, testing the temperature range of the PTC and NTC effects of CPCs has an important application value [23]. Moreover, there is a growing demand to design a CPC material possessing tunable PTC characteristics.

Up to now, research on regulating the PTC characteristics of CPCs has been reported [24–26]. It is known that the increasing temperature can induce the expansion of the polymer matrix and break the conductive links, exhibiting the PTC effect, whereas the re-agglomeration of the conductive additives can induce the NTC effect [27]. Most of the research published mainly focuses on the changes in temperature region in the vicinity of the melting point, thus, the PTC is considered an advantage to broaden the application area. Dai et al. [24] employed a segregated and double-percolated composite microstructure to develop a favorable NTC material by selectively distributing graphene in a polyamide 6 (PA6) phase between isolated ultra-high molecular weight polyethylene (UHMWPE) particles, achieving a relatively linear NTC effect through the whole heating process. Liu et al. [25] prepared high-density polyethylene (HDPE)/carbon fiber (CF) and isotactic polypropylene (iPP)/CF composites by melt blending, and only observed an abnormal PTC effect in an extremely narrow temperature range. Wang et al. [26] prepared a prototype of a multi-walled carbon nanotube (MWCNT)/epoxy resin flexible sensor, which shows obvious PTC and NTC phenomena. Nevertheless, the aforementioned methods were only feasible with their certain microstructure or chemical composition; importantly, most of the research mainly focuses on the one-component polymer, which would also lead to poor PTC reproducibility. It is still a challenge to tune the PTC efficiently.

Due to the outstanding characteristics of graphene nanoplatelets, such as a large aspect ratio and high electrical conductivity, the temperature response behaviors of CPCs filled with graphene attracted enormous attention in recent years. Pang et al. [27] constructed the segregated structure by controlling the migration of GNPs to the surface of ultra-high molecular weight polyethylene (UHMWPE), resulting in an increasing PTC intensity. Thus, the interaction of conductive fillers plays a pivotal role in the evolution of the conductive network in the temperature field. Although numerous research studies focused on investigating the morphology, electronic, and mechanical properties of the GNPs/polylactic acid (PLA) composites [28–30], the temperature response behaviors of the GNPs/PLA composites have rarely been studied so far. In particular, PLA as the biodegradable material combined with conductive fillers for temperature sensors is expected to have a potential multipurpose application in bio-nanomaterials and eco-friendly functional materials field [17,31].

In this paper, PLA/GNPs composites were prepared by an injection molding machine, and the internal conductive network structure of PLA/GNPs composites was adjusted by adding PBAT to form a double percolation conductive network structure inside the polymer. In addition, we explored the change in electrical conductivity of GNPs/PLA composites with time under different heat treatment temperatures, and the temperature-sensitive performance of PLA/GNPs composite in the temperature range of 37~140 °C was studied. At the same time, the influence of PBAT on the crystallization, electrical conductivity, and temperature sensitivity of PLA/PBAT/GNPs composites was systematically explored.

2. Materials and Methods

2.1. Materials

Polylactic acid (PLA), brand name 4032D, produced by Nature works; polybutylene adipate terephthalate (PBAT), brand name C1200, produced by BASF, Germany; graphene nanoplatelets (GNPs), thickness 4~20 nm, diameter 5~10 μm, Chengdu Institute of Organic Chemistry, Chinese Academy of Sciences.

2.2. Sample Preparation

Pure PLA pellets (100 g) were first blended with various contents of GNPs (0, 1, 2, 2.5, 3, 3.5, 4, and 4.5 wt%) and PBAT (5, 10, 20, 30, 40, and 50 wt%) using a twin-screw extruder (SHJ-20, Nanjing Haisi Extrusion Equipment Co., Ltd., Nanjing, China). The screw speed was 80 rpm and the temperature profile from hopper to die was from 160 to 195 °C. Then, the injection molding machine (HTF90W1, Ningbo Haitian Plastic Machine Group Co. LTD, Ningbo, China) was used to prepare standard dumbbell samples at an injection temperature profile of 190 to 200 °C from the hopper to the nozzle to simplify the name of the composites; $PLAGNP_xPB_y$ denotes the composite GNPs weight content as x and as PB weight content as y. For example, $PLAGNP_{3.5}PB_{10}$ composite contains 3.5 wt% GNPs and 10 wt% PBAT.

2.3. Characterization

2.3.1. Scanning Electron Microscopy (SEM)

To explore the dispersion state of GNPS in PLA and PBAT/PLA, the spline was first quenched using liquid nitrogen and then the section was observed using scanning electron microscopy (SEM). Scanning electron microscope (SEM) model: JSM-6390LV, Rigaku, Tokyo, Japan.

2.3.2. Differential Scanning Calorimetry (DSC)

To explore the influence of adding GNPs on the crystallinity of the composite, the samples were tested by DSC (model instrumentation: TGA/DSC1, Mettler-Toledo Instruments, Zurich, Switzerland). First, samples weighing 5~10 mg were placed in the crucible and were heated from 40 to 200 °C at 10 °C/min under a nitrogen atmosphere. They were then kept at 200 °C for 5 min to remove thermal history. Finally, the temperature was cooled from 200 to 40 °C at 3 °C/min.

The relative crystallinity (X_c) can be expressed by Equation (1) [32]:

$$X_c = \frac{\Delta H_m - \Delta H_{cc}}{\Delta H_f \omega_{PLA}} \times 100\% \tag{1}$$

where ΔH_m is the melting enthalpy of PLA, ΔH_{cc} is the cold crystallization enthalpy of PLA, and ΔH_f is standard melting enthalpy of 100%, being 93 J/g; ω_{PLA} is the mass fraction of PLA in the composite material.

2.3.3. Conductivity Testing

The electrical conductivity of samples was tested using an isolation resistance meter (TA2684A, Changzhou Tonghui Electronics, Changzhou, China).

The electrical conductivity (σ) for each sample was calculated [33]:

$$\sigma = L/RS \tag{2}$$

where σ is the conductivity of the material, R is the volume resistance, and S is the cross-sectional area.

2.3.4. Temperature Response Behavior Analyses

The temperature response behavior of PLA/PBAT/CNTs was investigated by self-designed equipment using an isolation resistance meter and temperature controller (WCY-SJ, Nanjing Sangli Electronic Equipment Factory, Nanjing, China). Based on the crystalline temperature of $PLAGNP_{3.5}$, the testing temperature was set from 25 to 100, 120 and 135 °C at 3 °C/min. In the meantime, the resistance changes of samples were online tested, as shown in Figure 1.

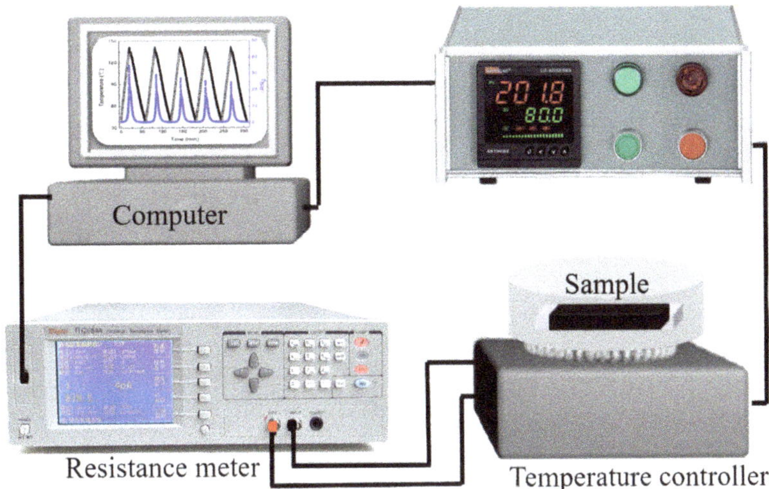

Figure 1. The self-designed equipment schematic diagram.

3. Results and Discussion

3.1. Electrical Conductivity of CPCs

To explore the influence of GNPs and PBAT addition on the electrical conductivity of PLA/PBAT/GNPs composites, all samples were tested as shown in Figure 2. Figure 2a shows that when GNPs content is in the range between 0 and 2.5 wt%, there is no change in the electrical conductivity of PLA/GNPs samples, showing the insulating behavior ($<10^{-8}$ S/m). The reason for this is that GNPs are not connected to form a conducting path in the PLA matrix, so electron transmission depends primarily on electrons hopping from adjacent fillers. With further increasing the GNPs content, the conductivity of the composites gradually increases. When the GNPs content exceeds 3 wt%, the conductivity of the samples increases rapidly from a value of 4.7×10^{-7} to 3.3×10^{-3} S/m, indicating that the internal conducting path of the composites is starting to form. Furthermore, when the GNPs content is greater than 4 wt%, the conductivity of the samples has no obvious change, showing that the inner conducting network of the composite is perfect, and the composite material is gradually converted into a conductor. This indicates that the filler content plays a vital role in constructing a conductive network path for the conductivity of composites. For comparison, the classical percolation theory shown in Equation (3) was cited to predict the conductive percolation value of the composite material in which the fitting value was 2.75 wt% [33].

$$\sigma(p) = \delta(p - p_c)^t \tag{3}$$

where $\sigma(p)$ is the conductivity of the composite material, δ is the conductivity of the filler, p is the content of the conductive filler, p_c is the percolation threshold of the composite, and t is the critical index.

According to the above analysis, the GNPs content of 3.5 wt% was chosen to explore the influence of PBAT content on the electric conductivity, as shown in Figure 2c. From Figure 2c, it can be seen that the curve of the PLAGNP$_{3.5}$PB$_y$ shows an "n" type change trend. When 5 wt% PBAT is added, the conductivity of the composite rises to 9.6×10^{-3} S/m, which is two orders of magnitude higher than that of PLAGNP$_{3.5}$. This is due to the low content of PBAT playing a repulsive role in the composite material, which promotes a part of GNPs migration from the PLA phase to the interface of PBAT and forms the conductive paths. When the amount of PBAT added is further increased from 5 to 20 wt%, the conductivity of PLAGNP$_{3.5}$PB$_y$ shows a trend of first increasing and then maintaining a plateau.

This is due to the migration of GNPs into the PBAT phase and a small amount of PBAT is not enough to agglomerate GNPs in a large amount. Therefore, the process of increasing PBAT content is the process of increasing GNPs migration capacity. However, with increasing PBAT content to 50 wt%, the conductivity of the composite drops to 3.04×10^{-8} S/m, which is about three orders of magnitude lower than that of PLAGNP$_{3.5}$. This is due to the higher melting viscosity of PBAT and a large number of GNPs migrating into the PBAT phase, resulting in the large-area agglomeration and no more conductive paths.

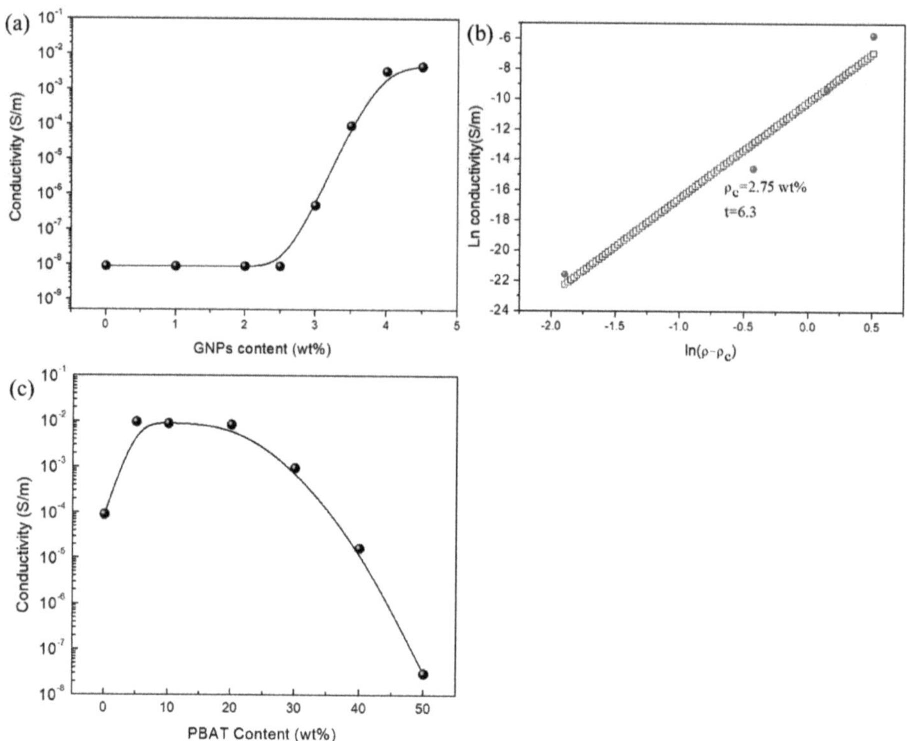

Figure 2. (**a**) Electric conductivity of PLA/GNPs composite, (**b**) the fitting curves of electric conductivity, and (**c**) electric conductivity curve of PLA/PBAT/GNPs composite as a function of PBAT content.

3.2. The Phase Morphology Analysis of CPCs

To better explain the effect of phase structure on the electrical conductivity of PLA/PBAT/GNPs composites, SEM was employed, as shown in Figure 3. As can be seen in Figure 3a, GNPs are distributed either flatly or vertically in the PLA matrix, and the surface structure of PLA is roughened, indicating that a large number of GNPs are overlapping on the surface of the quenched section. This phenomenon causes the PLA/GNPs composites to have a higher percolation threshold. With introducing PBAT into PLA/GNPs composites, a large number of white dots (sea-island structure) appear on the surface, which indicates low compatibility between PLA and PBAT. Moreover, the GNPs can be discerned on the surface. However, when further increasing the PBAT content from 30 to 40 wt%, the size of the dispersed phase is enlarged and the GNPs also vanish. When the PBAT content is increased to 50 wt%, the dispersed phase vanishes and forms co-continuous structures [14]. Thus, with increasing PBAT content, the changes in electric conductivity can be explained: (1) the high PBAT melting viscosity can induce the agglomeration of GNPs, thereby decreasing the formation of the conductive network; (2) it is easier for GNPs to migrate from the PLA phase to the

PBAT phase due to the intermolecular forces [34], thus, adding high PBAT content into PLA matrix can destroy the conductive network and reduce the electrical conductivity.

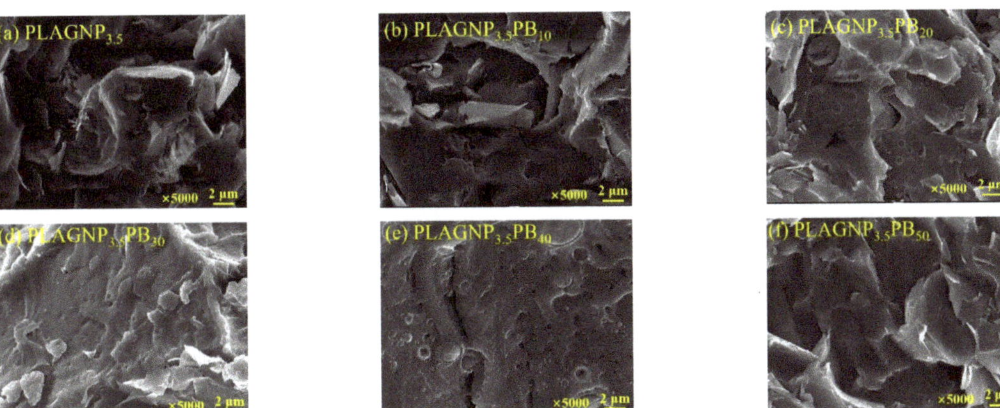

Figure 3. SEM images of PLA/PBAT/GNPs composites.

3.3. The Crystalline Properties of CPCs

To explore the influence of the addition of GNPs and PBAT on the crystallization properties of CPCs, a study of non-isothermal behaviors of CPCs was carried out. The results are shown in Figure 4. Table 1 shows the onset crystallization temperature (T_0), the peak crystallization temperature (T_p), and the half-time of crystallization ($T_{1/2}$). It can be seen from Figure 3a that the addition of GNPs shifts the T_0 and T_p of the composites to high temperature, in which the T_0 increases from 116.7 to 134.7 °C, and T_p increases from 93.2 to 120 °C. Moreover, introducing GNPs into the PLA matrix also significantly decreases $T_{1/2}$ from 35.8 to 27.3 min. These phenomena demonstrate that GNPs with a large surface can provide more nucleation sites and promote the crystalline process. However, introducing PBAT into PLA/GNPs composites displays an inverse phenomenon. When adding PBAT into PLA/GNPs composites, the T_0 and T_p decrease, while $T_{1/2}$ increases from 27.3 min to 36 min, as shown in Figure 4 and Table 1. The addition of PBAT results in the packaging of a large number of GNPs into PBAT, thereby reducing the number of nucleation sites and hindering the motion of PLA chains, which greatly inhibits the process of crystal growth.

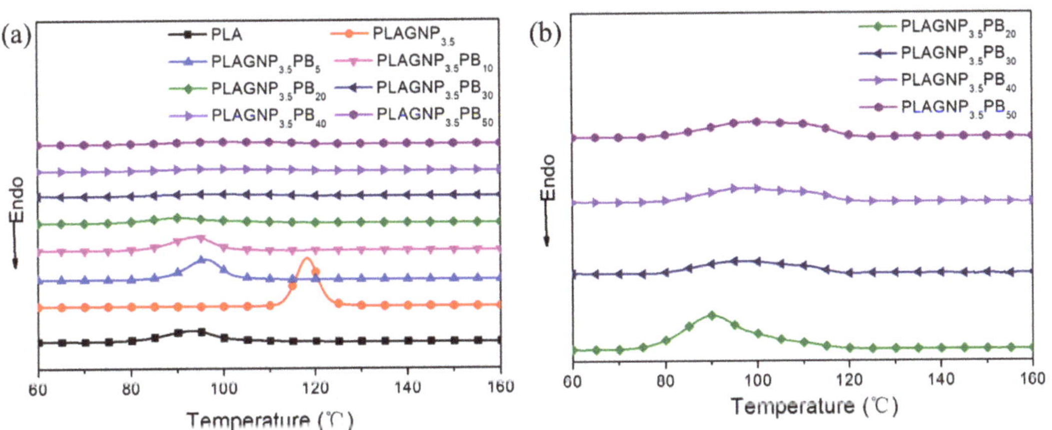

Figure 4. (**a**) DSC curve of PLAGNP$_{3.5}$PB$_y$ composite and (**b**) the magnification curve diagram.

Table 1. Comparison table of the crystallization process of $PLAGNP_{3.5}PB_y$ composite.

Sample	T_0 (°C)	T_p (°C)	$T_{1/2}$ (min)	Sample	T_0 (°C)	T_p (°C)	$T_{1/2}$ (min)
PLA	116.7	93.2	35.8	$PLAGNP_{3.5}PB_{20}$	120.9	90.3	36.0
$PLAGNP_{3.5}$	130.8	118.2	27.3	$PLAGNP_{3.5}PB_{30}$	120.6	97.6	34.1
$PLAGNP_{3.5}PB_5$	112.9	96.0	34.8	$PLAGNP_{3.5}PB_{40}$	125.3	97.2	33.5
$PLAGNP_{3.5}PB_{10}$	110.7	93.8	35.7	$PLAGNP_{3.5}PB_{50}$	125.3	98.8	33.2

3.4. The Isothermal Temperature Response of CPCs

Based on the above analysis, 20 wt% PBAT was chosen to explore the temperature response behaviors of CPCs. An online isothermal process was performed and the data are shown in Figure 5, in which the samples were treated at varying temperatures (T_{end} = 100, 120, and 135 °C). As shown in Figure 4a, the conductivity of $PLAGNP_{3.5}$ first decreases from 6.99×10^{-5} S/m to 1.37×10^{-6} S/m with an action time of 14.15 min and then gradually increases. Moreover, by increasing the heat-treated temperature from 100 to 135 °C, the action time is reduced to 7.77 min (seen in Table 2), indicating that the increase in heat treatment temperature can enhance the action time. Afterward, the electric conductivity shows a similar increase trend, in which with time passing, the electrical conductivity gradually increases. The reason for this is that during the isothermal process, the crystalline structure that appears within the polymer chains is progressively refined, and the "crystal repulsion" effect results in a large displacement of GNPs into the amorphous region, leading to a slow increase in electrical conductivity. At the end of the cooling process after 60 min, the conductivity initially rises sharply and then gradually stabilizes due to the rapid growth of the crystal at the beginning of the cooling step, which makes the repulsion effect of the crystal most obvious. It also can be seen from Figure 5b that during the subsequent isothermal process at different isothermal temperatures, the conductivity of $PLAGNP_{3.5}PB_{20}$ also tends to slowly increase. However, compared to the $PLAGNP_{3.5}$ composite, the conductivity of the $PLAGNP_{3.5}PB_{20}$ drops from 8.62×10^{-3} S/m to 4.60×10^{-4} S/m at 100 °C and action time is significantly shortened to 2.19 min.

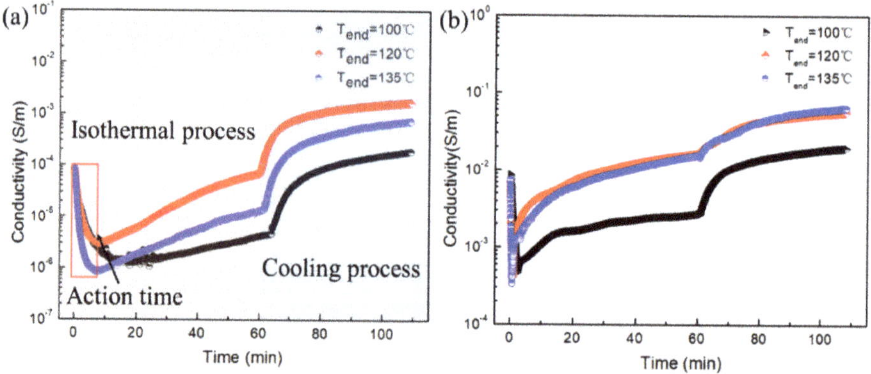

Figure 5. The temperature response behavior of $PLAGNP_{3.5}$ (a) and $PLAGNP_{3.5}PB_{20}$ composites (b).

The corresponding heat-treated samples were further tested by DSC to explore the crystallinity changes of PLA-based composites, as shown in Figure 6 and the related parameters are shown in Table 2. Figure 6 shows that $PLAGNP_{3.5}$ and $PLAGNP_{3.5}PB_{20}$ have similar changing trends, in which increasing the T_{end} can enhance the melting enthalpy and melting temperature, indicating that the crystalline structures become more perfect. Moreover, the relative crystallinity of samples, as shown in Table 2, is the highest when T_{end} is 120 °C, corresponding to 52.3% ($PLAGNP_{3.5}$) and 46.8% ($PLAGNP_{3.5}PB_{20}$), respectively. This is in agreement with the final conductivity of the composites when the isothermal temperature is 120 °C in the time plot of the electrical conductivity, which proves the effects of "crystalline rejection" on the conductivity of the composites.

Table 2. The relative conductive parameters and DSC data at different heat treatment temperatures.

Heat Treatment Temperature	Initial Conductivity (10^{-5} S/m)		Minimum Conductivity (10^{-6} S/m)		Action Time (min)		Relative Crystallinity (%)	
	a	b	a	b	a	b	a	b
T_{end} = 25 °C							27.4	32.5
T_{end} = 100 °C	6.99	862	1.37	460	14.15	2.19	44.3	35.0
T_{end} = 120 °C	8.42	711	2.97	410	7.25	1.00	52.3	46.8
T_{end} = 135 °C	8.72	700	8.61	332	7.77	0.69	50.3	46.0

a represents PLAGNP$_{3.5}$ and b represents PLAGNP$_{3.5}$PB$_{20}$.

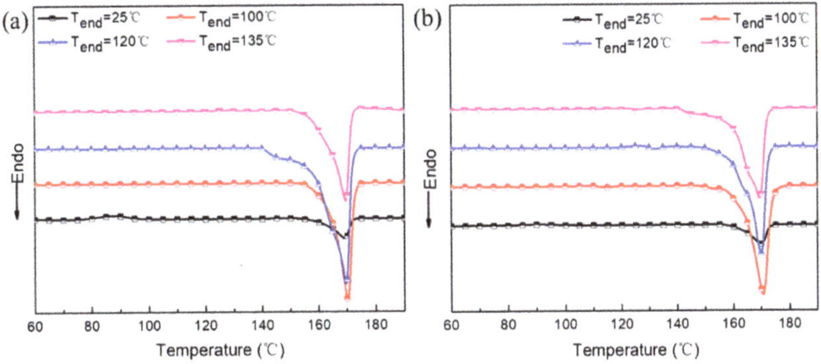

Figure 6. (**a**) The heating DSC curve of PLAGNP$_{3.5}$ composite and (**b**) DSC curve of PLAGNP$_{3.5}$PB$_{20}$ composite at different heat treatment temperatures.

3.5. The Non-Isothermal Temperature-Response Behaviors of CPCs

To explore the non-isothermal temperature response behavior of PLAGNP$_{3.5}$ and PLAGNP$_{3.5}$PB$_{20}$, non-isothermal temperature tests were performed and shown in Figure 7. The ratio $\Delta R/R_0$ ($\Delta R = R - R_0$, R is the real resistance and R_0 is the initial resistance) represents the resistance changes with the changing temperature, in which the maximum $\Delta R/R_0$ was used to characterize the PTC intensity. As can be seen from Figure 6a, when increasing the testing temperature from room temperature to 135 °C, the value of $\Delta R/R_0$ is also increased, demonstrating the PTC phenomenon under the single cycle process. Moreover, the maximum $\Delta R/R_0$ of the PLAGNP$_{3.5}$ composites (seen Figure 7c) show a gradual decrease with increasing cycle times. The reason for this is that PLA is an amorphous polymer when the temperature is increased and decreased for the PLAGNP$_{3.5}$ composite, and the temperature field, therefore, destroys some of the imperfect crystalline structure in the composites; so, in the slow cooling process, the imperfect crystal gradually achieves perfect crystallization so that the effect of "crystal repulsion" is increased, and the maximum $\Delta R/R_0$ is reduced in the following cycle. Furthermore, introducing the 20 wt% PBAT into PLA/GNPs composites has no obvious changes on the PTC phenomenon. However, the PTC value is enhanced and becomes more stable after the first cycle process compared with PLAGNP$_{3.5}$, in which the PTC value can be kept at 25. Thus, introducing PBAT into PLA/GNPs composites can provide a new method to prepare the temperature sensor with more sensibility and stability.

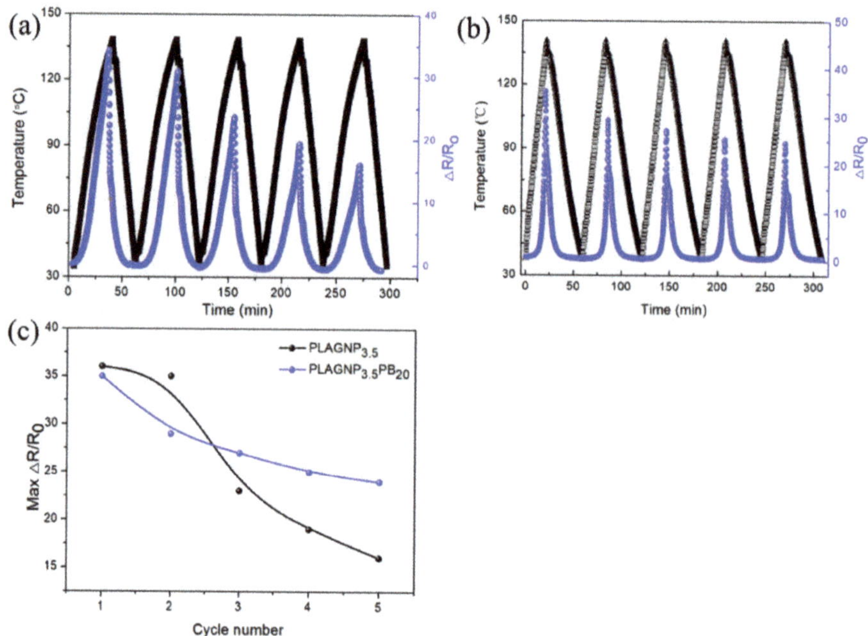

Figure 7. The electrical conductivity versus temperature curves of (**a**) PLAGNP$_{3.5}$, (**b**) PLAGNP$_{3.5}$PB$_{20}$ and (**c**) the changes of max $\Delta R/R_0$.

4. Conclusions

In this study, the PLA/GNPs/PBAT composites were prepared by injection molding and the electric conductivity, crystallization performance, and temperature response behavior were investigated in detail. Increasing GNPs content can gradually enhance the electrical conductivity and the percolation threshold is 2.75 wt% by fitting with a classic percolation theory. Moreover, the crystallization performance of PLAGNP$_{3.5}$ is improved by the addition of GNPs, which has a stronger ability to induce the crystallization of PLA due to their strong nucleation effect, including growth in crystallization rate and crystallinity. As for PLA/PBAT/GNPs composites, introducing a lower PBAT content can form a better conductive network, further improving the electrical conductivity, while decreasing crystalline temperatures, due to the high melting viscosity of PBAT and migration process of GNPs from PLA to PBAT. During the temperature-response testing, introducing PBAT can significantly shorten the action time from 14.15 min to 2.19 min and the cycle temperature-response stability is also gradually improved, in which the PTC value can be kept at 25 when compared with PLAGNP$_{3.5}$. The investigation of the microstructure evolution of a conductive network provides a guideline for the design and fabrication of temperature-sensing devices with high conductivity, high stability, and high-temperature sensitivity in a variety of applications.

Author Contributions: H.Y.: conceptualization, methodology, formal analysis, writing—original draft preparation, project administration and funding acquisition; R.X.: resources; C.W.: formal analysis; C.C.: investigation; X.X.: data curation; P.Z.: resources and funding acquisition; Z.Z.: conceptualization, writing—review and editing and supervision; Y.L.: data curation. All authors have read and agreed to the published version of the manuscript.

Funding: This research was funded by the Key Research and Development of Shaanxi Province (No. 2021GY-213) and Key Research and Development Program of Shaanxi (No. 2020ZDLGY13-08).

Institutional Review Board Statement: Not applicable.

Polymers **2023**, *15*, 138

Informed Consent Statement: Not applicable.

Data Availability Statement: The data presented in this study are available on request from the corresponding author.

Conflicts of Interest: The authors declare no conflict of interest. The funders had no role in the design of the study; in the collection, analyses, or interpretation of data; in the writing of the manuscript; or in the decision to publish the results.

References

1. Henry, A.; Plumejeau, S.; Heux, L.; Louvain, N.; Monconduit, L.; Stievano, L.; Boury, B. Conversion of Nanocellulose Aerogel into TiO$_2$ and TiO$_2$@C Nano-thorns by Direct Anhydrous Mineralization with TiCl4. Evaluation of Electrochemical Properties in Li Batteries. *ACS. Appl. Mater. Inter.* **2015**, *7*, 14584–14592. [CrossRef] [PubMed]
2. Hur, J.; Im, K.; Kim, S.W.; Kim, J.; Chung, D.Y.; Kim, T.H.; Jo, K.H.; Hahn, J.H.; Bao, Z.; Hwang, S. Polypyrrole/Agarose-based electronically conductive and reversibly restorable hydrogel. *Acs. Nano* **2014**, *8*, 10066–10076. [CrossRef] [PubMed]
3. Pang, H.; Yan, D.X.; Bao, Y.; Chen, J.B.; CHen, C.; Li, Z.M. Super-tough conducting carbon nanotube/ultrahigh-molecular-weight polyethylene composites with segregated and double-percolated structure. *J. Mater. Chem.* **2012**, *22*, 23568–23575. [CrossRef]
4. Wang, G.; Wang, L.; Mark, L.H.; Shaayegan, V.; Wang, G.; Li, H.; Zhao, G.; Park, C.B. Ultralow-Threshold and Lightweight Biodegradable Porous PLA/MWCNT with Segregated Conductive Networks for High-Performance Thermal Insulation and Electromagnetic Interference Shielding Applications. *ACS. Appl. Mater. Inter.* **2018**, *10*, 1195–1203. [CrossRef] [PubMed]
5. Fugetsu, B.; Sano, E.; Sunada, M.; Sambongi, Y.; Shibuya, T.; Wang, X.; Hiraki, T. Electrical conductivity and electromagnetic interference shielding efficiency of carbon nanotube/cellulose composite paper. *Carbon* **2008**, *46*, 1256–1258. [CrossRef]
6. Qi, X.D.; Jing, H.Y.; Nan, Z.; Ting, H.; Zuo, W.Z.; Kühnert, I.; Pötschke, P.; Wang, Y. Selective localization of carbon nanotubes and its effect on the structure and properties of polymer blends. *Prog. Polym. Sci.* **2021**, *123*, 101471. [CrossRef]
7. Korhonen, J.T.; Hiekkataipale, P.; Malm, J.; Karppinen, M.; Ikkala, O.; Ras, R. Inorganic hollow nanotube aerogels by atomic layer deposition onto native nanocellulose templates. *Acs. Nano* **2011**, *5*, 1967. [CrossRef]
8. Farjana, S.; Toomadj, F.; Lundgren, P.; Sanz-Velasco, A.; Naboka, O.; Enoksson, P. Conductivity-Dependent Strain Response of Carbon Nanotube Treated Bacterial Nanocellulose. *J. Sens.* **2013**, *2013*, 741248. [CrossRef]
9. Hamedi, M.; Karabulut, E.; Marais, A.; Herland, A.; Nystrom, G.; Wagberg, L. Nanocellulose aerogels functionalized by rapid layer-by-layer assembly for high charge storage and beyond. *Angew. Chem. Int. Ed. Engl.* **2013**, *52*, 12038–12042. [CrossRef]
10. Kang, Y.J.; Chun, S.J.; Lee, S.S.; Kim, B.Y.; Kim, J.H.; Chung, H.; Lee, S.Y.; Kim, W. All-solid-state flexible supercapacitors fabricated with bacterial nanocellulose papers, carbon nanotubes, and triblock-copolymer ion gels. *Acs. Nano* **2012**, *6*, 6400–6406. [CrossRef]
11. Joshi, G.M.; Deshmukh, K. Conjugated polymer/graphene oxide nanocomposite as thermistor. In *AIP Conference Proceedings*; AIP Publishing LLC: Melville, NY, USA, 2015; Volume 1667, p. 050017.
12. Yu, Y.X.; Huang, Q.F.; Rhodes, S.; Fang, J.Y.; An, L.N. SiCNO-GO composites with the negative temperature coefficient of resistance for high-temperature sensor applications. *J. Am. Ceram. Soc.* **2017**, *100*, 592–601. [CrossRef]
13. Feng, C.P.; Wei, F.; Sun, K.Y.; Wang, Y.; Lan, H.B.; Shang, H.J.; Ding, F.Z.; Bai, L.; Yang, J.; Yang, W. Emerging Flexible Thermally Conductive Films: Mechanism, Fabrication, Application. *Nano-Micro Lett.* **2022**, *14*, 24–57. [CrossRef] [PubMed]
14. Zhao, Z.; Zhou, S.; Quan, L.; Xia, W.; Fan, Q.; Zhang, P.; Chen, Y.; Tang, D. Polypropylene Composites with Ultrahigh Low-Temperature Toughness by Tuning the Phase Morphology, Ind. *Eng. Chem. Res.* **2022**, *61*, 15240–15248. [CrossRef]
15. Khosravani, M.R.; Berto, F.; Ayatollahi, M.R.; Reinicke, T. Characterization of 3D-printed PLA parts with different raster orientations and printing speeds. *Sci. Rep.* **2022**, *12*, 1016. [CrossRef] [PubMed]
16. Yu, B.; Zhao, Z.; Fu, S.; Meng, L.; Liu, Y.; Chen, F.; Wang, K.; Fu, Q. Fabrication of PLA/CNC/CNT conductive composites for high electromagnetic interference shielding based on Pickering emulsions method. *Compos. Part A Appl. Sci. Manuf.* **2019**, *125*, 105558. [CrossRef]
17. Kallannavar, V.; Kattimani, S. Effect of temperature and porosity on free vibration characteristics of a doubly-curved skew laminated sandwich composite structures with 3D printed PLA core. *Thin-Wall. Struct.* **2023**, *182*, 110263. [CrossRef]
18. Bai, B.C.; Kang, S.C.; Im, J.S.; Lee, S.H.; Lee, Y.S. Effect of oxy fluorinated multi-walled carbon nanotube additives on positive temperature coefficient/negative temperature coefficient behavior in high-density polyethylene polymeric switches. *Mater. Res. Bull.* **2011**, *46*, 1391–1397. [CrossRef]
19. Asare, E.; Evans, J.; Newton, M.; Peijs, T.; Bilotti, E. Effect of particle size and shape on positive temperature coefficient (PTC) of conductive polymer composites (CPC)-a model study. *Mater. Design.* **2016**, *97*, 459–463. [CrossRef]
20. Abazine, K.; Anakiou, H.; El Hasnaoui, M.; Graça, M.P.F.; Fonseca, M.A.; Costa, L.C.; Achour, M.E.; Oueriagli, A. Electrical conductivity of multiwalled carbon nanotubes/polyester polymer nanocomposites. *J. Compos. Mater.* **2016**, *50*, 3283–3290. [CrossRef]
21. Xiang, Z.D.; Chen, T.; Li, Z.M.; Bian, X.C. Negative Temperature Coefficient of Resistivity in Lightweight Conductive Carbon Nanotube/Polymer Composites. *Macromol. Mater. Eng.* **2009**, *294*, 91–95. [CrossRef]
22. Gong, S.; Zhu, Z.H.; Li, Z. Electron tunnelling and hopping effects on the temperature coefficient of resistance of carbon nanotube/polymer nanocomposites. *Phys. Chem. Chem. Phys.* **2017**, *19*, 5113–5120. [CrossRef] [PubMed]

23. Zhao, S.; Lou, D.; Zhan, P.; Li, G.; Dai, K.; Guo, J.; Zheng, G.; Liu, C.; Shen, C.; Guo, Z. Heating-induced negative temperature coefficient effect in conductive graphene/polymer ternary nanocomposites with a segregated and double-percolated structure. *J. Mater. Chem. C* **2017**, *5*, 8233–8242. [CrossRef]

24. Zhang, X.; Zheng, X.; Ren, D.; Liu, Z.; Yang, W.; Yang, M. Unusual positive temperature coefficient effect of polyolefin/carbon fiber conductive composites. *Mater. Lett.* **2016**, *164*, 587–590. [CrossRef]

25. Wang, F.X.; Liang, W.Y.; Wang, Z.Q.; Yang, B.; He, L.; Zhang, K. Preparation and property investigation of multi-walled carbon nanotube (MWCNT)/epoxy composite films as high-performance electric heating (resistive heating) element. *Express. Polym. Lett.* **2018**, *12*, 285–295. [CrossRef]

26. Li, G.; Hu, C.; Zhai, W.; Zhao, S.; Zheng, G.; Dai, K.; Liu, C.; Shen, C. Particle size induced tunable positive temperature coefficient characteristics in electrically conductive carbon nanotubes/polypropylene composites. *Mater. Lett.* **2016**, *182*, 314–317. [CrossRef]

27. Pang, H.; Zhang, Y.C.; Chen, T.; Zeng, B.Q.; Li, Z.M. Tunable positive temperature coefficient of resistivity in an electrically conducting polymer/graphene composite. *Appl. Phys. Lett.* **2010**, *96*, 251907. [CrossRef]

28. Wu, D.; Cheng, Y.; Feng, S. Crystallization Behavior of Polylactide/Graphene Composites. *Ind. Eng. Chem. Res.* **2013**, *52*, 6371–6379. [CrossRef]

29. Sabzi, M.; Jiang, L.; Liu, F.; Ghasemi, I.; Atai, M. Graphene nanoplatelets as poly(lactic acid) modifier: Linear rheological behavior and electrical conductivity. *J. Mater. Chem. A* **2013**, *1*, 8253–8261. [CrossRef]

30. Zhang, D.; Chi, B.; Li, B.; Gao, Z.W.; Du, Y.; Guo, J.B.; Wei, J. Fabrication of highly conductive graphene flexible circuits by 3D printing. *Synth. Met.* **2016**, *217*, 79–86. [CrossRef]

31. Kergariou, C.D.; Saidani-Scott, H.; Perriman, A.; Scarapa, F.; Le, D.A. The influence of the humidity on the mechanical properties of 3D printed continuous flax fibre reinforced poly(lactic acid) composites. *Compos. Part A Appl. Sci. Manuf.* **2022**, *155*, 106805. [CrossRef]

32. Zhao, Z.; Tang, D.; Jia, S.; Ai, T. Favorable formation of stereocomplex crystals in long-chain branched poly(L-lactic acid)/poly(D-lactic acid) blends: Impacts of melt effect and molecular chain structure. *J. Mater. Sci.* **2021**, *56*, 6514–6530. [CrossRef]

33. Zhao, Z.; Zhou, S.; Ai, T.; Li, Y.; Yang, Q.; Jia, S.; Tang, D. Fabrication of highly electrically conductive polypropylene/polyolefin elastomer/multiwalled carbon nanotubes composites via constructing ordered conductive network assisted by die-drawing process. *J. Appl. Polym. Sci.* **2022**, *139*, e52939. [CrossRef]

34. Kashi, S.; Kashi, S. Graphene Nanoplatelet-Based Nanocomposites: Electromagnetic Interference Shielding Properties and Rheology. Ph.D. Thesis, RMIT University, Melbourne, Australia, 2017.

polymers

Article

Improving Thermal Conductivity of Injection Molded Polycarbonate/Boron Nitride Composites by Incorporating Spherical Alumina Particles: The Influence of Alumina Particle Size

Chuxiang Zhou, Yang Bai, Huawei Zou * and Shengtai Zhou *

State Key Laboratory of Polymer Materials Engineering, Polymer Research Institute, Sichuan University, Chengdu 610065, China
* Correspondence: hwzou@163.com (H.Z.); qduzst@163.com (S.Z.)

Abstract: In this work, the influences of alumina (Al_2O_3) particle size and loading concentration on the properties of injection molded polycarbonate (PC)/boron nitride (BN)/Al_2O_3 composites were systematically studied. Results indicated that both in-plane and through-plane thermal conductivity of the ternary composites were significantly improved with the addition of spherical Al_2O_3 particles. In addition, the thermal conductivity of polymer composites increased significantly with increasing Al_2O_3 concentration and particle size, which were related to the following factors: (1) the presence of spherical Al_2O_3 particles altered the orientation state of flaky BN fillers that were in close proximity to Al_2O_3 particles (as confirmed by SEM observations and XRD analysis), which was believed crucial to improving the through-plane thermal conductivity of injection molded samples; (2) the presence of Al_2O_3 particles increased the filler packing density by bridging the uniformly distributed BN fillers within PC substrate, thereby leading to a significant enhancement of thermal conductivity. The in-plane and through-plane thermal conductivity of PC/50 μm-Al_2O_3 40 wt%/BN 20 wt% composites reached as high as 2.95 and 1.78 W/mK, which were 1183% and 710% higher than those of pure PC, respectively. The prepared polymer composites exhibited reasonable mechanical performance, and excellent electrical insulation properties and processability, which showed potential applications in advanced engineering fields that require both thermal conduction and electrical insulation properties.

Keywords: injection molding; polycarbonate; boron nitride; spherical alumina; thermal conductivity; microstructure; mechanical properties; electrical insulation

Citation: Zhou, C.; Bai, Y.; Zou, H.; Zhou, S. Improving Thermal Conductivity of Injection Molded Polycarbonate/Boron Nitride Composites by Incorporating Spherical Alumina Particles: The Influence of Alumina Particle Size. *Polymers* **2022**, *14*, 3477. https://doi.org/10.3390/polym14173477

Academic Editor: Marcelo Antunes

Received: 3 August 2022
Accepted: 24 August 2022
Published: 25 August 2022

Publisher's Note: MDPI stays neutral with regard to jurisdictional claims in published maps and institutional affiliations.

1. Introduction

Nowadays, the booming development in the fields of new energy sectors, cloud computing, the Internet of Things, high-speed communication, and artificial intelligence are imperceptibly changing our daily life. However, heat dissipation is a growing concern due to the integration of multifunctional components in confined areas, especially in the fields of microelectronics [1], battery units [2], and new energy sectors [3]. Thermally conductive polymer composites demonstrate an edge over metals and ceramic materials in terms of weight advantage, excellent resistance to corrosive environments, and most importantly, good processability and moldability [4,5], which can be scaled at large quantities without causing much additional costs [6]. Therefore, developing high performance thermally conductive polymer composites has become the core interest of researchers from both academic and industrial spheres.

It is known that the intrinsic thermal conductivity of polymers is very low (0.1–0.5 W/mK), which cannot meet the stringent requirements of industrial sectors [2,7]. Thus, a great effort has been devoted to developing highly thermally conductive polymers to suit the needs of the above-mentioned fields. The synthesis of intrinsically thermally conductive polymers is currently impractical, since very complex processes are involved, and the yields are always not satisfying [8–10]. In addition, synthesizing intrinsically thermally conductive polymers is costly, and only a few types of materials are available that are not cost effective and also not compatible for large scale industrial applications [11,12]. The commonly accepted approach is adding thermally conductive fillers to the polymer matrix using either solution [13,14] or melt blending methods [15–17]. Although solution mixing achieves better filler distribution and the prepared composites exhibit higher thermally conductive properties [18], the use of large amounts of solvents limits its wide use in industrial sectors [19]. Therefore, the melt blending method, which is industrially compatible and environmentally benign, has been widely adopted to prepare thermally conductive polymer composites [20,21].

Conventionally, different types of thermally conductive fillers such as carbonaceous fillers, metallic fillers, and ceramic fillers are adopted for preparing thermally conductive polymer composites [22]. However, the use of metallic [23] and carbonaceous [24,25] fillers impairs electrical insulation properties of polymer composites, which restricts their applications in fields that require electrical insulation performance. As a result, ceramic fillers such as boron nitride (BN) [26], alumina (Al_2O_3) [27], and aluminum nitride (AlN) [2] are commonly employed to prepare thermally conductive yet electrically insulative polymer composites. It has been accepted that the formation of a thermally conductive network and the increase of filler packing density are prerequisites for achieving thermally conductive polymer composites [28]. Under such circumstances, filler concentrations as high as 30 vol% are not rare in terms of preparing thermally conductive polymer composites [2]. Thus, great effort has been paid to constructing thermally conductive pathways and increasing filler packing density while not significantly impairing the mechanical and processing properties.

BN, which exhibits a planar structure alike flake graphite has been widely adopted to prepare thermally conductive polymer composites due to its intrinsically high thermal conductivity and thermal stability [29,30]. Injection molding, which is geared towards mass production at industrial scales, exerts a complex shearing influence on polymer melts that leads to a preferential orientation of planar fillers in injection molded articles [31–33]. In this scenario, a great discrepancy was noted in terms of the measured thermal conductivity with respect to melt flow direction, i.e., in-plane (along the flow direction) and through-plane (perpendicular to the flow direction) directions [34,35]. For example, there is a great possibility of forming intact filler conductive pathways along the melt flow direction due to the preferred orientation of planar fillers, whereas the properties in pathways perpendicular to the flow direction are significantly impaired [36,37]. As a result, there is a great anisotropy in the values of thermal conductivity for injection molded samples related to the melt flow direction [38].

Presently, numerous studies have indicated that hybrid filler loading is effective in improving the thermal conductivity of polymer composites by facial construction of thermally conductive pathways [39–41]. In a previous study [42], we found that the incorporation of spherical Al_2O_3 particles was effective in altering the orientation state of planar BN fillers in polycarbonate (PC)-based composites, thereby minimizing the difference between in-plane and through-plane thermal conductivity. In addition, Liu et al. [43] reported that the addition of a small amount (5 wt%) of spherical graphite was instrumental in building a more compact and denser filler packing structure in 45 wt% flake graphite (FG)-filled polypropylene (PP) composites, which was beneficial to improving the thermal conductivity of the resultant moldings. In another work [44], the same authors reported that the loading of spherical Al_2O_3 particles was able to affect the orientation state of FG in PP-based composites. As a result, both in-plane and through-plane thermal conductivity were enhanced for PP/FG composites. Moreover, they reported that the size of Al_2O_3 particles played a role in building thermally conductive pathways, and the addition of smaller size Al_2O_3 particles was proven to be more effective. However, the above studies were performed using the compression molding method, which exerts much lower shearing impact on the polymer melts when compared with the injection molding process [45]. To the best of our knowledge, the influence of spherical particle size on the distribution state of planar fillers under the influence of high shear rates has scarcely been studied.

To attempt to bridge this knowledge gap, spherical Al_2O_3 particles of different sizes were employed by using PC/BN composites as the model systems, and the properties such as thermal conductivity, morphological and mechanical properties, as well as rheological properties of ternary $PC/Al_2O_3/BN$ composites were systematically investigated. In this work, we reported that both the loading content and the size of spherical particles played a role in determining the distribution state of planar BN fillers. This work provided a new perspective in simultaneously improving the in-plane and through-plane thermal conductivities of injection molded polymer composites without significantly impairing the processability and mechanical properties, which show potential applications in the fields that require both thermal dissipation and excellent electrical insulation properties.

2. Experimental Section

2.1. Materials

Polycarbonate (PC, L-1225M) with a melt flow index of 28.8 g/10 min (300 °C @ 1.2 kg load) was produced by Teijin Polycarbonate China Ltd. (Jiaxing, China). Two-dimensional boron nitride (BN) fillers with an average size of 35 μm were purchased from Dandong Rijin Science and Technology Co., Ltd. (Dandong, China). Spherical alumina (Al_2O_3) with respective particle sizes of 5 and 50 μm were provided by Zhengzhou Sanhe New Materials Co., Ltd. (Zhengzhou, China).

2.2. Preparation of Samples

Briefly, PC, BN, and Al_2O_3 were thoroughly dried at 60 °C for at least 10 h prior to melt blending. Then a series of filler-containing PC-based composites (Table 1) was prepared using a twin-screw extruder (TSSJ-25/33, Chengdu Tarise Chemical Engineering Co. Ltd., Chengdu, China). The screw rotation speed was set at 30 rpm. The temperatures from the hopper to die zones were set from 240 to 260 °C. Then, the extrudates were pelletized, fully dried under the above-mentioned conditions and used for injection molding using an MA-2000 injection molding machine (Ningbo Haitian Machinery Inc., Ningbo, China). The melting and mold temperatures were set at 280 and 100 °C, respectively. The injection speed was set at 150 mm/s. The approximated shear rates were higher than 10^5 1/s. The larger size, i.e., 50 μm, Al_2O_3 particles were denoted as A_L, the smaller Al_2O_3 particles were named as A_S, and the BN was abbreviated as B. Thus, PC/50 μm-Al_2O_3 40 wt%/BN 20 wt% composites were termed as PC/A_L40B20. The same nomenclature system was applicable to the other systems, as listed in Table 1.

Table 1. Formulation of $PC/Al_2O_3/BN$ composites.

Samples	PC (wt%)	BN (wt%)	5 μm-Al_2O_3 (Small, wt%)	50 μm-Al_2O_3 (Large, wt%)
PC/B5	95	5	0	0
PC/B20	80	20	0	0
PC/A_L10-60	90-40	0	0	10-60
PC/A_S10-60	90-40	0	10-60	0
PC/A_L20B5	75	5	0	20
PC/A_L20B20	60	20	0	20
PC/A_L40B5	55	5	0	40
PC/A_L40B20	40	20	0	40
PC/A_S40B5	55	5	40	0
PC/A_S40B20	40	20	40	0

2.3. Characterization

The thermal conductivity (λ) of PC-based composites was measured using a LFA467 flash apparatus (NETZSCH, Selb, Germany). Samples with a diameter of 25 and a thickness of 0.2 mm were adopted. It should be noted that this technique reports λ in both the in-plane and through-plane directions. Five replicates were tested for each sample.

The morphology of filler-containing composites was observed by a scanning electronic microscope (SEM; Thermoscientific Apreo S, Oxford Instruments, Abingdon, UK). All samples were fractured in liquid nitrogen, and then the cryo-fractured samples were coated with gold to enhance image resolution.

XRD patterns of the composites were collected using an X-ray diffractometer (Ultima IV, Rigaku, Tokyo, Japan), and the scans were conducted in a 2θ range of 20–$60°$ at a scanning speed of $2°$/min.

The viscoelastic properties of samples with a diameter of 25 mm and a thickness of 1 mm were determined using a dynamic rheometer (MCR302, Anton Paar, Graz, Austria). The test was carried out under a constant-strain mode, where the applied strain was set at 1%. The scan frequency ranged from 100 to 0.01 Hz, and the test temperature was set at 260 °C.

The tensile strength was determined using a UTM4204 universal tester (Shenzhen Suns Company, Shenzhen, China) at 10 mm/min as per GB/T 1040.2-2006. The bending tests were carried out on a UTM4204 universal tester according to GB/T 9341-2008. The impact strength of specimen with a 2 mm V-notch was measured in accordance with GB/T 1843-2008. Five replicates were tested for each sample.

The resistance (R_x) of samples was measured using a high resistance meter (ZC-90F, Shanghai Taiou Electronics Co., Ltd., Shanghai, China), and the volume resistance (ρ_v) was obtained using the following equation:

$$\rho_v = R_x S / L \tag{1}$$

where L is the distance between the electrodes, and S is the cross-sectional area of the samples. Three replicates were tested for each sample.

3. Results and Discussion

3.1. Thermal Conductivity of PC/Al$_2$O$_3$ Composites

The thermal conductivity (λ) of PC/Al$_2$O$_3$ composites and their enhancement ratio to pure PC are presented in Figure 1a,b, respectively. Results showed that there was an increment on the λ with an increasing content of Al$_2$O$_3$. The addition of either larger size or smaller size Al$_2$O$_3$ particles contributed to an obvious increase of λ when compared with pure PC. However, the enhancement ratio became more appreciable when the filler content reached 40 wt%, regardless of the particle size. This was likely related to the formation of intact thermally conductive pathways at this certain filler concentration [46]. It is worth noting that the λ of PC/A$_S$ was higher than that of PC/A$_L$ counterparts when the filler concentration was less than 40 wt%; however, an opposite trend was observed when the filler concentration exceeded 40 wt%. The above observation further indicated that the thermally conductive network was likely constructed in the vicinity of 40 wt% Al$_2$O$_3$, and the difference in the values of λ between both PC/Al$_2$O$_3$ composites was likely related to the state of filler distribution. For example, a larger number of A$_s$ particles were present in PC composites when compared with the A$_L$-containing counterparts. According to Li and Shimizu [47], higher shearing conditions were effective for improving the state of distribution of inorganic fillers. Therefore, the improved distribution of spherical particles would be beneficial for improving the λ when the filler content was below 40 wt%. As shown in Figure 2, spherical Al$_2$O$_3$ particles exhibited a relatively uniform distribution in PC substrate, which was attributed to the high shearing conditions involved during the injection molding process [48]. Moreover, the mean distance between adjacent Al$_2$O$_3$ particles decreased significantly with increasing filler concentrations, which was critical for improving λ. When the concentration of Al_2O_3 reached 40 wt%, samples with larger size particles exhibited more compact filler packing structures and less filler/matrix interfacial thermal resistance [28,49], thereby leading to a higher increment in λ for corresponding PC-based moldings.

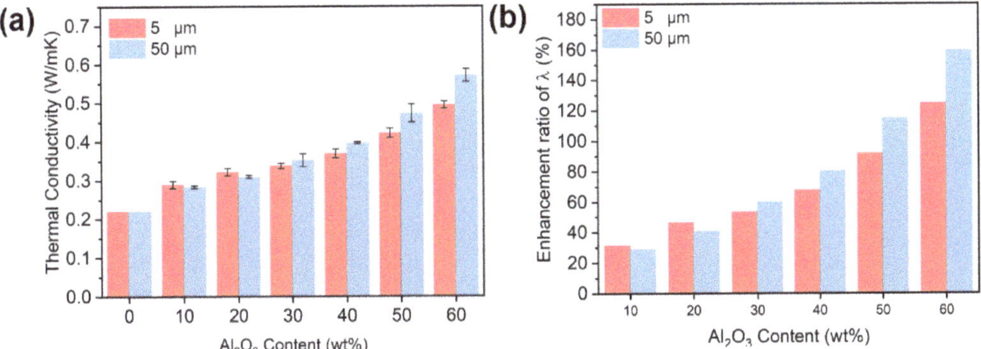

Figure 1. (**a**) Thermal conductivity (λ) and (**b**) enhancement ratio of λ of PC/Al_2O_3 composites in comparison with pure PC: influence of particle size and filler concentration.

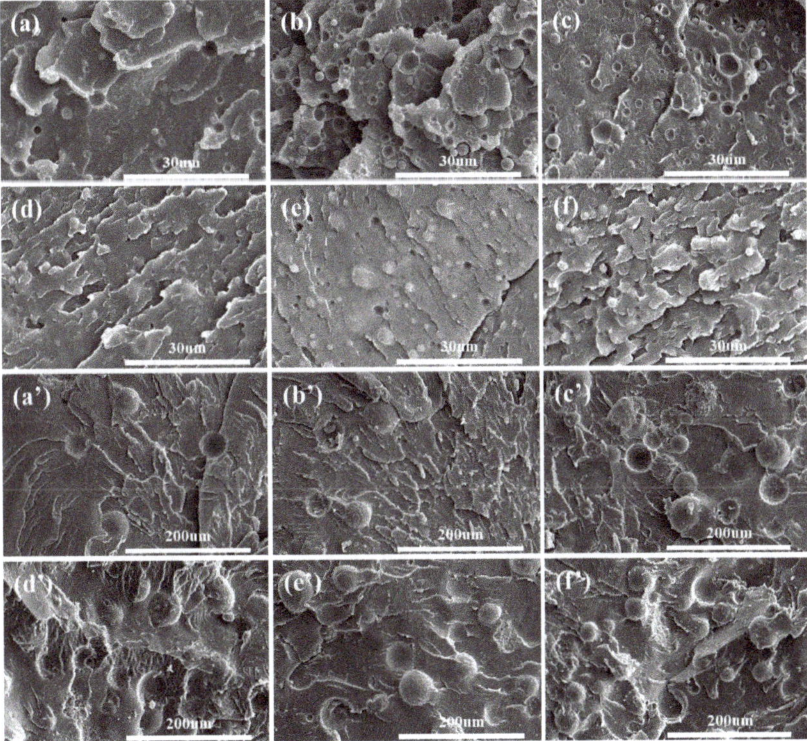

Figure 2. SEM images of PC/A_S composites: (**a**) PC/A_S10, (**b**) PC/A_S20, (**c**) PC/A_S30, (**d**) PC/A_S40, (**e**) PC/A_S50, and (**f**) PC/A_S60. SEM images of PC/A_L composites. (**a′**) PC/A_L10, (**b′**) PC/A_L20, (**c′**) PC/A_L30, (**d′**) PC/A_L40, (**e′**) PC/A_L50, and (**f′**) PC/A_L60.

Thus, it can be concluded that larger size Al_2O_3 particles exhibited a higher efficiency in forming thermally conductive pathways at higher filler concentrations (i.e., >40 wt%), which was likely related to the size effect of the fillers, i.e., Al_2O_3 particles. For example, less contact surface was likely formed between adjacent larger size particles when forming a thermally conductive network; the specific surface area of 50 μm Al_2O_3 particles was much smaller when compared with their 5 μm counterparts. As such, fewer matrix/filler interfacial defects were generated in PC-based composites with respect to larger size Al_2O_3 particles. Under such circumstances, less phonon scattering occurred in terms of 50 μm Al_2O_3 particle-containing PC-based composites, which led to a higher λ for corresponding composites when the filler concentration exceeded 40 wt% [49].

3.2. Thermal Conductivity of PC/Al$_2$O$_3$/BN Composites

As reported in previous studies [21,42], planar BN would exhibit a typical orientation along the flow direction due to the intense shearing effect, thereby leading to a higher anisotropy in thermal conduction properties in different directions, i.e., along the flow direction (i.e., in-plane λ) and perpendicular to the flow direction (i.e., through-plane λ). It was also found [42] that the orientation state of planar BN could be altered with the incorporation of spherical Al_2O_3 particles, thereby concurrently improving both the in-plane and through-plane λ of injection molded samples. Herein, the influence of the size of Al_2O_3 particles on the λ of PC/BN/Al_2O_3 composites was highlighted.

The λ values of PC/BN/Al_2O_3 composites, which were measured along the flow direction and perpendicular to the flow direction, are displayed in Figure 3. Figure 3a,b indicate that no significant enhancement in through-plane λ was observed for PC/BN composites and a great enhancement in in-plane λ when the BN concentration was increased from 5 to 20 wt%, which was related to the preferred orientation and increased addition of planar BN fillers in injection molded samples. Interestingly, there was a significant increase in both the in-plane and through-plane λ with the incorporation of spherical Al_2O_3 particles, as shown in Figure 3. For example, the through-plane λ and in-plane λ of PC/A_L40B20 composites reached as high as 1.78 and 2.95 W/(mK), which were 710% and 1183% higher than those of pure PC, respectively. This was definitely related to the formation of a more compact filler network that contributed to the significant increase of λ. Moreover, the addition of spherical Al_2O_3 particles would, to some extent, alter the orientation state of planar BN fillers, thereby contributing to a much greater enhancement in both in-plane and through-plane λ. The influence of particle size of Al_2O_3 on the λ of PC/BN/Al_2O_3 composites was studied as well. Results showed that samples with larger size Al_2O_3 exhibited a much greater increase in λ, which was attributed to the higher efficiency of forming thermally conductive pathways by using larger size spherical particles. In this scenario, the use of A_L particles not only reduced the contact thermal resistance between filler/matrix and filler/filler, but they also acted as bridges between adjacent fillers. Both factors contributed to a significant enhancement of λ for subsequent moldings.

The mechanism for thermal conductivity enhancement of subsequent composites is depicted in Figure 4. As shown in Figure 4a, planar BN fillers tended to preferentially align along the melt flow direction due to the predominant shearing effect induced by injection molding. Under such circumstances, injection molded PC/BN composites would exhibit a great discrepancy between in-plane λ and through-plane λ. The preferred orientation of BN was, to some extent, impaired with the incorporation of Al_2O_3 particles. As a result, in addition to the presence of Al_2O_3 particles, the partial deflection of planar BN would increase the likelihood of constructing thermally conductive pathways across the perpendicular to flow direction, thereby leading to an increase of through-plane λ without much influence on in-plane λ. Moreover, more planar BN fillers would be deflected with the increasing addition of Al_2O_3 particles. In this scenario, the difference between in-plane λ and through-plane λ was minimized at higher Al_2O_3 loading concentrations. In addition, the size of Al_2O_3 particles played a role in determining the deflection state of planar BN. Larger size Al_2O_3 particles tended to exert a greater influence on the orientation state of

BN fillers, which explained why the PC/BN/Al$_2$O$_3$ composites with 50 μm Al$_2$O$_3$ particles demonstrated higher λ than the 5 μm Al$_2$O$_3$-containing counterparts. As a result, the addition of larger size spherical Al$_2$O$_3$ particles had a more positive effect on improving the in-plane and through-plane λ of injection molded samples.

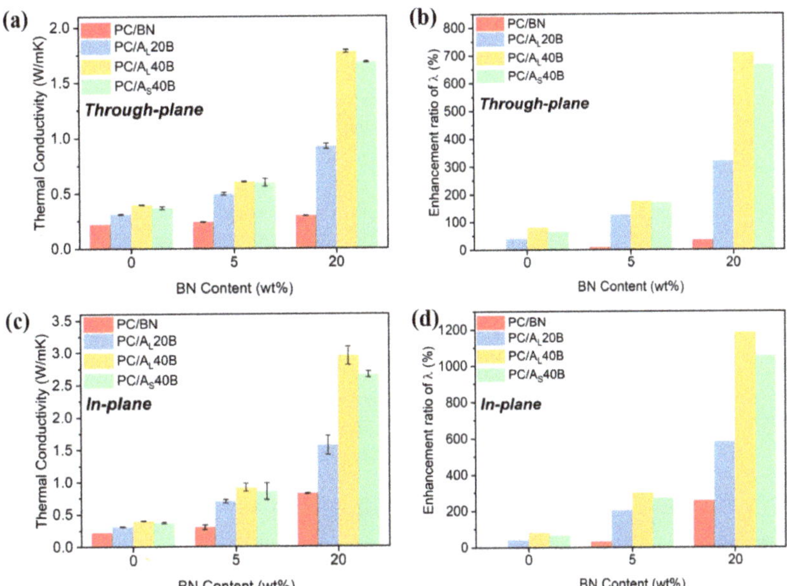

Figure 3. (**a**) Through-plane λ, (**b**) through-plane λ enhancement ratio, (**c**) in-plane λ, and (**d**) in-plane λ enhancement ratio of PC/Al$_2$O$_3$/BN composites with different Al$_2$O$_3$ particle sizes and contents.

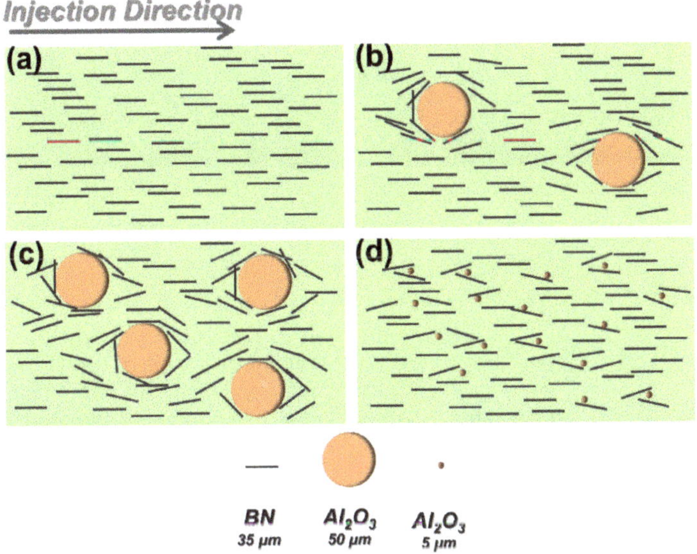

Figure 4. The schematic mechanism for thermal conductivity enhancement of PC/Al$_2$O$_3$/BN composites with different Al$_2$O$_3$ particle sizes and contents: (**a**) PC/BN, (**b**) PC/A$_L$20B, (**c**) PC/A$_L$40B, (**d**) PC/A$_S$40B.

3.3. Morphology of PC/Al$_2$O$_3$/BN Composites

The microstructure of PC/Al$_2$O$_3$/BN composites is displayed in Figure 5. As shown in Figure 5g,h, planar BN fillers exhibited preferential orientation along the melt flow direction, regardless of filler concentrations. In addition, both BN and Al$_2$O$_3$ particles showed a relatively uniform distribution within the PC substrate. The presence of BN fillers was labeled using red arrows. Figure 5a–f show that BN fillers, which were located far away from Al$_2$O$_3$ particles, remained in the preferential orientation state along the flow direction; however, planar BN fillers that were located in the vicinity of Al$_2$O$_3$ particles showed some deflection due to the steric hindrance effect imposed by Al$_2$O$_3$ particles. Under such circumstances, the possibility of constructing intact thermally conductive pathways was increased across the thickness direction, i.e., perpendicular to flow direction. As a result, the in-plane and through-plane λ would be concurrently enhanced for injection molded PC/BN/Al$_2$O$_3$ composites. Moreover, the deflection degree of BN fillers increased with increasing Al$_2$O$_3$ content and particle size, which is beneficial for building a denser thermally conductive network. Therefore, a larger size Al$_2$O$_3$ demonstrated an edge over smaller size particles in terms of altering the orientation state of BN fillers, thereby leading to a significant increase of both in-plane and through-plane λ for subsequent PC/BN/Al$_2$O$_3$ composites. The in-plane and through-plane λ of PC/A$_L$40B20 composites reached as high as 2.95 and 1.78 W/(mK), respectively, which were 94.1% and 63.3% higher than PC/A$_S$40B20 composites, and 1183% and 710% higher than pure PC, respectively. Thus, incorporating larger size spherical particles was proved effective in improving the λ of planar filler-containing injection molded composites.

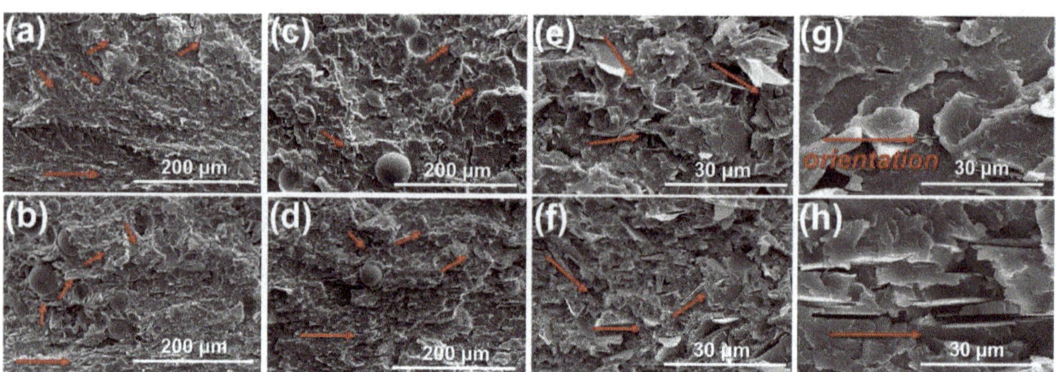

Figure 5. The microstructure of PC-based composites along the flow direction: (**a**) PC/A$_L$40B5, (**b**) PC/A$_L$40B20, (**c**) PC/A$_L$20B5, (**d**) PC/A$_L$20B20, (**e**) PC/A$_S$40B5, (**f**) PC/A$_S$40B20, (**g**) PC/B5; (**h**) PC/B20.

3.4. XRD Analysis

According to literature [50], the orientation degree of BN can be assessed using XRD analysis. Basically, the peak intensity ratio of the (002) plane to the (100) plane of BN was employed to reflect the orientation state of BN [41,51]. The XRD spectra and the peak intensity ratios for related polymer composites are presented in Figure 6. Results showed that the peak intensity ratio of the PC/B20 composite was the highest among the studied systems. The peak intensity ratio of the PC/B20 composite was higher than that of PC/B5, which was related to the reduced melt viscosity related to the incorporation of a larger fraction of planar fillers (see Figure 7c). However, the peak intensity ratios for PC/BN/Al$_2$O$_3$ composites decreased significantly when compared with binary PC/BN composites, which was ascribed to the filler-induced deflection of planar BN fillers [42].

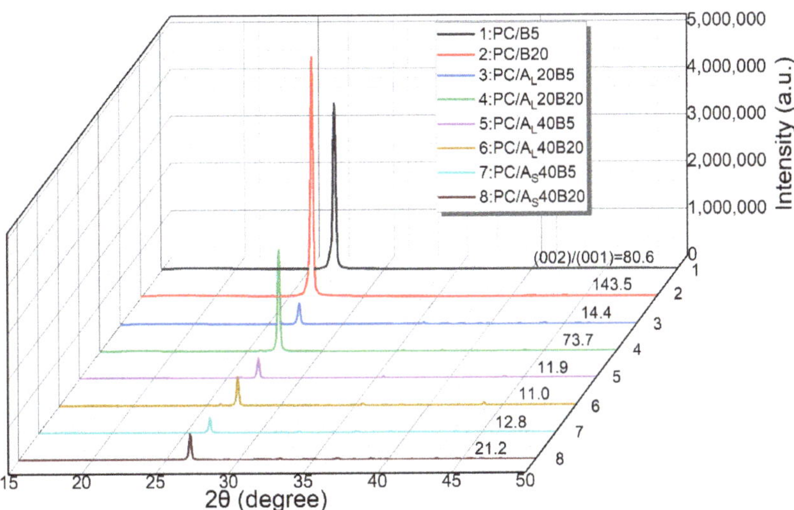

Figure 6. XRD spectra of PC/BN and PC/Al$_2$O$_3$/BN composites. The influence of particle size and filler content.

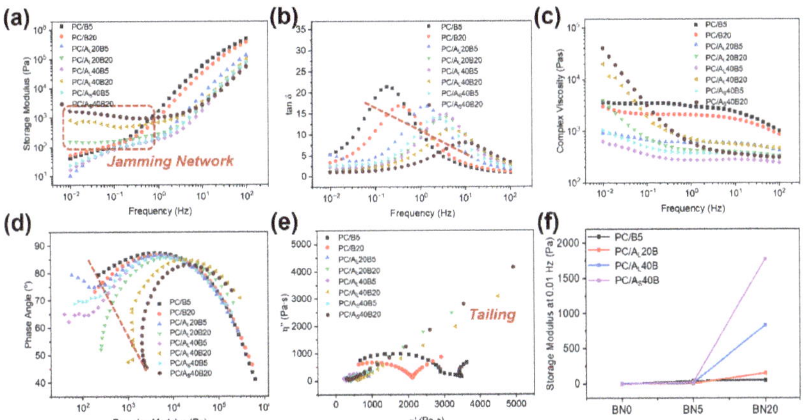

Figure 7. Rheological properties of PC/BN and PC/Al$_2$O$_3$/BN composites: (**a**) storage modulus vs. frequency, (**b**) tan δ vs. frequency, (**c**) complex viscosity vs. frequency, (**d**) phase angle vs. complex modulus, (**e**) η'' vs. η', and (**f**) storage modulus at 0.01 Hz for different filler-containing composites.

Specifically, the values of peak intensity ratio decreased with increasing content of Al$_2$O$_3$, provided that the content of BN and the particle size of Al$_2$O$_3$ remained the same. In addition, the peak intensity ratios of PC/A$_L$/BN composites were higher than those of PC/A$_S$/BN counterparts when the concentrations of Al$_2$O$_3$ particles and BN were the same. Thus, it can be concluded that both increasing the concentration of Al$_2$O$_3$ particles and using larger size Al$_2$O$_3$ led to a higher degree of deflection for planar fillers in injection molded samples, which was crucial to improving the through-plane λ of related polymer composites.

3.5. Rheological Properties

The viscoelastic properties of PC/BN and PC/Al$_2$O$_3$/BN composites were studied using a dynamic rheometer, as displayed in Figure 7. The values of storage modulus (G′) as a function of sweeping frequency are given in Figure 7a. Results showed that the values of G' increased with increasing sweeping frequency and total filler loading concentrations. In particular, the values of G' at the lowest frequency, i.e., $G'_{0.01Hz}$, increased with total filler concentrations, which was related to the reinforcement effect of the added fillers [52]. A plateau was not visible for PC/BN and PC/Al$_2$O$_3$/BN 5 wt% composites, suggesting that an intact filler network structure was absent in the above systems. Moreover, a plateau was visible for PC/Al$_2$O$_3$/BN 20 wt% composites, signifying the existence of an intact filler network that consisted of spherical Al$_2$O$_3$ and planar BN, which was crucial for improving the λ. Figure 7b,c show that the peak of tan δ shifted to higher frequency regions coupling with a reduction of peak value, and a shear thinning behavior became more noticeable with increasing filler concentrations, which was attributed to the transition from a viscous to solid state of polymer melts arising from the presence of inorganic fillers [36,53].

The microstructural changes of polymer composites were further evaluated using the van Gurp–Palmen and Cole–Cole plots, as given in Figure 7d,e, respectively. Figure 8d showed that the values of phase angle for PC/Al$_2$O$_3$/BN composites decreased greatly at lower complex modulus regions when the concentration of BN reached 20 wt%. Moreover, the Cole–Cole plot changed from a semicircular to a linear shape with an obvious tailing effect, suggesting that the PC/Al$_2$O$_3$/BN composites exhibited typical solid-like behavior owing to the formation of three-dimensional thermally conductive pathways [42,54]. Furthermore, Figure 7f shows that there were negligible changes in the values of $G'_{0.01Hz}$ when the filler content of BN was increased from 5 to 20 wt% for binary PC/BN composites, suggesting that no microstructural change (i.e., the formation of a filler network) was detected in PC/BN composites. Moreover, the changes in $G'_{0.01Hz}$ were also insignificant for PC/Al$_2$O$_3$/BN 5 wt% composites, and such changes became more noticeable when the concentration of BN reached 20 wt%, especially for PC/Al$_2$O$_3$ 40 wt%/BN 20 wt% composites, which was related to the microstructural change related to the increasing addition of inorganic fillers [52]. It was worth noting that the values of $G'_{0.01Hz}$ at lower frequency regions of the PC/A$_S$40B20 composite were higher than its PC/A$_L$40B20 counterparts, which suggested that the filler network structure consisting of 5 μm Al$_2$O$_3$ and BN exerted a much stronger confinement effect on polymer chains, which was likely related to the higher specific surface area of smaller Al$_2$O$_3$ particles [49]. However, this did not mean that such a filler network structure had a higher efficiency in dissipating heat because the particle size played a positive role in determining the λ of polymer composites [38,49,55].

3.6. Mechanical and Electrical Properties

The mechanical properties, including tensile strength, elongation at break, flexural strength, and notched impact strength of PC-based composites are presented in Figure 8. Results showed that the mechanical properties, particularly for elongation at break and notched impact strength, of PC/Al$_2$O$_3$/BN composites deteriorated significantly with the incorporation of BN and Al$_2$O$_3$ fillers, which was likely attributed to the poor interfacial interactions between the host substrate and inorganic fillers, as well as the formation of filler network structures that acted as structural defects [16]. Figure 8 shows that the mechanical properties of PC/A$_L$40B20 composites were higher than those of PC/A$_S$40B20 composites. For example, the tensile strength, elongation at break, flexural strength, and notched impact strength of the former were 49.8 MPa, 1.83%, 81.4 MPa, and 3.66 kJ/m^2, respectively, whereas those of the latter were 46.5 MPa, 1.59%, 69.2 MPa, and 3.48 kJ/m^2, respectively. The above results were likely related to the fact that smaller Al$_2$O$_3$ particles had a tendency to form agglomerates due to their higher specific surface areas and the presence of more filler/matrix interfacial defects [49,56], thereby leading to a much lower mechanical performance for smaller Al$_2$O$_3$ particle-containing polymer composites. In general, PC/A$_L$/BN composites showed reasonable mechanical properties, which demonstrated potential applications in industrial sectors.

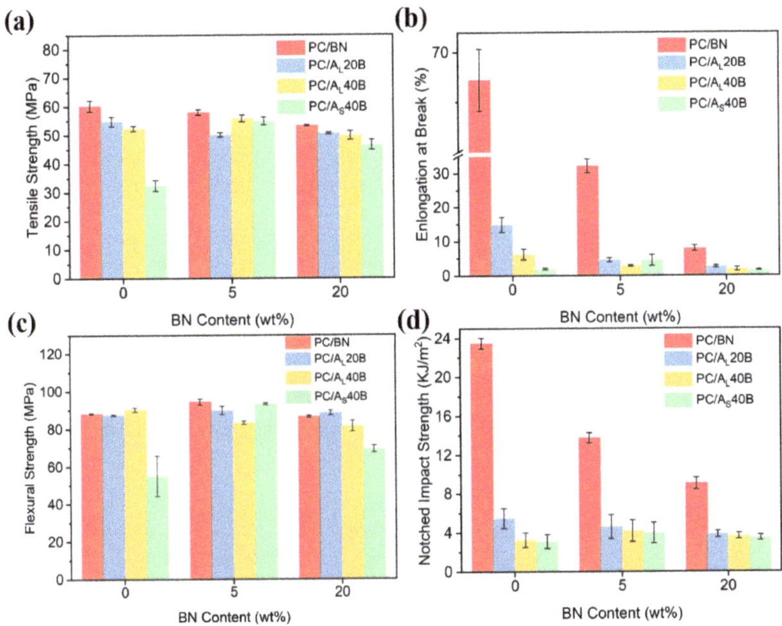

Figure 8. The mechanical properties of PC/BN and PC/Al$_2$O$_3$/BN composites: (**a**) tensile strength, (**b**) elongation at break, (**c**) flexural strength, and (**d**) notched impact strength.

The volume resistivity of PC/BN and PC/Al$_2$O$_3$/BN composites is shown in Figure 9. It can be seen that all samples demonstrated excellent electrical insulation properties, since the volume resistivity of all PC/BN and PC/Al$_2$O$_3$/BN composites exceeded 10^{15} Ω·cm. Therefore, the prepared polymer composites demonstrated exceptional electrical insulation properties, excellent thermal dissipation properties in both in-plane and through-plane directions of injection molded parts, and reasonable mechanical properties that can be targeted for applications in the fields of electronic devices and battery units, where both thermal conduction and electrical insulation are primary concerns.

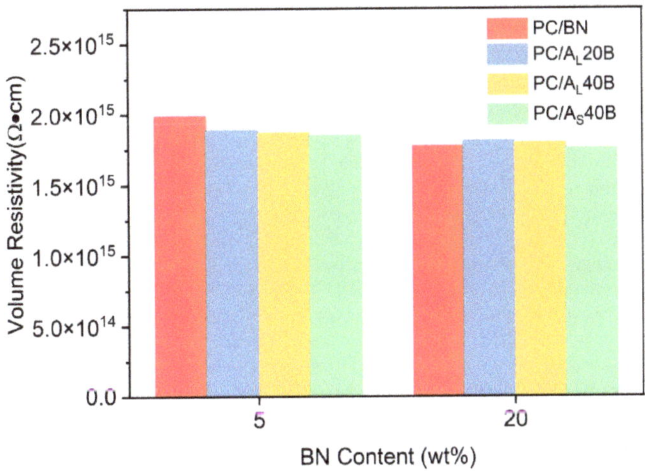

Figure 9. The volume resistance of PC/BN and PC/Al$_2$O$_3$/BN composites.

4. Conclusions

In this work, incorporating spherical alumina (Al_2O_3) particles was found to be effective in altering the orientation state of planar boron nitride (BN) fillers in injection molded PC-based composites. Both the in-plane and through-plane thermal conductivity of PC/Al_2O_3/BN composites were concurrently enhanced with the addition of spherical Al_2O_3 particles. Both increased particle size and loading content of Al_2O_3 were found to be crucial in affecting the orientation state of BN fillers, especially for those in the vicinity of Al_2O_3 particles, as corroborated by SEM observations and XRD analysis. The deflection state of planar BN fillers and the presence of larger size Al_2O_3 particles were beneficial for constructing intact thermally conductive pathways, thereby leading to a significant improvement in through-plane thermal conductivity without much influence on in-plane thermal conductivity. For example, the through-plane and in-plane thermal conductivity of PC/A_L40B20 composites reached as high as 1.78 and 2.95 W/mK, which were 710% and 1183% higher than those of pure PC, respectively. In addition, the prepared PC/Al_2O_3/BN composites demonstrated reasonable mechanical and exceptional electrical insulation properties, which show potential applications in the fields of electronics, battery units, and so forth.

Author Contributions: C.Z.: Investigation, visualization, writing—original draft. Y.B.: Investigation, data curation, formal analysis. H.Z.: Resources, funding acquisition, supervision. S.Z.: Conceptualization, funding acquisition, supervision, writing—review and editing. All authors have read and agreed to the published version of the manuscript.

Funding: This work was funded by the National Natural Science Foundation of China (52103040) and the China Postdoctoral Science Foundation (2020M673217).

Institutional Review Board Statement: Not applicable.

Informed Consent Statement: Not applicable.

Data Availability Statement: The data presented in this study are available on request from the corresponding author.

Conflicts of Interest: There are no competing financial interests or personal relationships that could have appeared to influence the work reported in this paper.

References

1. Ma, H.; Gao, B.; Wang, M.; Yuan, Z.; Shen, J.; Zhao, J.; Feng, Y. Strategies for enhancing thermal conductivity of polymer-based thermal interface materials: A review. *J. Mater. Sci.* **2020**, *56*, 1064–1086. [CrossRef]
2. Han, Z.; Fina, A. Thermal conductivity of carbon nanotubes and their polymer nanocomposites: A review. *Prog. Polym. Sci.* **2011**, *36*, 914–944. [CrossRef]
3. Coetzee, D.; Venkataraman, M.; Militky, J.; Petru, M. Influence of Nanoparticles on Thermal and Electrical Conductivity of Composites. *Polymers* **2020**, *12*, 742. [CrossRef] [PubMed]
4. Zhou, S.; Chen, Y.; Zou, H.; Liang, M. Thermally conductive composites obtained by flake graphite filling immiscible Polyamide 6/Polycarbonate blends. *Thermochim. Acta* **2013**, *566*, 84–91. [CrossRef]
5. Zhou, S.; Yu, L.; Song, X.; Chang, J.; Zou, H.; Liang, M. Preparation of highly thermally conducting polyamide 6/graphite composites via low-temperature in situ expansion. *J. Appl. Polym. Sci.* **2014**, *131*, 131. [CrossRef]
6. Vadivelu, M.A.; Kumar, C.R.; Joshi, G.M. Polymer composites for thermal management: A review. *Compos. Interfaces* **2016**, *23*, 847–872. [CrossRef]
7. Wei, B.; Chen, X.; Yang, S. Construction of a 3D aluminum flake framework with a sponge template to prepare thermally conductive polymer composites. *J. Mater. Chem. A* **2021**, *9*, 10979–10991. [CrossRef]
8. Gu, J.; Ruan, K. Breaking Through Bottlenecks for Thermally Conductive Polymer Composites: A Perspective for Intrinsic Thermal Conductivity, Interfacial Thermal Resistance and Theoretics. *Nanomicro Lett.* **2021**, *13*, 110. [CrossRef]
9. Li, Y.; Gong, C.; Li, C.; Ruan, K.; Liu, C.; Liu, H.; Gu, J. Liquid crystalline texture and hydrogen bond on the thermal conductivities of intrinsic thermal conductive polymer films. *J. Mater. Sci. Technol.* **2021**, *82*, 250–256. [CrossRef]
10. Ruan, K.; Zhong, X.; Shi, X.; Dang, J.; Gu, J. Liquid crystal epoxy resins with high intrinsic thermal conductivities and their composites: A mini-review. *Mater. Today Phys.* **2021**, *20*, 100456. [CrossRef]

11. Yang, X.; Liang, C.; Ma, T.; Guo, Y.; Kong, J.; Gu, J.; Chen, M.; Zhu, J. A review on thermally conductive polymeric composites: Classification, measurement, model and equations, mechanism and fabrication methods. *Adv. Compos. Hybrid Mater.* **2018**, *1*, 207–230. [CrossRef]

12. Zhang, F.; Feng, Y.; Feng, W. Three-dimensional interconnected networks for thermally conductive polymer composites: Design, preparation, properties, and mechanisms. *Mater. Sci. Eng. R Rep.* **2020**, *142*, 100580. [CrossRef]

13. Liu, B.; Li, Y.; Fei, T.; Han, S.; Xia, C.; Shan, Z.; Jiang, J. Highly thermally conductive polystyrene/polypropylene/boron nitride composites with 3D segregated structure prepared by solution-mixing and hot-pressing method. *Chem. Eng. J.* **2020**, *385*, 123829. [CrossRef]

14. Cui, X.; Ding, P.; Zhuang, N.; Shi, L.; Song, N.; Tang, S. Thermal Conductive and Mechanical Properties of Polymeric Composites Based on Solution-Exfoliated Boron Nitride and Graphene Nanosheets: A Morphology-Promoted Synergistic Effect. *ACS Appl. Mater. Interfaces* **2015**, *7*, 19068–19075. [CrossRef]

15. Zhang, X.; Wu, K.; Liu, Y.; Yu, B.; Zhang, Q.; Chen, F.; Fu, Q. Preparation of highly thermally conductive but electrically insulating composites by constructing a segregated double network in polymer composites. *Compos. Sci. Technol.* **2019**, *175*, 135–142. [CrossRef]

16. Zhou, S.; Shi, Y.; Bai, Y.; Liang, M.; Zou, H. Preparation of thermally conductive polycarbonate/boron nitride composites with balanced mechanical properties. *Polym. Compos.* **2020**, *41*, 5418–5427. [CrossRef]

17. Zhou, S.; Lei, Y.; Zou, H.; Liang, M. High thermally conducting composites obtained via in situ exfoliation process of expandable graphite filled polyamide 6. *Polym. Compos.* **2013**, *34*, 1816–1823. [CrossRef]

18. Agari, Y.; Ueda, A.; Nagai, S. Thermal-Conductivities of Composites in Several Types of Dispersion-Systems. *J. Appl. Polym. Sci.* **1991**, *42*, 1665–1669. [CrossRef]

19. Al-Saleh, M.H.; Sundararaj, U. A review of vapor grown carbon nanofiber/polymer conductive composites. *Carbon* **2009**, *47*, 2–22. [CrossRef]

20. Leung, S.N. Thermally conductive polymer composites and nanocomposites: Processing-structure-property relationships. *Compos. Part B Eng.* **2018**, *150*, 78–92. [CrossRef]

21. Bai, Y.; Shi, Y.; Zhou, S.; Zou, H.; Liang, M. Highly Thermally Conductive Yet Electrically Insulative Polycarbonate Composites with Oriented Hybrid Networks Assisted by High Shear Injection Molding. *Macromol. Mater. Eng.* **2021**, *307*, 2100632. [CrossRef]

22. He, X.; Wang, Y. Recent Advances in the Rational Design of Thermal Conductive Polymer Composites. *Ind. Eng. Chem. Res.* **2021**, *60*, 1137–1154. [CrossRef]

23. Krupa, I.; Cecen, V.; Boudenne, A.; Prokeš, J.; Novák, I. The mechanical and adhesive properties of electrically and thermally conductive polymeric composites based on high density polyethylene filled with nickel powder. *Mater. Des.* **2013**, *51*, 620–628. [CrossRef]

24. Li, X.-H.; Liu, P.; Li, X.; An, F.; Min, P.; Liao, K.-N.; Yu, Z.-Z. Vertically aligned, ultralight and highly compressive all-graphitized graphene aerogels for highly thermally conductive polymer composites. *Carbon* **2018**, *140*, 624–633. [CrossRef]

25. Xu, F.; Bao, D.; Cui, Y.X.; Gao, Y.Y.; Lin, D.; Wang, X.; Peng, J.W.; Geng, H.L.; Wang, H.Y. Copper nanoparticle-deposited graphite sheets for highly thermally conductive polymer composites with reduced interfacial thermal resistance. *Adv. Compos. Hybrid Mater.* **2021**. [CrossRef]

26. Yu, C.; Zhang, J.; Tian, W.; Fan, X.; Yao, Y. Polymer composites based on hexagonal boron nitride and their application in thermally conductive composites. *RSC Adv.* **2018**, *8*, 21948–21967. [CrossRef]

27. Ouyang, Y.; Bai, L.; Tian, H.; Li, X.; Yuan, F. Recent progress of thermal conductive ploymer composites: Al2O3 fillers, properties and applications. *Compos. Part A Appl. Sci. Manuf.* **2022**, *152*, 106685. [CrossRef]

28. Zhou, S.; Luo, W.; Zou, H.; Liang, M.; Li, S. Enhanced thermal conductivity of polyamide 6/polypropylene (PA6/PP) immiscible blends with high loadings of graphite. *J. Compos. Mater.* **2015**, *50*, 327–337. [CrossRef]

29. Guerra, V.; Wan, C.; McNally, T. Thermal conductivity of 2D nano-structured boron nitride (BN) and its composites with polymers. *Prog. Mater. Sci.* **2019**, *100*, 170–186. [CrossRef]

30. Lei, Y.; Liang, M.; Chen, Y.; Zhou, S.; Zou, H. Crystallization and thermal conductivity of poly(vinylidene fluoride)/boron nitride nanosheets composites. *Polym. -Plast. Technol. Mater.* **2020**, *59*, 1552–1561. [CrossRef]

31. Wieme, T.; Duan, L.; Mys, N.; Cardon, L.; D'Hooge, R.D. Effect of Matrix and Graphite Filler on Thermal Conductivity of Industrially Feasible Injection Molded Thermoplastic Composites. *Polymers* **2019**, *11*, 87. [CrossRef] [PubMed]

32. Fiske, T.; Gokturk, H.S.; Yazici, R.; Kalyon, D.M. Effects of flow induced orientation of ferromagnetic particles on relative magnetic permeability of injection molded composites. *Polym. Eng. Sci.* **1997**, *37*, 826–837. [CrossRef]

33. Zhou, S.; Hrymak, A.; Kamal, M. Electrical and morphological properties of microinjection molded polypropylene/carbon nanocomposites. *J. Appl. Polym. Sci.* **2017**, *134*, 45462. [CrossRef]

34. Yoo, Y.; Lee, H.L.; Ha, S.M.; Jeon, B.K.; Won, J.C.; Lee, S.-G. Effect of graphite and carbon fiber contents on the morphology and properties of thermally conductive composites based on polyamide 6. *Polym. Int.* **2014**, *63*, 151–157. [CrossRef]

35. Mazov, I.; Burmistrov, I.; Il'inykh, I.; Stepashkin, A.; Kuznetsov, D.; Issi, J.-P. Anisotropic thermal conductivity of polypropylene composites filled with carbon fibers and multiwall carbon nanotubes. *Polym. Compos.* **2015**, *36*, 1951–1957. [CrossRef]

36. Bai, Y.; Zhou, S.; Lei, X.; Zou, H.; Liang, M. Enhanced thermal conductivity of polycarbonate-based composites by constructing a dense filler packing structure consisting of hybrid boron nitride and flake graphite. *J. Appl. Polym. Sci.* **2022**, *139*, e52895. [CrossRef]

37. Zhou, S.; Wu, Y.; Zou, H.; Liang, M.; Chen, Y. Tribological properties of PTFE fiber filled polyoxymethylene composites: The influence of fiber orientation. *Compos. Commun.* **2021**, *28*, 100918. [CrossRef]

38. Chen, H.; Ginzburg, V.V.; Yang, J.; Yang, Y.; Liu, W.; Huang, Y.; Du, L.; Chen, B. Thermal conductivity of polymer-based composites: Fundamentals and applications. *Prog. Polym. Sci.* **2016**, *59*, 41–85. [CrossRef]

39. Lee, G.-W.; Park, M.; Kim, J.; Lee, J.I.; Yoon, H.G. Enhanced thermal conductivity of polymer composites filled with hybrid filler. *Compos. Part A Appl. Sci. Manuf.* **2006**, *37*, 727–734. [CrossRef]

40. Wang, Z.-G.; Gong, F.; Yu, W.-C.; Huang, Y.-F.; Zhu, L.; Lei, J.; Xu, J.-Z.; Li, Z.-M. Synergetic enhancement of thermal conductivity by constructing hybrid conductive network in the segregated polymer composites. *Compos. Sci. Technol.* **2018**, *162*, 7–13. [CrossRef]

41. Pan, C.; Kou, K.; Zhang, Y.; Li, Z.; Wu, G. Enhanced through-plane thermal conductivity of PTFE composites with hybrid fillers of hexagonal boron nitride platelets and aluminum nitride particles. *Compos. Part B Eng.* **2018**, *153*, 1–8. [CrossRef]

42. Bai, Y.; Shi, Y.; Zhou, S.; Zou, H.; Liang, M. A Concurrent Enhancement of Both In-Plane and Through-Plane Thermal Conductivity of Injection Molded Polycarbonate/Boron Nitride/Alumina Composites by Constructing a Dense Filler Packing Structure. *Macromol. Mater. Eng.* **2021**, *306*, 2100267. [CrossRef]

43. Liu, H.; Gu, S.; Cao, H.; Li, X.; Li, Y. A dense packing structure constructed by flake and spherical graphite: Simultaneously enhanced in-plane and through-plane thermal conductivity of polypropylene/graphite composites. *Compos. Commun.* **2020**, *19*, 25–29. [CrossRef]

44. Cao, H.; Gu, S.; Liu, H.; Li, Y. Disordered graphite platelets in polypropylene (PP) matrix by spherical alumina particles: Increased thermal conductivity of the PP/flake graphite composites. *Compos. Commun.* **2021**, *27*, 100856. [CrossRef]

45. Abbasi, S.; Carreau, P.J.; Derdouri, A. Flow induced orientation of multiwalled carbon nanotubes in polycarbonate nanocomposites: Rheology, conductivity and mechanical properties. *Polymer* **2010**, *51*, 922–935. [CrossRef]

46. Luo, W.; Cheng, C.; Zhou, S.; Zou, H.; Liang, M. Thermal, electrical and rheological behavior of high-density polyethylene/graphite composites. *Iran. Polym. J.* **2015**, *24*, 573–581. [CrossRef]

47. Li, Y.; Shimizu, H. High-shear processing induced homogenous dispersion of pristine multiwalled carbon nanotubes in a thermoplastic elastomer. *Polymer* **2007**, *48*, 2203–2207. [CrossRef]

48. Zhou, S.; Hrymak, A.N.; Kamal, M.R. Microinjection molding of polypropylene/multi-walled carbon nanotube nanocomposites: The influence of process parameters. *Polym. Eng. Sci.* **2018**, *58*, E226–E234. [CrossRef]

49. Wang, Y.; Zhang, W.; Feng, M.; Qu, M.; Cai, Z.; Yang, G.; Pan, Y.; Liu, C.; Shen, C.; Liu, X. The influence of boron nitride shape and size on thermal conductivity, rheological and passive cooling properties of polyethylene composites. *Compos. Part A Appl. Sci. Manuf.* **2022**, *161*, 107117. [CrossRef]

50. Hamidinejad, M.; Zandieh, A.; Lee, J.H.; Papillon, J.; Zhao, B.; Moghimian, N.; Maire, E.; Filleter, T.; Park, C.B. Insight into the Directional Thermal Transport of Hexagonal Boron Nitride Composites. *ACS Appl. Mater. Interfaces* **2019**, *11*, 41726–41735. [CrossRef]

51. Sun, N.; Sun, J.; Zeng, X.; Chen, P.; Qian, J.; Xia, R.; Sun, R. Hot-pressing induced orientation of boron nitride in polycarbonate composites with enhanced thermal conductivity. *Compos. Part A Appl. Sci. Manuf.* **2018**, *110*, 45–52. [CrossRef]

52. Kasgoz, A.; Akın, D.; Durmus, A. Rheological behavior of cycloolefin copolymer/graphite composites. *Polym. Eng. Sci.* **2012**, *52*, 2645–2653. [CrossRef]

53. Pötschke, P.; Abdel-Goad, M.; Alig, I.; Dudkin, S.; Lellinger, D. Rheological and dielectrical characterization of melt mixed polycarbonate-multiwalled carbon nanotube composites. *Polymer* **2004**, *45*, 8863–8870. [CrossRef]

54. Zheng, Q.; Zhang, X.-W.; Pan, Y.; Yi, X.-S. Polystyrene/Sn-Pb alloy blends. I. Dyn. Rheol. Behav. *J. Appl. Polym. Sci.* **2002**, *86*, 3166–3172.

55. Kim, H.S.; Bae, H.S.; Yu, J.; Kim, S.Y. Thermal conductivity of polymer composites with the geometrical characteristics of graphene nanoplatelets. *Sci. Rep.* **2016**, *6*, 26825. [CrossRef]

56. Osman, M.A.; Atallah, A. Effect of the particle size on the viscoelastic properties of filled polyethylene. *Polymer* **2006**, *47*, 2357–2368. [CrossRef]

Article

Development of an Injection Molding Process for Long Glass Fiber-Reinforced Phenolic Resins

Robert Maertens [1,2,*], Wilfried V. Liebig [1], Kay A. Weidenmann [3] and Peter Elsner [1,2,†]

1   Karlsruhe Institute of Technology (KIT), Institute for Applied Materials—Materials Science and Engineering (IAM-WK), Engelbert-Arnold-Str. 4, 73161 Karlsruhe, Germany; wilfried.liebig@kit.edu (W.V.L.); peter.elsner@ict.fraunhofer.de (P.E.)

2   Fraunhofer Institute for Chemical Technology ICT, Joseph-von-Fraunhofer-Str. 7, 76327 Pfinztal, Germany

3   Institute for Materials Resource Management MRM, Hybrid Composite Materials, Augsburg University, Am Technologiezentrum 8, 86159 Augsburg, Germany; kay.weidenmann@mrm.uni-augsburg.de

*   Correspondence: robert.maertens@kit.edu

†   The authors dedicate this work to the memory of our friend and colleague Prof. Dr.-Ing. Peter Elsner.

Abstract: Glass fiber-reinforced phenolic resins are well suited to substitute aluminum die-cast materials. They meet the high thermomechanical and chemical demands that are typically found in combustion engine and electric drive train applications. An injection molding process development for further improving their mechanical properties by increasing the glass fiber length in the molded part was conducted. A novel screw mixing element was developed to improve the homogenization of the long fibers in the phenolic resin. The process operation with the mixing element is a balance between the desired mixing action, an undesired preliminary curing of the phenolic resin, and the reduction of the fiber length. The highest mixing energy input leads to a reduction of the initial fiber length $L_0 = 5000$ μm to a weighted average fiber length of $L_p = 571$ μm in the molded part. This is an improvement over $L_p = 285$ μm for a short fiber-reinforced resin under comparable processing conditions. The mechanical characterization shows that for the long fiber-reinforced materials, the benefit of the increased homogeneity outweighs the disadvantages of the reduced fiber length. This is evident from the increase in tensile strength from $\sigma_m = 21$ MPa to $\sigma_m = 57$ MPa between the lowest and the highest mixing energy input parameter settings.

Keywords: thermoset injection molding; phenolic molding compound; fiber length measurement; long fiber processing; plasticizing work; injection work; screw mixing element; process data acquisition

Citation: Maertens, R.; Liebig, W.V.; Weidenmann, K.A.; Elsner, P. Development of an Injection Molding Process for Long Glass Fiber-Reinforced Phenolic Resins. *Polymers* **2022**, *14*, 2890. https://doi.org/10.3390/polym14142890

Academic Editors: Shengtai Zhou and Andrew N. Hrymak

Received: 4 July 2022
Accepted: 12 July 2022
Published: 16 July 2022

Publisher's Note: MDPI stays neutral with regard to jurisdictional claims in published maps and institutional affiliations.

1. Introduction

1.1. Phenolic Molding Compounds and Their Applications

Phenolic resins are versatile polymers that are used in a variety of industrial and consumer applications. For example, they are used as binding systems for wood composites, paper, abrasives, and friction materials [1]. They also serve as matrix systems in fiber-reinforced composite materials, such as continuous fiber-reinforced laminates [2], long fiber-reinforced compression molding materials [3], and short fiber-reinforced injection molding compounds [4].

Typical applications for phenolic molding compounds that are established in the state of the art are oil pump housings, intake manifolds, valve blocks [4,5], and other small automotive parts, for example in the air condition systems [6] or turbochargers [7]. In these applications, the beneficial properties of the phenolic resin, such as high heat resistance, chemical resistance, and an overall excellent dimensional accuracy and stability come into play. Usually, the mechanical strength of the material plays a subordinate role for these parts. In the latest developments, phenolic molding compounds are used in electric motors [8–10], larger parts in combustion engines (such as camshaft modules) [11], and entire parts of crank cases [12]. In addition to the beneficial properties mentioned above, the mechanical requirements become more important due to the size and the structural nature of such large parts.

The most significant mechanical disadvantage of parts made from phenolic molding compounds are their low elongation at break and their high brittleness compared to thermoplastic polymers [4]. Long fiber reinforcement is especially beneficial for increasing the impact strength of a fiber-reinforced polymer material. This was proven by Gupta et al. [13], Thomason and Vlug [14], Rohde et al. [15], and Kim et al. [16] for glass fiber-reinforced polypropylene. Boroson et al. [17] conducted a study with glass fiber-reinforced phenolic resins with initial fiber length values in the molding compound ranging between $L = 3.5$ mm and $L = 12.7$ mm and found out that longer fibers significantly reduce the notch sensitivity during impact testing. However, they did not measure the residual fiber length in the molded part. Based on the literature data, it can be concluded that an attractive development aim is increasing the fiber length in parts manufactured from phenolic molding compounds to improve the impact toughness.

Typically, there are two possibilities for increasing the fiber length in molded parts: First, using a semi-finished material, such as a long fiber granulate; or second, using a direct process in which longer fibers are incorporated into the final part. This decision-making process is a trade-off between the higher material costs of a semi-finished long fiber material and the more complex manufacturing processes of a direct process, most of which also involve a higher capital investment [18,19]. For fiber-reinforced thermoplastics, both process routes are well established and are available from multiple material and machinery equipment suppliers. However, in the case of phenolic molding compounds, neither long fiber-reinforced injection molding compounds nor established long fiber injection molding processes exist. Within this research paper, the development and the validation of a long fiber injection molding process for thermosetting phenolic resins is described.

1.2. Long Fiber Injection Molding Materials and Processes

Thermoplastic long fiber granulates are available in a variety of different lengths, typically ranging between $s = 6$ mm and $s = 25$ mm pellet length. Due to the pellet manufacturing process, the maximum initial fiber length, i.e., the fiber length before taking into account any process-induced fiber shortening, is limited to the size of the granulate [18]. The granulate size, in turn, is limited by the available dosing and feeding technology. For phenolic resin matrix systems, no long fiber granulates designed for injection molding are available on the market today. However, there are long fiber phenolic molding compounds for compression molding applications. They have a plate-like shape and are available in length classes of 5 mm, 12 mm, and 24 mm [20]. For compression molding applications, it is claimed that impact strength values that are 10 ... 20 times superior to conventional short fiber phenolic molding [4], but these values were never reached in injection molding trials by Saalbach et al. [21] and Raschke [22].

Several process variants with direct fiber feeding were developed for thermoplastic materials [19,23–33]. In the injection molding compounder (IMC{ XE "IMC" \t *"injection molding compounder"*}), a co-rotating twin screw extruder is combined with an injection unit. The process was patented by Putsch [34] and first industrialized by KraussMaffei Technologies GmbH. Continuous roving strands are pulled into the extruder, which is coupled with the discontinuous injection process using a melt buffer. Besides the increase in fiber length, the main advantage of the inline compounding process is lowering the material costs. Due to the high capital investment costs and the responsibility for the material formulation, it is mostly only used for high-volume applications [26]. Another inline compounding process was developed by Composite Products, Inc. (CPI{ XE "CPI" \t *"Composite Products, Inc."*}). It combines a continuous compounding process with a discontinuous injection process using a melt buffer [29]. The melting and compounding tasks are divided between two single-screw extruders. In contrast to the IMC process, the melt buffer and the injection unit are combined in one component. By using a check valve in the piston head, the compounding extruder can fill the backside of the piston head during the injection and holding phase. After the holding phase, the material can flow through the check valve to the other side of the piston head.

By conducting the melting and compounding discontinuously, matching the injection molding cycle, no melt buffer is required. This reduces the capital investment costs and makes the direct compounding feasible for lower-volume applications. The direct compounding injection molding (DCIM{ XE "DCIM" \t "*direct compounding injection molding*"}) process, invented by Exipnos and KraussMaffei Technologies, couples a single screw compounding extruder with a traditional injection molding machine [30]. In the compounding extruder, the molder can tailor the material according to his needs by adding fibers, fillers, and other additives to the thermoplastic polymer. In the DIF{ XE "DIF" \t "*direct incorporation of continuous fibers*"} process (direct incorporation of continuous fibers), invented by Truckenmüller at the University of Stuttgart, continuous fibers are directly pulled into the screw of the injection molding machine [23,25]. Mixing elements on the injection molding screw are required for obtaining good fiber dispersion. The mechanical properties of the produced samples are comparable to conventional long fiber granulate. Another direct process for the injection molding of long fiber-reinforced thermoplastics was developed by Arburg GmbH & Co. KG in cooperation with SKZ Kunststofftechnik GmbH. In this process, called fiber direct compounding (FDC{ XE "FDC" \t "*fiber direct compounding*"}), the unreinforced thermoplastic granulate is passively pulled into the screw and melted, such as in a conventional injection molding machine [31–33]. In contrast to the DIF process, the continuous fibers are cut to a selectable length of $L = 2$ mm ... 100 mm using a fiber chopper and are fed to the injection molding machine via a twin screw sidefeed. Since the injection molding is a discontinuous process, the fiber feed is coordinated with the screw movement via the machine control system. At the position of the fiber feed, the screw core diameter is reduced to facilitate the incorporation of the fibers.

1.3. Fiber Shortening and Fiber Length

Using the adhesion between fiber and matrix τ_{int}, the fiber diameter D, and the tensile strength of the fiber σ_F, the critical fiber length L_c can be calculated according to Equation (1) [35].

$$L_c = \frac{D\sigma_F}{2\,\tau_{int}}, \tag{1}$$

which is considered the minimum fiber length that is required for fully utilizing the reinforcement potential of the fibers. A fiber length $L < L_c$ still leads to a reinforcing effect, but does not fully utilize the available potential. Literature values for the critical fiber length L_c in glass fiber-reinforced phenolic molding compounds vary between $L_c = 2$ mm [36] and $L_c = 8$ mm [37].

During the injection molding process, the fibers are subjected to high mechanical loads, causing fiber damage and breaking. Three distinct mechanisms for fiber shortening are identified [15,38–40]. First, fluid–fiber interactions are caused by viscous forces transferred from the polymer matrix into the fibers. For example, Gupta et al. [41] found in their study on the fiber length reduction of glass fiber-reinforced polypropylene, that a thin polymer film is initially formed on the surface of the screw and barrel wall when the matrix is melted. In this region, fibers that are anchored on one side in solid granulate are exposed to the shear flows of the molten polymer, which can lead to flexural failure of the fibers. According to their calculations, forces can occur that lead to fiber damage by buckling. Second, fiber–fiber interactions can be caused by fiber overlap. The amount of fiber–fiber interactions increases with increasing fiber content and increasing fiber length [42]. At the junction points of two overlapping fibers, the contact forces cause bending deformation of the fibers, which might lead to fiber breakage. Third, fiber–wall interactions happen at contact locations to machine parts. This is visible by the abrasive wear that can be found on the screw, the barrel, and other machine parts.

Agglomerations and fiber bundles reduce the overall extent of the fiber shortening, resulting in a higher average fiber length compared to well-homogenized parts. Opening the fiber bundles works in the same way as breaking the fibers. Truckenmüller [43] investigated the opening of fiber bundles in the DIF process and concluded that fiber bundles can be treated as a single fiber with a larger fiber diameter and therefore a smaller L/D aspect ratio. This underlines the conclusion that fiber bundle opening is not possible without fiber shortening: Once the fiber bundle is opened, the aspect ratio of the individual fiber is significantly larger than the aspect ratio of the bundle from which the fiber originated. If the fluid forces are high enough for opening the fiber bundles, they likely will be high enough for shortening the individual fiber. An indicator for judging the existence of agglomerations and the degree of dispersion quality is the FLD ratio (fiber length distribution{ XE "FLD" \t "*fiber length distribution*"}) defined by Meyer et al. according to Equation (2) [44].

$$\text{FLD} = \frac{L_p}{L_n} \tag{2}$$

The average fiber length L_n and the weighted average fiber length L_p can be calculated according to Equations (3) and (4). L_i is the length of the individual fiber i. The weighted average fiber length is the second moment of the fiber length distribution and is generally considered to be more descriptive because it has a higher emphasis on long fibers [45].

$$L_n = \frac{\sum_{i=1}^n n_i L_i}{\sum_{i=1}^n n_i} \tag{3}$$

$$L_p = \frac{\sum_{i=1}^n n_i L_i^2}{\sum_{i=1}^n n_i} \tag{4}$$

Meyer et al. calculated a theoretical value of FLD = 1.44 for a fiber break in the middle due to viscous forces on the fibers. Once this value is reached, no further breakdown of the fibers due to fiber–fluid interactions shall occur.

1.4. Data Acquisition during the Injection Molding Process

With modern data acquisition technologies, the quantification of energy input into the polymer during plasticization and injection is possible. This is particularly important for reactive thermoset materials, such as phenolic resins. The screw torque during the plasticizing process M_{Plast} was monitored and analyzed by several authors. According to Rauwendaal [46], it is a good measure to quantify the mechanical power consumed by the extrusion process. For hydraulic injection molding machines, this plasticizing torque is typically calculated by measuring the pressure drop Δp_{Hydr} over the screw drive according to Equation (5).

$$M_{\text{Plast}} = \frac{\Delta p_{\text{Hydr}} \eta_{\text{Hydr}} V_{\text{Drive}}}{20\pi}, \tag{5}$$

Using the hydraulic efficiency η_{Hydr} and the hydraulic volume of the drive V_{Drive}, Scheffler et al. identified an initial decrease in plasticizing torque with rising moisture content for phenolic molding compounds, followed by an increase towards very high moisture content values [47]. The fundamental softening effect of water in the polymer is the same for thermoplastics and thermosets, which explains the initial decrease in plasticizing torque. However, due to the lack of a non-return valve, further increasing the moisture content leads to a higher backflow during the injection phase for the thermoset molding compounds, and consequently a higher number of fully filled screw flights. In those fully filled screw flights, the molding compound is agitated and mixed during the screw rotation, leading to the rise in plasticizing torque [47]. In general, Scheffler [48] concludes that the plasticizing torque for thermosetting molding compounds is influenced by multiple factors, but has a strong correlation to the backflow during the injection phase. Several authors [49,50] used the injection work W_{Plast}, which is the integral of the plasticizing power P_{Plast}, as a measure for the total energy input into the polymer during the plasticization phase, see Equation (6).

$$W_{\text{Plast}} = \int_{\text{Pl}_{\text{St}}}^{\text{Pl}_{\text{End}}} P_{\text{Plast}}\mathrm{d}t = \int_{\text{Pl}_{\text{St}}}^{\text{Pl}_{\text{End}}} (M_{\text{Plast}} \times \omega)\mathrm{d}t$$
$$= 2\pi \int_{\text{Pl}_{\text{St}}}^{\text{Pl}_{\text{End}}} (M_{\text{Plast}} \times n)\mathrm{d}t \tag{6}$$

In Equation (6), M_{Plast} is the plasticizing torque and n is the screw rotational speed. For a standard injection molding process using thermoplastic materials, Kruppa [49] observed an increase in the plasticizing work with increasing screw speed. This increased energy input leads to a stronger shortening of glass fibers, as described by Truckenmüller [43]. With increasing plasticizing work, fiber length asymptotically approaches a threshold value, which appears to be independent of initial fiber length and glass fiber content. A similar approach to quantify the energy input into the material during the injection phase of the process is the calculation of the injection work W_{Inj}, which is the integral of the injection force F_{Inj} over the injection distance s according to Equation (7).

$$W_{\text{Inj}} = \int_{\text{Inj}_{\text{St}}}^{\text{Inj}_{\text{End}}} F_{\text{Inj}}\mathrm{d}s = A_{\text{Piston}} \int_{\text{Inj}_{\text{St}}}^{\text{Inj}_{\text{End}}} p_{\text{Hydr.,Inj}}\mathrm{d}s \tag{7}$$

Lucyshyn et al. [51], as well as Schiffers [52], use the injection work as a measure for viscosity changes of thermoplastic polymers during the process, e.g., due to a change in moisture content. A higher moisture content leads to a lower viscosity and consequently to a lower injection work. The injection work is also used as a control parameter for the injection process by several authors. Woebcken [53] described a method to compensate for changes in the material and/or the machine and mold setup by adjusting the screw movement during injection to reach a specific, previously defined injection work value. Cavic [54] used the injection work for judging the reproducibility of the injection molding process. All cited works deal with thermoplastic materials. The usage of the injection work to evaluate the curing state of the material in the thermoset injection molding processes is not yet reported.

As outlined above, no process for the direct feeding of long glass fibers into the injection molding process for thermoset resins in general and for phenolic resins in particular exists in the state of the art. In this paper, the development of such a process is described. An essential part of the process development and a significant addition to the state of the art is the application of a novel screw mixing element for the injection molding of fiber-reinforced phenolic resins. The machine process data are analyzed and used for the process development by calculating the injection and plasticizing work. By means of the structural and mechanical properties of the molded parts, the long fiber injection molding process is evaluated and compared to the state-of-the-art processing of short glass fiber-reinforced phenolic molding compounds.

The methods and the results that are presented within this paper were partially published in earlier publications by the authors. In publication [55], the method development for the fiber length measurement is described in detail. For the present paper, this measurement method is used to generate the fiber length distribution results. The results themselves are not published yet. The publication [56] describes the process development for the twin screw extruder compounding of the short glass fiber-reinforced phenolic molding compounds. The compounds that were manufactured according to this method constitute the basis for the new experimental investigations and process data analyses that are presented here. In small extracts, the results of the material characterization were presented at conferences [57,58]. This paper is a comprehensive presentation of both the process development and the material characterization results.

2. Materials and Methods

2.1. Materials

The phenolic molding compound used within this work is based on the Vyncolit® X6952 short glass fiber-reinforced compound by Sumitomo Bakelite (Gent, Belgium). The material has a tailored composition of short glass fibers (SGF) and long glass fibers (LGF). The SGFs of the type DS5163-13P with a diameter of $D = 13$ µm were sourced from 3B fibreglass (Hoeilaart, Belgium) and added to the molding compound in fractions between $\phi = 0$ wt. $-$ % and $\phi = 30$ wt. $-$ % by a twin screw extruder compounding on a lab scale extruder with a screw diameter of $d = 27$ mm (Leistritz Extrusionstechnik GmbH, Nürnberg, Germany). The powdery resin components were melted in the first zones of the extruder by the barrel heating and by a screw kneading zone. Further downstream, the glass fibers were fed into the molten resin by using a sidefeed. A second kneading zone was used for opening the chopped fiber bundles. After leaving the extruder, the compound was cooled and granulated by using a cutting mill (Hosokawa Alpine, Augsburg, Germany). The method for controlling the energy input into the phenolic resin during the compounding, which was developed by the authors, is described in detail in the publication [56]. In publication [56], the detailed screw layout and the barrel temperature profile are presented. For the LGF, the $Tt = 2400$ tex direct roving 111AX11 with a filament diameter of $D = 17$ µm by 3B Fibreglass was used. The nomenclature of the material formulations follows the scheme PF-SGFx-LGFx. The variable x indicates the fiber weight content ϕ of the short or long glass fibers. Figure 1 gives an overview of the variations that were conducted. For selected material formulations, additional process and material variations were conducted. They are marked by the symbol and line styles in Figure 1.

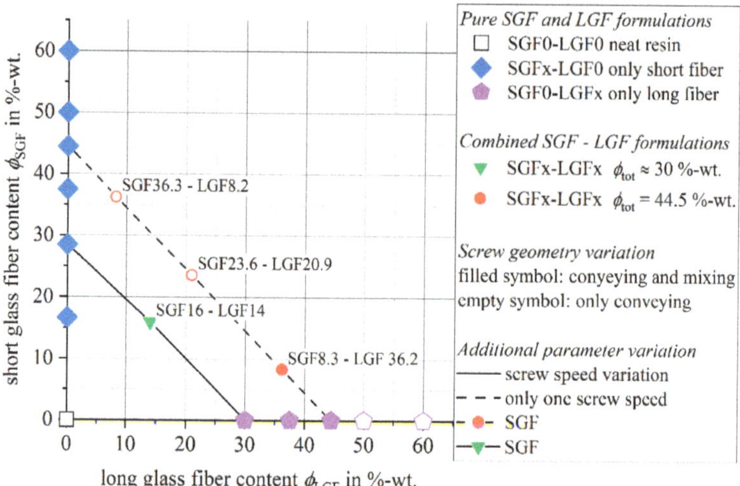

Figure 1. Material and process variations.

For this paper, the focus will be on material formulations with a total fiber content of $\phi_{tot} \approx 30$ wt.$-$%, which are positioned on the diagonal in the bottom left of Figure 1.

2.2. Long Fiber Thermoset Injection Molding Process

The long fiber thermoset injection molding process enables a flexible combination of SGF and LGF by separating the two mass flows (see Figure 2).

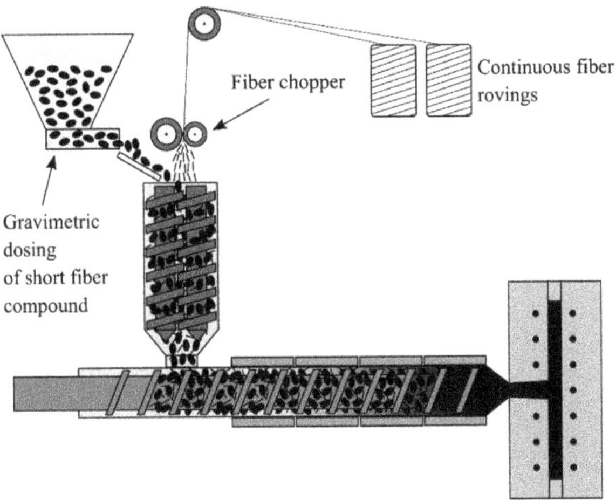

Figure 2. Process scheme of the long fiber thermoset injection molding process [57].

The SGF are gravimetrically fed as a part of the phenolic molding compound, whereas the LGF are chopped from the continuous rovings. Both mass flows are fed into the plasticizing unit with a twin screw sidefeed. The injection molding screw is a conveying screw with an interchangeable screw tip, which allows for the adaption of either a conventional conveying geometry or a newly designed, thermoset specific Maddock mixing element, which is shown in Figure 3a.

(a) thermoset-specific Maddock mixing screw tip

(b) cross-section drawing of the thermoset-specific Maddock mixing screw tip

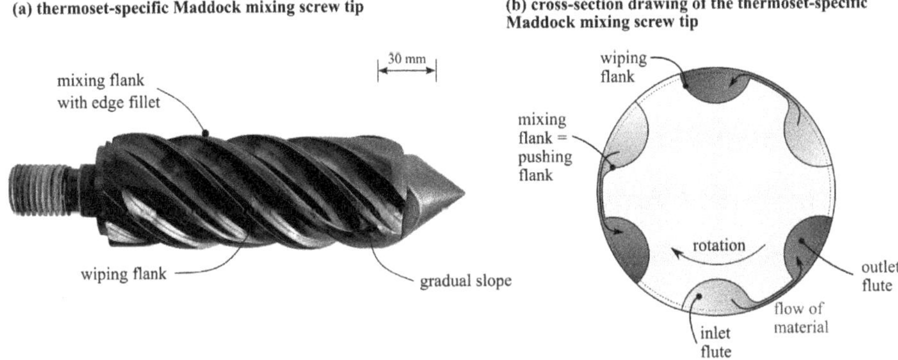

Figure 3. Thermoset-specific Maddock mixing element image (**a**) and cross section drawing (**b**).

The mixing element is set apart from a conventional Maddock mixing element by three distinct geometrical features. First, the inlet channels have a gradual slope at their ends to facilitate the flow of material and to avoid material accumulations. Second, the thermoset Maddock mixing element has an edge fillet on the mixing flight, which results in additional elongational stresses on the material when passing through the shear gap. The third main feature is the reversed positioning of the mixing flight and wiping flight compared to the state of the art; traditionally, the pushing flank is also the wiping flank of the mixing element. Since thermoset molding compounds only start to melt in the foremost screw flights under the influence of screw flank pressure, the material close to the pushing screw flanks is molten, whereas the material distant from the screw flanks might still be granular [59]. If the molding compound entered a traditional mixing element in such a state, the granular fraction would be pushed through the shear gap, possibly blocking it. The thermoset-specific design ensures that only molten material enters the shear gap.

2.3. Injection Molding Parameters

During all injection molding trials with the long fiber process variants, rectangular plates with a size of 190 mm × 480 mm and a thickness of h = 4 mm were molded. The plates were filled via a central sprue with a diameter of $d = 15$ mm by using a KraussMaffei 550/2000 GX injection molding machine (KraussMaffei Technologies GmbH, Munich, Germany), which has a screw diameter of $d = 60$ mm and a maximum clamping force of $F = 5500$ kN. The basic specifications are given in Table 1. After molding, all plates were post-cured according to the temperature cycle in Figure 4.

Table 1. Specifications of the KM 550/2000 GX injection molding machine.

Specification	Value	Unit
Screw diameter	60	mm
Max. plasticizing volume	792	cm^3
Number of cylinder heating zones	4	-
Max. injection pressure	2420	bar
Max. injection speed	848	cm^3/s
Clamping force	5500	kN

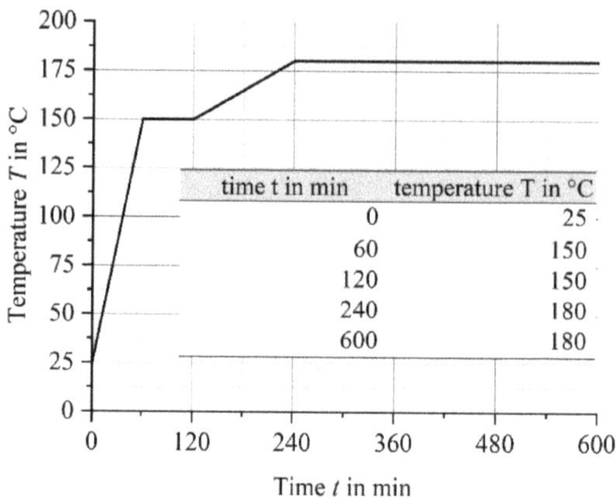

Figure 4. Post-cure cycle for molded plates.

2.4. Material Characterization

The test specimens were cut out of the molded and post-cured plates by waterjet cutting according to the cutting patterns shown in Figure 5a,b. A waterjet cutting machine, iCUT water SMART (imes-icore GmbH, Eiterfeld, Germany), was used.

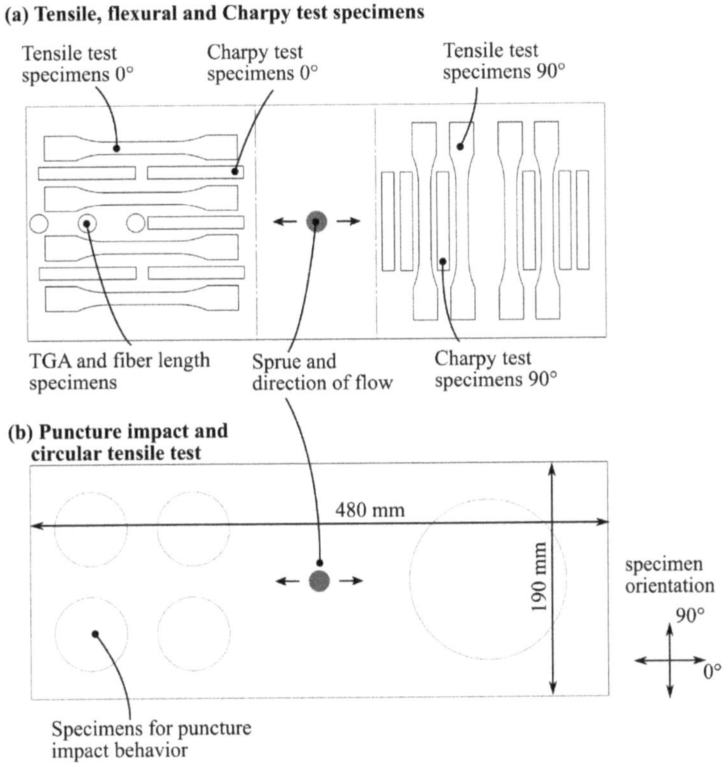

Figure 5. Cutting pattern for waterjet cutting of test specimens.

The quasistatic mechanical testing was carried out according to the technical standards of DIN EN ISO 527-2 (tensile testing) [60] and DIN EN ISO 6603-2 [61] (instrumented puncture impact testing). After the mechanical testing, the fracture surfaces were analyzed by using a Zeiss Supra 55VP instrument. For the overview images on the left side of Figure 13, an acceleration voltage of $U \approx 10$ kV and a working distance of WD ≈ 28 mm were used. The detail images on the left side of Figure 13 were obtained with $U \approx 3$ kV ... 5 kV and WD ≈ 7 mm.

The fiber length measurement procedure, as well as the validation investigations regarding fiber damaging and selectivity towards longer or shorter fibers, are described in detail by the authors in the publication [55]. In the first step of the measurement method, a circular sample with a diameter of $d = 25$ mm was extracted from the molded plate using waterjet cutting. A typical weight for such a sample is approximately $m = 3$ g. Subsequently, the phenolic matrix was removed by means of pyrolysis at $T = 650\,°C$ for a duration of $t = 36$ h under air atmosphere by using a LECO TGA 701 (St. Joseph, MI, USA). The ash residue, which solely consists of the dry glass fibers, was transferred into $V = 1.5$ L distilled water, and a small amount of acetic acid was added to support the fiber dispersion. The suspension was subjected to $t = 2$ min in an ultrasonic bath to open the fiber bundles. The fiber concentration in this suspension was too high for obtaining an analyzable image, which is why further dilution was necessary. By transferring the suspension into a dilution device for further down-sampling, this process can be conducted in a repeatable and controlled manner. The dilution device consists of a beaker glass with a capacity of $V = 4$ L and an outlet tap with a diameter of $d = 10$ mm attached to its side. A propeller stirrer keeps the fibers distributed homogeneously within the suspension.

The dilution and sample taking process steps are accomplished by opening the outlet tap and refilling the beaker with distilled water. Once the desired degree of dilution was reached, measurement samples were taken through the outlet tap and transferred to a Petri dish, which was then analyzed using the FASEP { XE "FASEP" \t *"fiber length measurement system (no abbreviation)"*} system by IDM systems (Darmstadt, Germany). The cropping of the image and thresholding were done manually, but the fiber detection was done automatically using the algorithms provided by the FASEP system. Per Petri dish, approximately $n = 3000$ (long fiber molding compound) to $n = 6000$ (short fiber molding compound) fibers were measured. To reduce the influence of the variation in the sample taking, it was repeated at least four times per specimen.

3. Results

3.1. Process Development

Figure 6 shows the injection pressure for a PF-SGF0-LGF30 formulation on the left side (a), and for a comparable short fiber formulation (PF-SGF28.5-LGF0) on the right side (b). Both materials were molded with both screw geometries. Using the conveying screw produces a pronounced pressure peak at the beginning of the injection stroke. The pressure requirement for the rest of the injection stroke is rather constant or slightly decreasing.

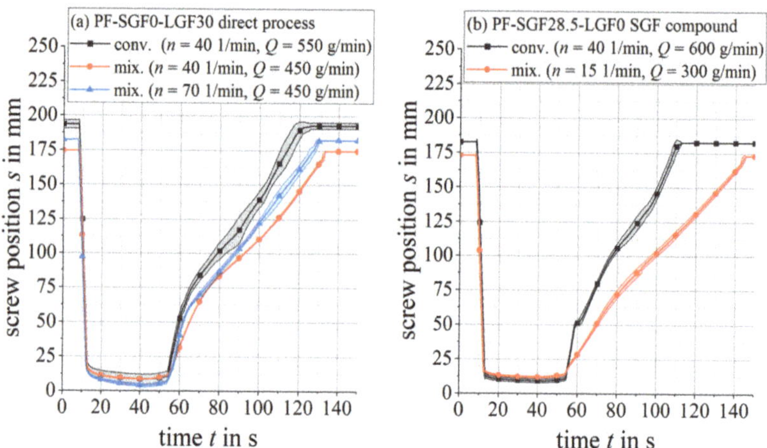

Figure 6. Injection pressure for LGF and SGF formulations with conveying and mixing screw.

In contrast, the mixing element has a significantly lower initial pressure peak and a lower pressure requirement during the filling phase. Towards the end of the injection stroke, the pressure rises sharply until the switchover point to the holding pressure is reached. When comparing the two different plasticizing screw speeds $n = 40$ 1/min and $n = 70$ 1/min for the PF-SGF0-LGF30 material formulation and the mixing element (Figure 7a), the higher screw speed results in a tendentially lower initial pressure peak. During the remaining injection stroke, no clear distinction between the two screw speeds can be observed.

For the same material formulations as above, Figure 7 shows the screw position during injection and plasticizing. The total material throughput Q, which is adjusted by the peripheral devices (gravimetric loss-in-weight feeder and fiber chopper), approaches as close to the maximum possible feeding rate as possible, so that the plasticizing time is minimized.

For the long fiber formulation PF-SGF0-LGF30 in Figure 7a, the mixing element leads to a smoother screw movement with less scattering than the conveying geometry. To a less pronounced extent, this is also valid for the short fiber formulation PF-SGF28.5-LGF0.

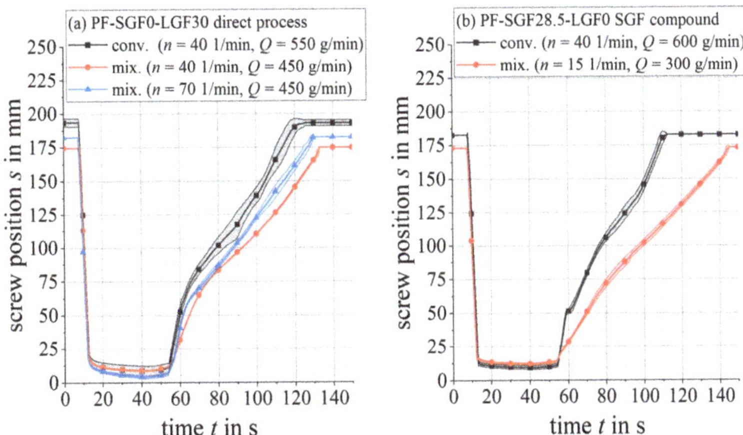

Figure 7. Screw position for LGF and SGF formulations with conveying and mixing screw.

The characteristic values plasticizing work and injection work are used for the evaluation of the process stability and to determine the process limits of the long fiber direct injection molding process. In contrast to a conventional injection molding process, in which the screw is flood fed by pulling the granulate out of the material hopper, the long fiber direct injection molding process offers the possibility to starve feed the screw due to the adjustability of the material feeding rate. The material throughput, and therefore the plasticizing time, is defined by the mass flow provided by the gravimetric dosing of the granulate and the cutting speed of the fiber chopper. It is independent of the screw speed, which means that the screw speed can be used as a parameter for influencing the mixing quality and the energy input into the material. Figure 8 shows the plasticizing and injection work for a parameter study by using a PF-SGF0-LGF30 material formulation and the screw mixing element. Over the course of 13 injection molding cycles, the screw speed was increased from $n = 30\ 1/\mathrm{min}$ to $n = 120\ 1/\mathrm{min}$.

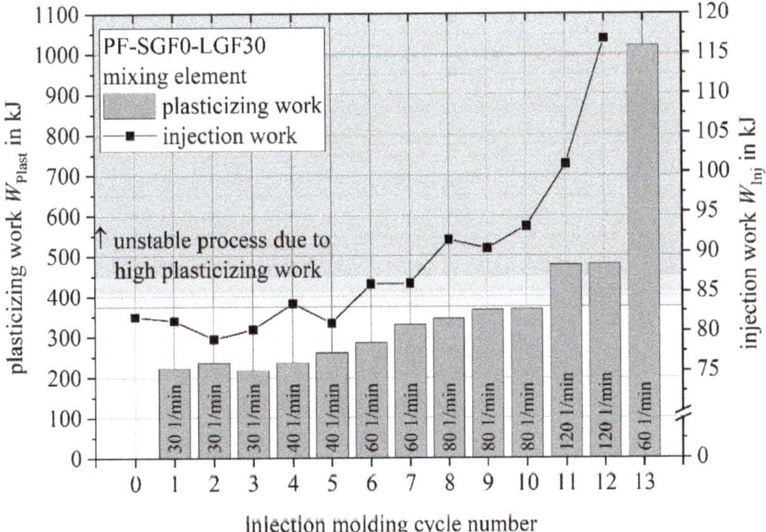

Figure 8. Plasticizing work and injection work for PF-SGF0-LGF30 screw speed study (usage and extension of the results presented in [57]).

For each injection molding cycle, Figure 8 shows the plasticizing work and the corresponding injection work. For the first cycle 0, the material was plasticized in manual mode, which is why no plasticizing work was recorded by the machine. The increase in plasticizing work with increasing screw speed is clearly visible. Up to a screw speed of n = 80 1/min, both work integrals remain stable at the respective screw speed increments. For the highest screw speed value n = 120 1/min, the injection work rises despite a constant plasticizing work. The cycle 12 was the last moldable part of this parameter study. Despite reducing the screw speed to n = 60 1/min after recognizing the instability of the process, the plasticizing work increased dramatically, and no injection was possible due to a curing of the material on the mixing element.

Figure 9 shows the plasticizing work and the injection work of a process stability study using screw speeds of n = 40 1/min and n = 70 1/min. For both parameter combinations, respectively, 10 (trial number 1) and 9 (trial number 2) injection molding cycles were performed after a stable process was established.

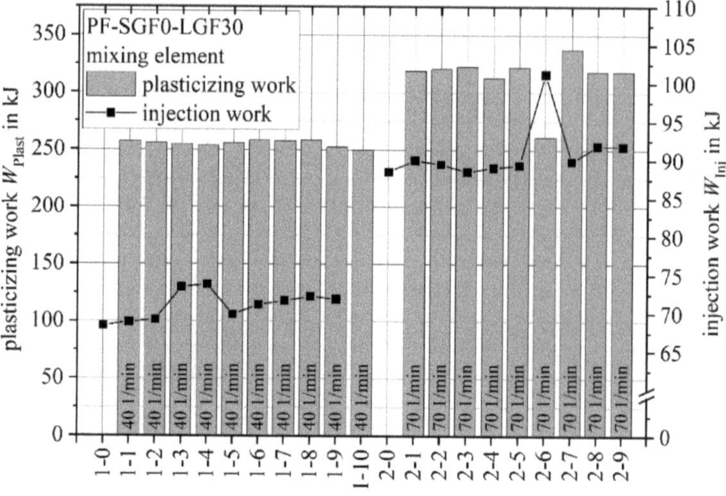

Figure 9. Plasticizing work and injection work for PF SGF0 LGF30 process stability study.

For both screw speeds, the plasticizing work is stable. The effect of the increased screw speed on the plasticizing work and the injection work is in accordance with the values measured during the parameter study shown in Figure 8.

The plasticizing work was analyzed for both the mixing element and the conveying screw for several material formulations (see Figure 10).

For the conveying screw geometry, a clear increase in plasticizing work with increasing fiber content is visible. This is valid for short glass fiber (SGF), long glass fiber (LGF) and combined (PF-SGFx-LGFx) material formulations. In contrast to the conveying screw geometry, no clear correlation between the fiber content and the plasticizing work can be drawn for the mixing element. The mixing element causes an overall significantly higher plasticizing work. Formulations containing long glass fibers require a significantly higher plasticizing work compared to formulations with only short glass fibers.

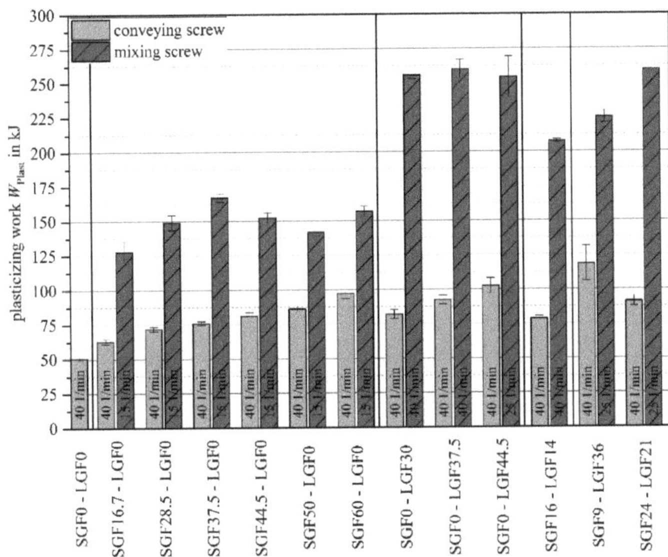

Figure 10. Plasticizing work for screw layout and material variations.

As noted above, the allowable material throughput had to be adjusted based on the material formulation. Lowering the throughput was required for higher fiber contents and longer fiber lengths. With both factors, the apparent density of the granulate–fiber dry blend decreases and the shear energy input during plasticizing increases. This means that less material is pulled into the screw per screw rotation (apparent density), and at the same time, the screw rotational speed must be decreased (to keep the shear energy input of the mixing element in a controllable range and to avoid overheating). Both aspects result in an increase in plasticizing time. For long glass fiber materials with $L = 5$ mm fiber length, up to $\phi = 60$ wt. $- \%$ is possible with the conveying element, whereas only $\phi = 44.5$ wt. $- \%$ can be molded with the mixing element before the plasticizing time exceeds the heating time. Plasticizing times that exceed the heating time are undesirable because of the long contact time of the machine nozzle to the hot mold.

3.2. Mechanical Properties

For evaluating the effect screw mixing element on the mechanical proper, the focus of this section will be on the formulations with a fiber content of $\phi = 30$ wt. $- \%$ and an initial long glass fiber length of $L = 5$ mm. Figure 11 shows the tensile strength parallel (0°) and perpendicular (90°) to the flow of material.

Switching from the conveying screw geometry to the Maddock mixing element significantly increases the tensile strength for all formulations and for both specimen orientations. Increasing the plasticizing screw speed when using the mixing element leads to a further increase in tensile strength for the PF-SGF0-LGF30 formulation in 0° orientation. For the other formulations and orientations, the change in tensile strength with increasing screw speed is within the standard deviation of the measurement. For most material formulation and process parameter combinations, the scattering of the measurement results also increases when using the mixing element. While the positive effect of the mixing element on the tensile strength is clearly visible from the measurement results, it must be noted that the overall highest absolute strength value for the formulations with a fiber content of $\phi = 30$ wt. $- \%$ is still reached by the short fiber material PF-SGF28.5-LGF0.

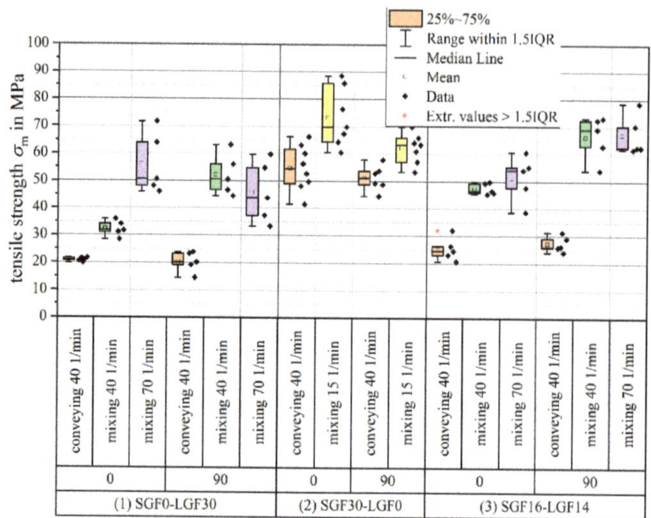

Figure 11. Tensile strength of 30 wt.-% specimens (usage and extension of the results presented in [57,58]).

The direct comparison of samples with a total fiber content of $\phi = 30$ wt. $-$ % shows that the formulations containing LGF profit the most from using the mixing element (see Figure 12).

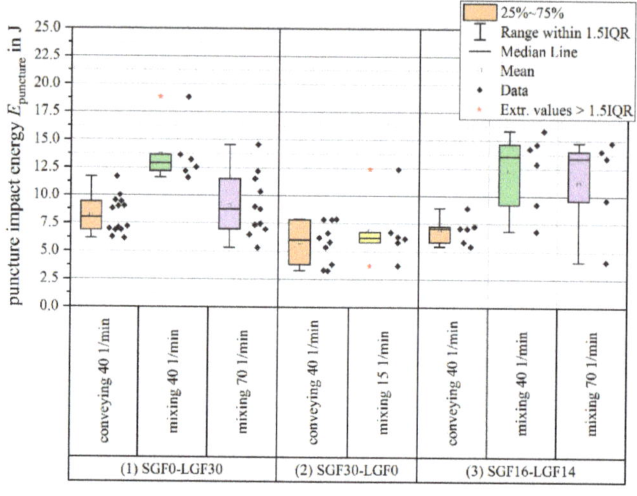

Figure 12. Puncture impact energy of 30 wt.-% specimens (usage and extension of the results presented in [57,58]).

Both for the PF-SGF0-LGF30 and the PF-SGF16-LGF14 material, the puncture impact energy increases significantly when using the mixing element. An increase in screw speed with the mixing element has no significant effect; the average value of puncture impact energy decreases, but within the scattering of the measurement. Compared to the SGF material, both formulations that contain LGF have significantly higher puncture impact energy when using the mixing element.

3.3. Scanning Electron Microscopy

Figure 13 shows a comparison between the PF-SGF28.5-LGF0 sample (conveying screw) and the PF-SGF0-LGF30 sample (with conveying screw and mixing element). The images show the fracture surface of unnotched Charpy impact test specimens with a specimen orientation parallel to the flow of material, e.g., the flow of material is oriented perpendicular to the plane of the image.

(a) PF-SGF28.5-LGF0 granulate conveying screw

(b) PF-SGF0-LGF30 direct process conveying screw

(c) PF-SGF0-LGF30 direct process mixing screw 70 1/min

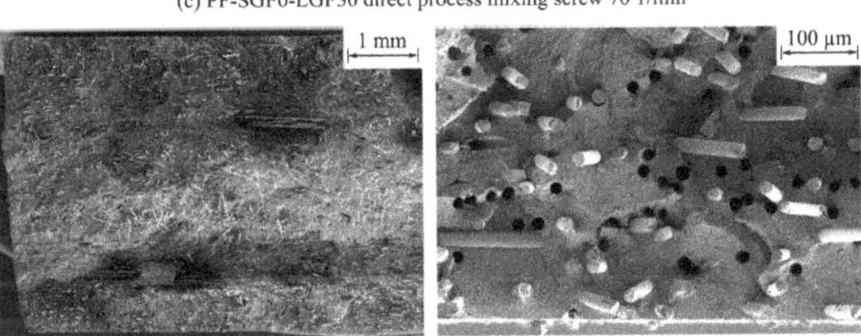

Figure 13. SEM for SGF granulate and LGF direct process mechanical testing specimens.

The SGF material in Figure 13a shows a skin and core layer structure with a predominant fiber orientation in the specimen direction (e.g., 0° to the flow) on the two surfaces of the specimen and a predominantly perpendicular fiber orientation (e.g., 90° to the flow of material) in the core layer. The fibers are pulled out of the fracture surface with some resin residues on the face sides of the fibers. The holes in the matrix created by pulling out the fibers are frayed and irregular.

Comparing the LGF specimens manufactured with the conveying screw in Figure 13b to the specimens molded with mixing element in Figure 13c shows a strong reduction in the number of fiber bundles. The specimen manufactured with the conveying screw has a very inhomogeneous fracture surface with fiber-rich bundle regions and resin-rich regions where almost no fibers are present. By using the mixing element, the number of bundles is reduced significantly and a more homogeneous distribution of the fibers across the sample is achieved. Additionally, a skin and core layer structure becomes visible, which is not detectable with the conveying screw setup. The overall visual impression suggests that the fibers were shortened by using the mixing element.

Analyzing the detail SEM images on the right side of Figure 13b,c shows that the fiber surfaces are blank and smooth. Fibers that are pulled out of the fracture surface leave sharp and well-defined holes in the matrix. The mixing element improves the dispersion of the fibers and reduces the number of bundles, but no difference is observed regarding the resin residue on the fibers. It is remarkable that a high number of fibers have resin residues on their face sides; this indicates that the resin adhesion on the face sides is significantly stronger than on the fiber circumference.

3.4. Fiber Length Measurement

The fiber length measurement results are shown in Figures 14 and 15. For all measurements, the initial fiber length L_0, the weighted average fiber length in the part L_p, and the quotient FLD $= L_p / L_n$ are given.

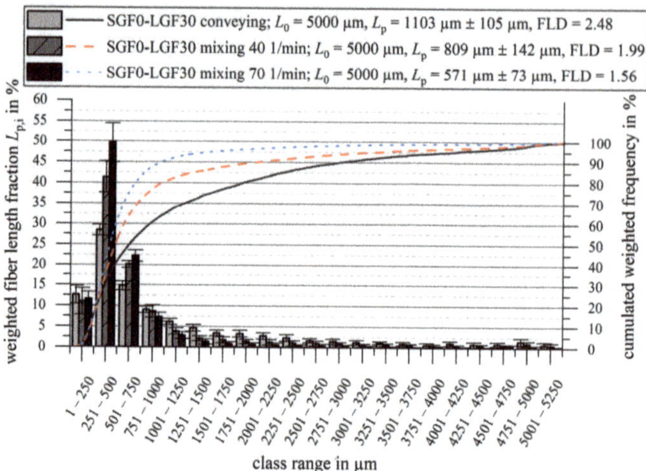

Figure 14. Fiber length measurement results for PF-SGF0-LGF30 (usage and extension of the results presented in [57,58]).

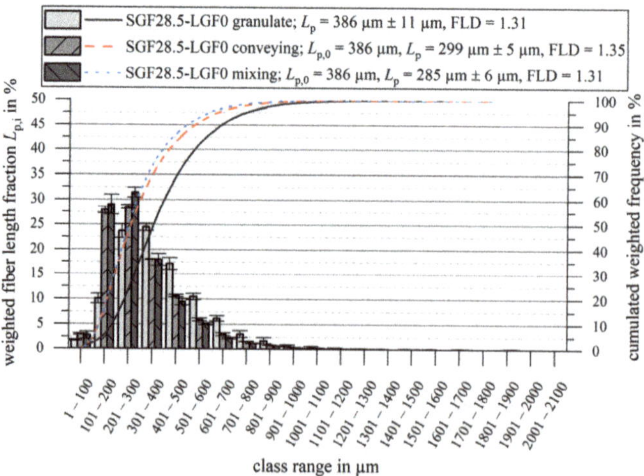

Figure 15. Fiber length measurement results for PF-SGF28.5-LGF0.

For PF-SGF0-LGF30, the weighted average fiber length in the molded part is reduced from $L_p = 1103$ µm for the conveying screw tip to $L_{p,\,40\ 1/\text{min}} = 809$ µm and to $L_{p,\,70\ 1/\text{min}} = 571$ µm for the mixing element at $n = 40\ 1/\text{min}$ and $n = 70\ 1/\text{min}$, respectively. The fiber length measurement results show that the frequency of fibers in the length classes from $L = 251$ µm to $L = 750$ µm is the highest for the mixing element with the high plasticizing screw speed. For all classes $L > 751$ µm, the conveying screw tip results in the highest fraction. The cumulative frequency curves confirm these results. The ratio $\text{FLD} = L_p/L_n$ is also reduced when using the mixing element.

When using a pure SGF compound, the fiber shortening is less pronounced. For both screw geometries, a fiber shortening of $\Delta L_p \approx 80$ µm$\ldots 100$ µm compared to the granulate takes place. No significant change in the ratio $\text{FLD} = L_p/L_n$ occurs during the injection molding process.

4. Discussion

4.1. Process Development

It was possible to run a stable injection molding process with the developed thermoset-specific mixing element. In comparison to the conventional conveying screw geometry, the process value scatter, both during the injection and the plasticizing phase of the process, is reduced. The reduced process value scatter is especially visible for material formulations containing long fibers. It is assumed that the shear gaps of the mixing element reduce the backflow during the injection phase due to the flow resistance being higher, as compared to the standard conveying screw geometry. This assumption is in accordance with the findings of Kruppa et al. [62,63], who found that the small gaps of their mixing element act comparable or superior to a standard non-return valve.

The increased plasticizing work input by the mixing element is visible from the reduced pressure peak at the beginning of the injection stroke, which is typically attributed to the cold plug in the nozzle. The lack of this pressure peak indicates that the material in front of the screw is hotter and consequently has a lower viscosity and a less pronounced cold plug. The increase in injection pressure requirement towards the end of the stroke can be attributed to the advanced chemical reaction progress, which in turn results in an increased viscosity.

It is deducted that the effect of backflow prevention accomplished by the mixing element outweighs the backflow-increasing effect of the lower material viscosity. To investigate the lower process scattering with the mixing element during the plasticizing phase of the injection molding process, the back pressure and the screw position during plasticizing is analyzed (see Figure 16).

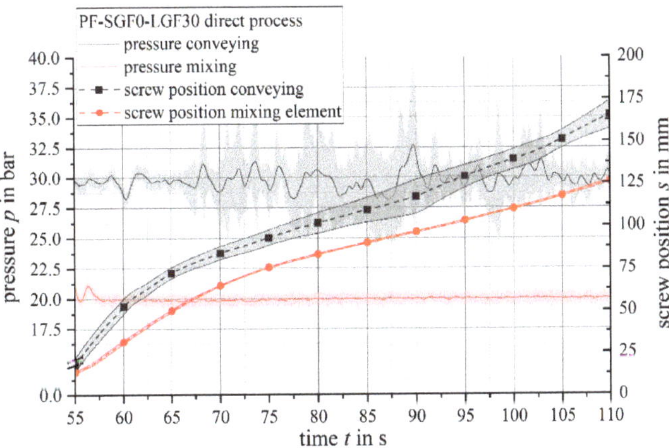

Figure 16. Pressure and screw position during plasticization of PF-SGF0-LGF30 material.

With the conveying screw geometry, the pressure that is required for melting and homogenizing the compound is applied by the injection molding machine's hydraulic system. Especially for long fiber materials with a low apparent density, the injection molding machine's hydraulic system had difficulties maintaining a constant back pressure. It is assumed that this scatter in back pressure resulted in the unsteady backward screw movement. When using the mixing element, the pressure for melting is applied geometrically by the reduction of the flow channel cross section in the shear gap of the element [64]. The back pressure of the injection molding machine is only applied to compact the already molten material in front of the screw. Consequently, the machine was able to maintain a much more stable back pressure level and a steadier backward screw movement.

Based on the parameter study findings, an upper plasticizing work limit of $W_{\mathrm{Plast}} \approx 400$ kJ was set. The available data does not allow for a sharp distinction between a stable and an unstable process. This plasticizing work limit corresponds to a screw speed limit of $n \approx 80$ 1/min ... 90 1/min for a PF-SGF0-LGF30 formulation.

From the plasticizing times that were recorded with the mixing element and the conveying screw, it is concluded that the short glass fiber formulations PF-SGFx-LGF0 can be molded up to a fiber content of $\phi = 60$ wt. $-$ % with the mixing element. The underlying assumption for this conclusion is that a plasticizing time that exceeds the heating time is not acceptable. For long glass fiber materials with $L = 5$ mm fiber length, up to $\phi = 60$ wt. $-$ % is possible with the conveying element, whereas only $\phi = 44.5$ wt. $-$ % can be molded with the mixing element.

4.2. Structure and Properties

Switching from the conveying screw to the mixing element increases the plasticizing work input, which leads to a stronger shortening of the fibers. In this regard, no difference to thermoplastics was found. Most studies of thermoplastics state that the fiber length in the molded part decreases with increasing screw speed, which is represented by the increased plasticizing work. Moritzer and Bürenhaus [65] confirmed this for PP-GF, Lafranche et al. drew the same conclusions for PA66-GF [66,67] { XE "PA66" \t *"polyamide 6.6 with glass fibers"*}. As an exception to the general consensus, Rohde et al. [15] only found a slight, but not statistically significant shortening effect of the screw speed for PP-GF. While a stronger fiber shortening for higher plasticizing work values is observed for the long fiber formulations, this is not the case for the pure short fiber formulation PF-SGF28.5-LGF0. No significant additional shortening of the fibers compared to the conventional conveying screw is observed. Since the ratio FLD $= L_p / L_n$ is also unaffected, it is concluded that the slight fiber shortening is caused by abrasive wear on the machine surfaces and not by breakage due to fluid forces.

In the SEM fracture surface images, a reduction of fiber bundle size and count is observed when using the mixing element. The very small amount of resin residue on the fibers indicates a very weak fiber–matrix adhesion for the specimens manufactured in the long fiber direct process. For manufacturing those samples, the LGF of the type 111AX11 [68], with a 2400 tex roving size, were used. The fiber type was chosen based on recommendations by the fiber and the resin suppliers. In fact, the same fiber type 111AX in a 1200 tex size is used for manufacturing the commercially available long fiber granulate PF-SGF0-LGF55 [69]. For this reason, the weak fiber–matrix adhesion for all the long fiber specimens is surprising. The SEM investigations also show the typical three-layer setup of the parts, consisting of two skin layers with fibers oriented in the direction of the flow and a core layer with perpendicular fiber orientation [70]. This structure could be observed for the materials with a high homogeneity. If many bundles are present, no distinct skin and core structure is visible. Instead, fiber bundles with a predominantly perpendicular orientation characterize the structure.

It is remarkable that using the screw mixing element leads to a higher fraction of the skin layer, which consequently leads to a higher fraction of fibers oriented parallel to the direction of the material flow. The process development results show that the mixing element and the screw speed both increase the plasticizing work, i.e., the energy input into the resin during plasticization. This likely results in a further advance in the resin's curing process. The formation of the skin layer happens by the incremental curing of the resin on the surface [71]. Englich [70] found out that higher mold temperatures and longer injection times result in a thicker skin layer, because the incremental curing on the mold surface is either quicker (higher mold temperature) or has more time (longer injection time). For this reason, the conclusion that a higher plasticizing work also causes a thicker skin layer is drawn, because the resin's curing is progressed further, which consequently facilitates the incremental final curing on the hot mold surface. The skin layer fraction is also plausibly the reason why the mechanical properties of the SGF and LGF profit more in $0°$ orientation than in $90°$ orientation when using the mixing element.

The results of the puncture impact testing clearly show that the absorbed impact energy increases with increasing LGF content and is further improved by using the mixing element. The puncture impact specimens have a diameter of $d = 60$ mm, which means that a large area is mechanically stressed during the testing. The LGF are likely better capable of distributing the load into this bigger area than the SGF.

5. Conclusions

With the developed long fiber direct thermoset injection molding process, a significant increase in the fiber length in the molded parts was achieved. The average weighted fiber length was up to four times higher than the state-of-the-art short fiber-reinforced phenolic resins, resulting in values in the range of $L_p = 500$ μm ... 1100 μm. While this is a significant improvement, it remains significantly below the literature values for the critical fiber length of $L_c = 2$ mm ... 8 mm for the combination of phenolic resin and glass fibers. A strong fiber shortening in the long fiber direct thermoset injection molding process from the initial fiber length $L_0 = 5000$ μm down to a weighted average fiber length in the range of $L_p = 500$ μm ... 1100 μm was observed.

In terms of the specific processing challenges presented for the long fiber thermoset injection molding process, the plasticizing work input by the screw mixing element must be carefully controlled to avoid overheating and curing. The characterization results show that in the conflicting field between fiber dispersion and fiber length, the focus must be on the good fiber dispersion in the phenolic matrix. Every mechanical property characteristic that was investigated was increased by an improved material homogeneity, despite the accompanying fiber shortening. Besides the glass fiber length, other structural characteristics, such as the distinctiveness of the skin and core layer structure, are influenced by the mixing element. The additional mixing energy input leads to a clearer development of this structure, as well as to a tendentially higher skin layer fraction, which in turn leads to a higher fraction of fibers oriented parallel to the direction of the flow.

The poor fiber–matrix adhesion reduces the expressiveness of some results. Despite those restrictions, it is concluded that the central key to good mechanical properties is a homogeneous distribution of the fibers in the phenolic resin matrix. This homogenization is more important than the fiber length. With the successful process development and the invention of the screw mixing element for thermoset molding compounds, future research studies will now focus on more detailed material and process parameter variations.

Author Contributions: Conceptualization, R.M., W.V.L. and K.A.W.; methodology, R.M., W.V.L. and K.A.W.; investigation, R.M.; resources, W.V.L., K.A.W. and P.E.; data curation, R.M.; writing—original draft reparation, R.M.; writing—review and editing, R.M., W.V.L., K.A.W. and P.E.; visualization, R.M.; supervision, W.V.L., K.A.W. and P.E.; project administration, K.A.W. and P.E.; funding acquisition, K.A.W. and P.E. All authors have read and agreed to the published version of the manuscript.

Funding: This research was funded by the Deutsche Forschungsgemeinschaft (DFG, German Research Foundation), grant number EL 470/9-1 and WE 4273/10-1.

Institutional Review Board Statement: Not applicable.

Informed Consent Statement: Not applicable.

Polymers **2022**, *14*, 2890

Data Availability Statement: The data presented in this study are available in the article.

Acknowledgments: The phenolic molding compound was provided by Sumitomo Bakelite Co., Ltd. (Vyncolit N.V., Gent, Belgium). We acknowledge support by the KIT-Publication Fund of the Karlsruhe Institute of Technology.

Conflicts of Interest: The authors declare no conflict of interest. The funders had no role in the design of the study, in the collection, analyses, or interpretation of data, in the writing of the manuscript, or in the decision to publish the results.

References

1. Pilato, L. Introduction. In *Phenolic Resins: A Century of Progress*; Pilato, L.A., Ed.; Springer: Berlin/Heidelberg, Germany, 2010; pp. 1–8. ISBN 978-3-642-04713-8.
2. Chawla, K.K. *Composite Materials: Science and Engineering*, 4th ed.; Springer: New York, NY, USA, 2019; ISBN 978-3-030-28983-6.
3. Swentek, I.; Ball, C.A.; Greydanus, S.; Nara, K.R. Phenolic SMC for Fire Resistant Electric Vehicle Battery Box Applications. In *SAE Technical Paper Series, Proceedings of the WCX SAE World Congress Experience, Detroit, MI, USA, 21–23 April 2020*; SAE International: Warrendale, PA, USA, 2020; pp. 1–5.
4. Koizumi, K.; Charles, T.; Keyser, H.D. Phenolic Molding Compounds. In *Phenolic Resins: A Century of Progress*; Pilato, L.A., Ed.; Springer: Berlin/Heidelberg, Germany, 2010; pp. 383–437. ISBN 978-3-642-04713-8.
5. Ball, C. Phenolic molding compounds in automotive powertrain applications. In Proceedings of the 17th Annual Automotive Conference & Exhibition, Novi, MI, USA, 6–8 September 2017; pp. 1–11.
6. Beran, T.; Hübel, J.; Maertens, R.; Reuter, S.; Gärtner, J.; Köhler, J.; Koch, T. Study of a polymer ejector design and manufacturing approach for a mobile air conditioning. *Int. J. Refrig.* **2021**, *126*, 35–44. [CrossRef]
7. Keyser, H.D. Carbon fibre/phenolic parts for engines. *Adv. Compos. Bull.* **2004**, *12*, 12–13.
8. Reuter, S.; Berg, L.F.; Doppelbauer, M. Performance evaluation of a high-performance motor with thermoset molded internal cooling. In Proceedings of the 2021 11th International Electric Drives Production Conference (EDPC), Erlangen, Germany, 7–9 December 2021; IEEE: New York, NY, USA, 2021; pp. 1–5, ISBN 978-1-6654-1809-6.
9. Reuter, S.; Sorg, T.; Liebertseder, J.; Doppelbauer, M. Design and Evaluation of a Houseless High-Performance Machine with Thermoset Molded Internal Cooling. In Proceedings of the 2021 11th International Electric Drives Production Conference (EDPC), Erlangen, Germany, 7–9 December 2021; IEEE: New York, NY, USA; 2021; pp. 1–6, ISBN 978-1-6654-1809-6.
10. Reuter, S.; Keyser, H.D.; Doppelbauer, M. Performance study of a PMSM with molded stator and composite housing with internal slot cooling–resulting in high continuous power. In Proceedings of the Dritev-Drivetrain for Vehicles, Bonn, Germany, 27–28 June 2018; pp. 17–31.
11. Schindele, K.; Sorg, T.; Hentschel, T.; Liebertseder, J. Lightweight camshaft module made of high-strength fiber-reinforced plastic. *MTZ Worldw.* **2020**, *81*, 26–31. [CrossRef]
12. Jauernick, M.; Pohnert, D.; Kujawski, W.; Otte, R. Hybrid lightweight cylinder crankcase: Challenges and feasibility. *MTZ Worldw.* **2019**, *80*, 70–75. [CrossRef]
13. Gupta, V.B.; Mittal, R.K.; Sharma, P.K.; Mennig, G.; Wolters, J. Some Studies on Glass Fiber-Reinforced Polypropylene: Part II: Mechanical Properties and Their Dependence on Fiber Length, Interfacial Adhesion, and Fiber Dispersion. *Polym. Compos.* **1989**, *10*, 16–27. [CrossRef]
14. Thomason, J.L.; Vlug, M.A. The Influence of fibre length and concentration on the properties of glass fibre-reinforced polypropylene: 4. Impact properties. *Compos. Part A Appl. Sci. Manuf.* **1997**, *28*, 277–288. [CrossRef]
15. Rohde, M.; Ebel, A.; Wolff-Fabris, F.; Altstädt, V. Influence of Processing Parameters on the Fiber Length and Impact Properties of Injection Molded Long Glass Fiber Reinforced Polypropylene. *IPP* **2011**, *26*, 292–303. [CrossRef]
16. Kim, Y.; Park, O.O. Effect of Fiber Length on Mechanical Properties of Injection Molded Long-Fiber-Reinforced Thermoplastics. *Macromol. Res.* **2020**, *28*, 433–440. [CrossRef]
17. Boroson, M.L.; Fitts, B.B.; Rice, B.A. Advances in Toughening of Phenolic Composites. In *SAE Technical Paper Series, Proceedings of the International Congress & Exposition, Detroit, MI, USA, 25 February 1991*; SAE International: Worrendale, PA, USA, 1991; pp. 1–10.
18. Ning, H.; Lu, N.; Hassen, A.A.; Chawla, K.K.; Selim, M.; Pillay, S. A review of Long fibre thermoplastic (LFT) composites. *Int. Mater. Rev.* **2020**, *65*, 164–188. [CrossRef]
19. Schemme, M. LFT–Development status and perspectives. *Reinf. Plast.* **2008**, *52*, 32–39. [CrossRef]
20. Sumitomo Bakelite Europe n.v. Technical Data Sheet Porophen®GF9201L12a, Gent. 2019. Available online: https://www.sbhpp.com/products-applications/catalog/item/porophen-gf-9202 (accessed on 11 July 2022).
21. Saalbach, H.; Maenz, T.; Englich, S.; Raschke, K.; Scheffler, T.; Wolf, S.; Gehde, M.; Hülder, G. *Faserverstärkte Duroplaste Für Die Großserienfertigung Im Spritzgießen: Ergebnisbericht des BMBF-Verbundprojektes FiberSet*; KraussMaffei Technologies GmbH: Munich, Germany, 2015.
22. Raschke, K. Grundlagenuntersuchungen zur Prozess-und Struktursimulation von Phenolharzformmassen mit Kurz-und Lang-glasfaserverstärkung. Original Language: German. Translated Title: Fundamental Investigations on Process and Structure Simulation of Phenolic Resin Molding Compounds with Short And Long Glass Fiber Reinforcement. Ph.D. Dissertation, Technische Universität Chemnitz, Chemnitz, Germany, 2017.

23. Truckenmüller, F.; Fritz, H.-G. Injection Molding of Long Fiber-Reinforced Thermoplastics: A Comparison of Extruded and Pultruded Materials with Direct Addition of Roving Strands. *Polym. Eng. Sci.* **1991**, *31*, 1316–1329. [CrossRef]
24. McLeod, M.; Baril, É.; Hétu, J.-F.; Deaville, T.; Bureau, M.N. Morphological and mechanical comparision of injection and compression moulding in-line compounding of direct long fibre thermoplastics. In Proceedings of the ACCE 2010, Troy, NY, USA, 15–16 September 2010; pp. 109–118, ISBN 9781618390240.
25. Truckenmüller, F. Direct Processing of Continuous Fibers onto Injection Molding Machines. *J. Reinf. Plast. Comp.* **1993**, *12*, 624–632. [CrossRef]
26. Markarian, J. Long fibre reinforced thermoplastics continue growth in automotive. *Plast. Add. Comp.* **2007**, *9*, 20–24. [CrossRef]
27. Roch, A.; Huber, T.; Henning, F.; Elsner, P. LFT foam: Lightweight potential for semi-structural components through the use of long-glass-fiber-reinforced thermoplastic foams. In Proceedings of the PPS-29 29th International Conference of the Polymer Processing Society, Nürnberg, Germany, 15–19 July 2013; Altstadt, V., Keller, J.-H., Fathi, A., Eds.; AIP Publishing: Melville, NY, USA, 2014; pp. 471–476.
28. Lohr, C.; Beck, B.; Henning, F.; Weidenmann, K.A.; Elsner, P. Mechanical properties of foamed long glass fiber reinforced polyphenylene sulfide integral sandwich structures manufactured by direct thermoplastic foam injection molding. *Compos. Struct.* **2019**, *220*, 371–385. [CrossRef]
29. Weber, C.; Ledebuhr, S.; Enochs, R.; Busch, J. A Novel, New Direct Injection Technology for In-Line Compounding and Molding of LFT Automotive Structures. In Proceedings of the 2002 SAE International Body Engineering Conference and Automotive & Transportation Technology Congress, Paris, France, 9 July 2002; pp. 1–4.
30. Hirsch, P.; Menzel, M.; Klehm, J.; Putsch, P. Direct Compounding Injection Molding and Resulting Properties of Ternary Blends of Polylactide, Polybutylene Succinate and Hydrogenated Styrene Farnesene Block Copolymers. *Macromol. Symp.* **2019**, *384*, 1–6. [CrossRef]
31. Heidenmeyer, P.; Deubel, C.; Kretschmer, K.; Schink, K. Spritzgießanlage und Spritzgießverfahren zur Herstellung von Faserverstärkten Kunststoffteilen. Original language: German. Translated Title: Injection Molding Machine and Injection Molding Method for the Production of Fiber-Reinforced Plastic Parts. Patent number DE102012217586A1, 27 March 2014.
32. Holmes, M. Expanding the market for long fiber technology. *Reinf. Plast.* **2018**, *62*, 154–158. [CrossRef]
33. Keck, B. Economical Alternative to High-Performance Polymer: Ros Reduces Costs and Part Weight with Fiber Direct Compounding. *Kunststoffe Int.* **2017**, *141*, 39–41.
34. Putsch, P. Compounding-Injection Moulding Process and Device. Patent WO1992000838A1, 23 January 1992.
35. Kelly, A.; Tyson, W.R. Tensile Properties of Fibre-Reinforced Metals: Copper/Tungsten and Copper/Molybdenum. *J. Mech. Phys. Solids* **1965**, *13*, 329–350. [CrossRef]
36. Chen, F.; Tripathi, D.; Jones, F.R. Determination of the Interfacial Shear Strength of Glass-Fibre-Reinforced Phenolic Composites by a Bimatrix Fragmentation Technique. *Compos. Sci. Technol.* **1996**, *56*, 609–622. [CrossRef]
37. Gore, C.R.; Cuff, G. Long-Short Fiber Reinforced Thermoplastics. In Proceedings of the ANTEC 1986, Boston, MA, USA, 28 April–1 May 1986; pp. 47–50.
38. von Turkovich, R.; Erwin, L. Fiber Fracture in Reinforced Thermoplastic Processing. *Polym. Eng. Sci.* **1983**, *23*, 743–749. [CrossRef]
39. Mittal, R.K.; Gupta, V.B.; Sharma, P.K. Theoretical and Experimental Study of Fibre Attrition During Extrusion of Glass-Fibre-Reinforced Polypropylene. *Compos. Sci. Technol.* **1988**, *31*, 295–313. [CrossRef]
40. Sasayama, T.; Inagaki, M.; Sato, N. Direct simulation of glass fiber breakage in simple shear flow considering fiber fiber interaction. *Compos. Part A Appl. Sci. Manuf.* **2019**, *124*, 1–10. [CrossRef]
41. Gupta, V.B.; Mittal, R.K.; Sharma, P.K.; Mennig, G.; Wolters, J. Some studies on glass fiber-reinforced polypropylene.: Part I: Reduction in fiber length during processing. *Polym. Compos.* **1989**, *10*, 8–15. [CrossRef]
42. Servais, C.; Maånson, J.-A.E.; Toll, S. Fiber–fiber interaction in concentrated suspensions: Disperse fibers. *J. Rheo.* **1999**, *43*, 991–1004. [CrossRef]
43. Truckenmüller, F. Direktverarbeitung von Endlosfasern auf Spritzgießmaschinen. Original Language: German. Translated Title: Direct Processing of Continuous Fibers on Injection Molding Machines: Possibilities and Limitations: Möglichkeiten und Grenzen. Ph.D. Dissertation, Universität Stuttgart, Stuttgart, Germany, 1996.
44. Meyer, R.; Almin, K.E.; Steenberg, B. Length reduction of fibres subject to breakage. *Br. J. Appl. Phys.* **1966**, *17*, 409–416. [CrossRef]
45. Goris, S.; Back, T.; Yanev, A.; Brands, D.; Drummer, D.; Osswald, T.A. A novel fiber length measurement technique for discontinuous fiber-reinforced composites: A comparative study with existing methods. *Polym. Compos.* **2018**, *39*, 4058–4070. [CrossRef]
46. Rauwendaal, C. *Polymer Extrusion*, 5th ed.; Hanser: Munich, Germany, 2014; ISBN 978-1-56990-516-6.
47. Scheffler, T.; Gehde, M.; Späth, M.; Karlinger, P. Determination of the Flow and Curing Behavior of Highly Filled Phenolic Injection Molding Compounds by Means of Spiral Mold. In Proceedings of the Europe/Africa Conference Dresden 2017, Polymer Processing Society PPS, Dresden, Germany, 27–29 June 2017; Wagenknecht, U., Pötschke, P., Wiessner, S., Gehde, M., Eds.; AIP Publishing: Melville, NY, USA, 2019; pp. 1–5.
48. Scheffler, T. Werkstoffeinflusse auf den Spritzgussprozess von Hochgefüllten Phenol-Formaldehydharz-Formmassen. Original Language: German. Translated Title: Material Influences on the Injection Molding Process of Highly Filled Phenol-Formaldehyde Resin Molding Compounds. Ph.D. Dissertation, Technische Universität Chemnitz, Chemnitz, Germany, 2018.

49. Kruppa, S. Adaptive Prozessführung und Alternative Einspritzkonzepte beim Spritzgießen von Thermoplasten. Original Language: German. Translated Title: Adaptive Process Control and Alternative Injection Concepts in the Injection Molding of Thermoplastics. Ph.D. Dissertation, Universtität Duisburg-Essen, Duisburg, Germany, 2015.

50. Fischbach, G.B.M. Prozessführung beim Spritzgießen Härtbarer Formmassen. Original Language: German. Translated Title: Process Control in Injection Molding of Thermosetting Molding Compounds. Ph.D. Dissertation, RWTH Aachen, Aachen, Germany, 1988.

51. Lucyshyn, T.; Kipperer, M.; Kukla, C.; Langecker, G.R.; Holzer, C. A physical model for a quality control concept in injection molding. *J. Appl. Polym. Sci.* **2012**, *124*, 4927–4934. [CrossRef]

52. Schiffers, R. Verbesserung der Prozessfähigkeit beim Spritzgießen durch Nutzung von Prozessdaten und eine Neuartige Schneckenhubführung. Original Language: German. Translated Title: Improving Process Capability in Injection Molding by Using Process Data and a New Type of Screw Stroke Guide. Ph.D. Dissertation, Universtität Duisburg-Essen, Duisburg, Germany, 2009.

53. Woebcken, W. Processes and Devices for Controlled Injection Molding on Different Plastic Injection Molding Machines with the Same Molded Part Properties and the Same Masses. Patent DE3524310C1, 19 June 1986.

54. Cavic, M. Kontinuierliche Prozeßüberwachung beim Spritzgießen unter Einbeziehung von Konzepten zur Verbesserung der Schmelzequalität. Original Language: German. Translated Title: Continuous Process Monitoring in Injection Molding, with Incorporation of Concepts for Improving Melt Quality. Ph.D. Dissertation, Universität Stuttgart, Stuttgart, Germany, 2005.

55. Maertens, R.; Hees, A.; Schöttl, L.; Liebig, W.V.; Elsner, P.; Weidenmann, K.A. Fiber shortening during injection molding of glass fiber-reinforced phenolic molding compounds: Fiber length measurement method development and validation. *Polym.-Plast. Tech. Mat.* **2021**, *60*, 872–885. [CrossRef]

56. Maertens, R.; Liebig, W.V.; Elsner, P.; Weidenmann, K.A. Compounding of Short Fiber Reinforced Phenolic Resin by Using Specific Mechanical Energy Input as a Process Control Parameter. *J. Compos. Sci.* **2021**, *5*, 127. [CrossRef]

57. Maertens, R. Long Fiber Thermoset Injection Molding: Process Development and Characterization of Material Properties. In Proceedings of the 11th International Thermoset Conference, Iserlohn, Germany, 25 November 2021.

58. Maertens, R. Development of a Direct Injection Molding Process for the Production of Long Glass Fiber-Reinforced Phenolic Resins Components. In Proceedings of the 20th European Conference on Composite Materials (ECCM20), Lausanne, Switzerland, 28 June 2022.

59. Singh, R.; Chen, F.; Jones, F.R. Injection molding of glass fiber reinforced phenolic composites. 2: Study of the injection molding process. *Polym. Compos.* **1998**, *19*, 37–47. [CrossRef]

60. *ISO 527-1*; Plastics—Determination of Tensile Properties. International Organization for Standardization: Geneva, Switzerland, 2012.

61. *ISO/DIS 6603-2*; Plastics—Determination of Puncture Impact Behaviour of Rigid Plastics: Part 2: Instrumented Puncture Test. International Organization for Standardization: Geneva, Switzerland, 2002.

62. Kruppa, S.; Karrenberg, G.; Wortberg, J.; Schiffers, R.; Holzinger, G.P. Backflow compensation for thermoplastic injection molding. In Proceedings of the ANTEC 2016, Indianapolis, IN, USA, 23–25 May 2016; Society of Plastics Engineers: Indianapolis, IN, USA, 2016; pp. 1034–1039.

63. Kruppa, S.; Wortberg, J.; Schiffers, R. Injection Molding Is Getting More Continuous. *Kunststoffe Int.* **2016**, *129*, 81–84.

64. Potente, H.; Többen, W.H. Improved Design of Shearing Sections with New Calculation Models Based on 3D Finite-Element Simulations. *Macromol. Mater. Eng.* **2002**, *287*, 808–814. [CrossRef]

65. Moritzer, E.; Bürenhaus, F. Influence of processing parameters on fiber length degradation during injection molding. In Proceedings of the ANTEC, Online, 10–21 May 2021; pp. 219–223.

66. Lafranche, E.; Krawczak, P.; Ciolczyk, J.-P.; Maugey, J. Injection moulding of long glass fiber reinforced polyamide 66: Processing conditions/microstructure/flexural properties relationship. *Adv. Polym. Tech.* **2005**, *24*, 114–131. [CrossRef]

67. Lafranche, E.; Krawczak, P.; Ciolczyk, J.P.; Maugey, J. Injection moulding of long glass fibre reinforced polyamide 6-6: Guidelines to improve flexural properties. *Express Polym. Lett.* **2007**, *1*, 456–466. [CrossRef]

68. 3B Fibreglass. *Technical Data Sheet Direct Roving 111AX11*; 3B Fibreglass: Hoeilaart, Belgium, 2020.

69. Keyser, H.D.; Sumitomo Bakelite High Performance Plastics; Muynck, M.D. Personal Communication, 2019.

70. Englich, S. Strukturbildung bei der Verarbeitung von Glasfasergefüllten Phenolformaldehydharzformmassen. Original Language: German. Translated Title: Structure Formation during Injection Molding of Glass Fiber Filled Phenolic Formaldehyde Resin Molding Compounds. Ph.D. Dissertation, Technische Universität Chemnitz, Chemnitz, Germany, 2015.

71. Tran, N.T.; Englich, S.; Gehde, M. Visualization of Wall Slip During Thermoset Phenolic Resin Injection Molding. *Int. J. Adv. Manuf. Tech.* **2018**, *3*, 1–7. [CrossRef]

MDPI AG
Grosspeteranlage 5
4052 Basel
Switzerland
Tel.: +41 61 683 77 34

Polymers Editorial Office
E-mail: polymers@mdpi.com
www.mdpi.com/journal/polymers